River Flowing from the Sunrise

An Environmental History of the Lower San Juan

A. R. Raplee's camp on the San Juan in 1893 and 1894. (Charles Goodman photo, Manuscripts Division, Marriott Library, University of Utah)

River Flowing from the Sunrise

An Environmental History of the Lower San Juan

James M. Aton
Robert S. McPherson

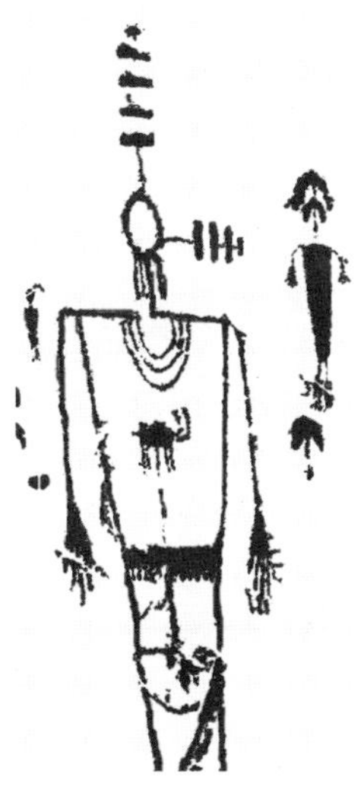

Utah State University Press
Logan, Utah

Utah State University Press
Logan, Utah 84322-7800

Manfactured in the United States of America
Printed on acid-free paper

6 5 4 3 2 1 00 01 02 03 04 05

Library of Congress Cataloging-in-Publication Data

Aton, James M., 1949–
River flowing from the sunrise : an environmental history of the lower San Juan / James M. Aton, Robert S. McPherson.
p. cm.
Includes bibliographical references and index.
ISBN 0-87421-404-1 (alk. paper) — ISBN 0-87421-403-3 (pbk. : alk. paper)
1. Nature—Effect of human beings on—San Juan River Valley (Colo.-Utah) 2. Human ecology—San Juan River Valley (Colo.-Utah)—History. 3. San Juan River Valley (Colo.-Utah)—Environmental conditions. I McPherson, Robert S., 1947– II. Title.
GF504.S35 A76 2000
304.2'09762'59—dc21
00-010229

For Jennifer,

My daughter and fellow traveller on the river of life

—JMA

And to Betsy and the children

—RSM

Contents

Illustrations

Foreword A River in Time

Donald Worster

St. John the Divine ended his Book of Revelation with "a pure river of water of life, clear as crystal, proceeding out of the throne of God and of the Lamb." On either side of that river grew the tree of life, bearing all manner of fruits every month of the year and shiny green leaves that could heal all the nations. He would not have liked the San Juan, the river of the American Southwest named by Spanish missionaries in his honor. Only cottonwood and tamarisk trees grow along its banks. Its water is dark with silt and has been polluted by oil. It flows not from a heavenly throne but from the state of Colorado, where gold miners have sought wealth more than spirituality. Native Americans, to be sure, have deeply religious feelings about this river. So do Mormon settlers in river towns like Bluff. But they have not lived together in peace; on the contrary, this river has experienced bitter conflict, fierce competition for its scarce resources, and not a few deaths. In other words, it has been a *real* river, not some phantasm in a dream, and how much more interesting that fact makes it.

James Aton and Robert McPherson have given us a splendid history of this harshly beautiful place. Heretofore it has been neglected by historians and other scholars, though they have written a surprising number of books and articles on the various peoples, the colorful individuals, who have passed along the river. Aton and McPherson have drawn on that literature extensively, while adding prodigious archival research of their own. But they have done more than sit in a library turning over brittle pages from the past. They have experienced this river firsthand. And they have completely reconceptualized the place and its history so that the whole stands forth, with a new clarity and integrity that it has not had before. They have done this by putting the river at the center of the story and then watching the civilizations come and go. The San Juan becomes the main character; it is no longer merely incidental to human endeavors.

We call this radical new perspective *environmental history*. It begins with the premise that the natural and human worlds are not totally separate but intertwined and interdependent. What nature does affects human beings in the most profound way; vice versa, what people do can influence the patterns and processes of nature profoundly, especially in the modern period, when technology gives us so much more power than we have ever had before. Often that impact has been felt not only by other species who share the place but also, through the intricacies of ecological feedback, by human communities as well. Because early Clovis hunters, the first people to leave their mark on the place, may have exterminated the local population of Columbian mammoths, both hunters and hunted suffered. Later, when the Navajos acquired sheep from the Spanish, they overgrazed the scanty vegetation and created an environmental disaster. The whites who crowded in with their large cattle herds during the late nineteenth century have followed an age-old pattern of land exploitation that likewise has brought serious economic and social problems. If this phenomenon of interdependence has been hard for people to learn, it has seldom entered the apprehension of historians—until the rise of environmental history, so well exemplified in this book.

Most dramatically, the river has been a powerful force over time. Study the canyon walls it has carved through ancient limestone, and you cannot miss that power of running water. What the river has done to the hard materiality of rock it has also done to the tangible dreams of human society: flooded, eroded, and washed them away. Although the federal Bureau of Reclamation has constructed Navajo Dam to control flooding, any historian of long view knows that such control is bound to be imperfect and temporary. Even the mighty Glen Canyon Dam downstream, just below the old confluence of the San Juan and Colorado Rivers, must one day become a man-made waterfall and its reservoir a vast plain of alluvial mud drying in the sun.

The history of the San Juan River stretches back millions of years, while the verified history of human beings dates only to between eleven and twelve thousand years ago and that of Euro-Americans only to 1765, when Juan Maria Antonio de Rivera came looking for the source of a silver ingot. From the perspective of the environmental historian, what happened can be divided into periods called Pleistocene, Anasazi, Ute, or American; these periods vary in length, but they all form one history.

Aton and McPherson are too wise to reduce that history to an oversimplified chronicle of progress or decline. Their perspective is more cyclical and multiple. The San Juan and its peoples pass through cycles of development in which expansion is regularly followed by stasis, even depopulation. And what looks like a time of progress to the whites may look like decline to the Utes or Navajos. Even now, as the authors show in the later chapters, change is coming to the river and its watershed. The old extractive economy created by the whites, which included lumbering, mining, and ranching, is failing, and its place is being taken by urban refugees looking for solitude and white-water rafting enthusiasts lining up like customers at an amusement park. In these changes lie many new problems as well as possible solutions to older ones. Neither a shallow optimism nor a shallow pessimism is supported by the always-tangled history of this place.

It is time that we got to know this river a little better. For too long it has been ignored as a mere tributary of the much larger and more celebrated Colorado, with its Grand Canyon and famous artists and explorers. Yet the San Juan has an amazing story to tell, too. Louis L'Amour found inspiration (and a home) here, and so has Tony Hillerman. But neither of them is a historian, working carefully through the records to tell the underlying story of this place. Aton and McPherson have brought together impressive talent, insight, perspective, and wisdom to write the environmental history of one of the most spectacular parts of the American continent. They are river guides in the fullest and best sense: boatmen who inspire the imagination and inform the mind as well as safely navigate the rapids.

Acknowledgments

Writing an environmental history is much like setting afloat for a trip on the river. Indeed this project began as we sat beside the San Juan under the yellow cottonwood leaves of fall, savoring peanut butter and jam sandwiches. It has taken a long time and many "miles" since that afternoon to bring us to this point in the journey. As we look back at the distance traveled and events along the way, there are a number of people and institutions that deserve thanks and recognition for making the entire tour possible.

Traditionally, the acknowledgments section in a book is the shortest but represents the greatest effort and assistance from others. This one is no different. In this case, length is not an indicator of gratitude, since without help from the following individuals and agencies, this book would not have been possible. The authors also recognize that although an agency has provided financial support or expertise, it is really people who make things happen. On the other hand, we have tried to compile a balanced recounting of the history of the Utah portion of the San Juan River, but if errors have crept in, we accept full responsibility for them.

The outfitters for our journey have been extremely helpful. Among the most prominent in launching and sustaining this work were the Utah Humanities Council, the Charles Redd Center for Western Studies, Southern Utah University's Faculty Development Fund, and the Manti-La Sal National Forest Service in Monticello, Utah. They provided financial support and/or assisted in the collection of Native American and other materials used throughout the text. In addition, the White Mesa Ute Council and the Navajo Nation Museum clarified traditional perspectives and, in the latter case, contributed photographs. Other agencies that supplied expertise and/or pictures are the San Juan County Historical Commission, the Utah State Historical Society, the Bureau of Land Management (San Juan Resource Area), the LDS Church History Archives, University of Utah Special Collections, the Huntington Library, the Museum of Northern Arizona, Northern Arizona University, Brigham Young University Special Collections, the California Academy of Sciences, and the Denver Federal Records Center. Southern Utah University and the College of Eastern Utah also offered each of us timely sabbaticals.

Many individuals also journeyed with us through parts of the manuscript, and their expertise as guides proved invaluable. Their names are sprinkled throughout the endnotes and encountered along the way. Collectively, thanks are due to members of the Navajo Nation and the White Mesa Utes for sharing their culture and history. Ray Hunt, trader and friend, who passed from this life as this manuscript was in progress, shared his many years of experience along the San Juan. He has left a legacy in his thoughts and words for future generations. Archaeologist Winston Hurst read and commented on parts of the manuscript and shared a knowledge of the land and its people that was extremely helpful. Gary Topping has been an endless source of information, friendship, and laughter over the years. Other readers who helped with all or parts of the manuscript are Charles S. Peterson, Mark W. T. Harvey, Rachel M. Gates, and Jill Wilks.

SUU Interlibrary Loan staff members Lorraine Warren and Loralyn Felix made much of the off-river research possible. Various SUU colleagues gave assistance: Rodney Decker, David Lee, Michael P. Cohen, S. S. Moorty, and Thomas Cunningham. SUU students Robert Sidford and Leann Walston helped with compiling the bibliography and scanning pictures. Tim Hatfield was a true artist developing black and white photographs. Special thanks go to Donald Worster, who commented on aspects of the work and wrote the book's excellent foreword. His knowledge of environmental history is well known and has played an important part in shaping our own thinking.

On a more personal level, we appreciate the patience and love extended by our families and friends as we worked on this project. Worthy of special note are Steve, Sue, and Emily Lutz. They opened their beautiful "Avenues" home during numerous research trips to Salt Lake City and also shared many wonderful river trips. All the float trips over the years were fun, and we hope that our children and friends understand now why some of those stops along the shore took longer than they thought necessary. This book is as much a testimony to their patience as it is to our perseverance. And like those trips that ended with sand-filled shoes and sunburned necks, there is a glow that comes with completion. We hope readers feel the same sense of accomplishment upon exiting the river as we do.

It is often called River Flowing from the Sunrise.

—Chester Cantsee
Weeminuche Ute Tribal Elder
1994

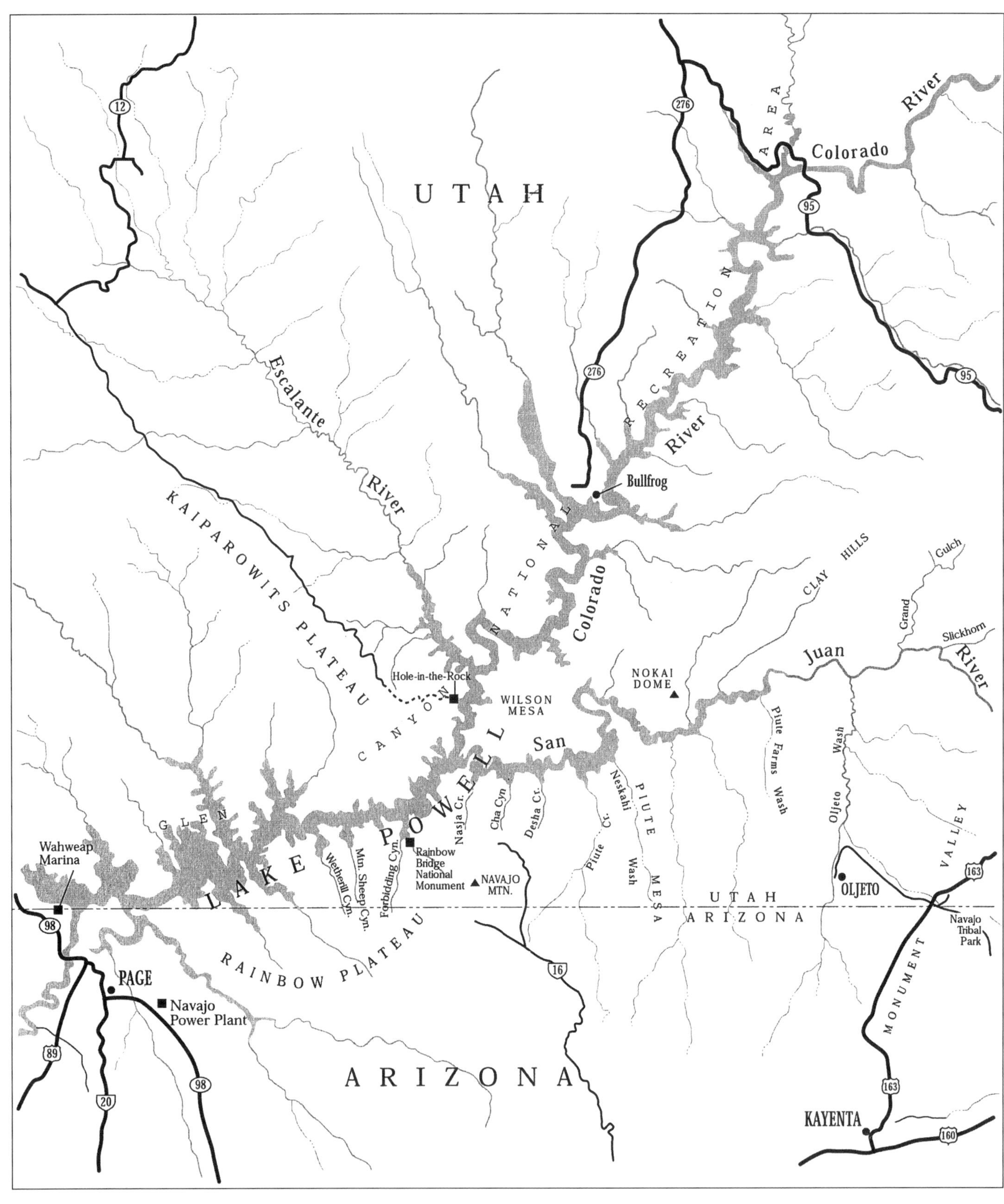
UTAH
ARIZONA
Colorado River
RECREATION AREA
River
Bullfrog
Escalante River
KAIPAROWITS PLATEAU
NATIONAL
Colorado
CLAY HILLS
Grand Gulch
Slickhorn
San Juan River
NOKAI DOME
Hole-in-the-Rock
WILSON MESA
CANYON
GLEN
LAKE POWELL
Piute Farms Wash
Oljeto Wash
MONUMENT VALLEY
Nasja Cr.
Cha Cyn
Desha Cr.
Neskahi Wash
Piute Cr.
PIUTE MESA
Wahweap Marina
Wetherill Cyn.
Mtn. Sheep Cyn.
Forbidding Cyn.
Rainbow Bridge National Monument
NAVAJO MTN.
UTAH
ARIZONA
OLJETO
Navajo Tribal Park
RAINBOW PLATEAU
PAGE
Navajo Power Plant
KAYENTA
12
276
95
98
89
20
16
163
160

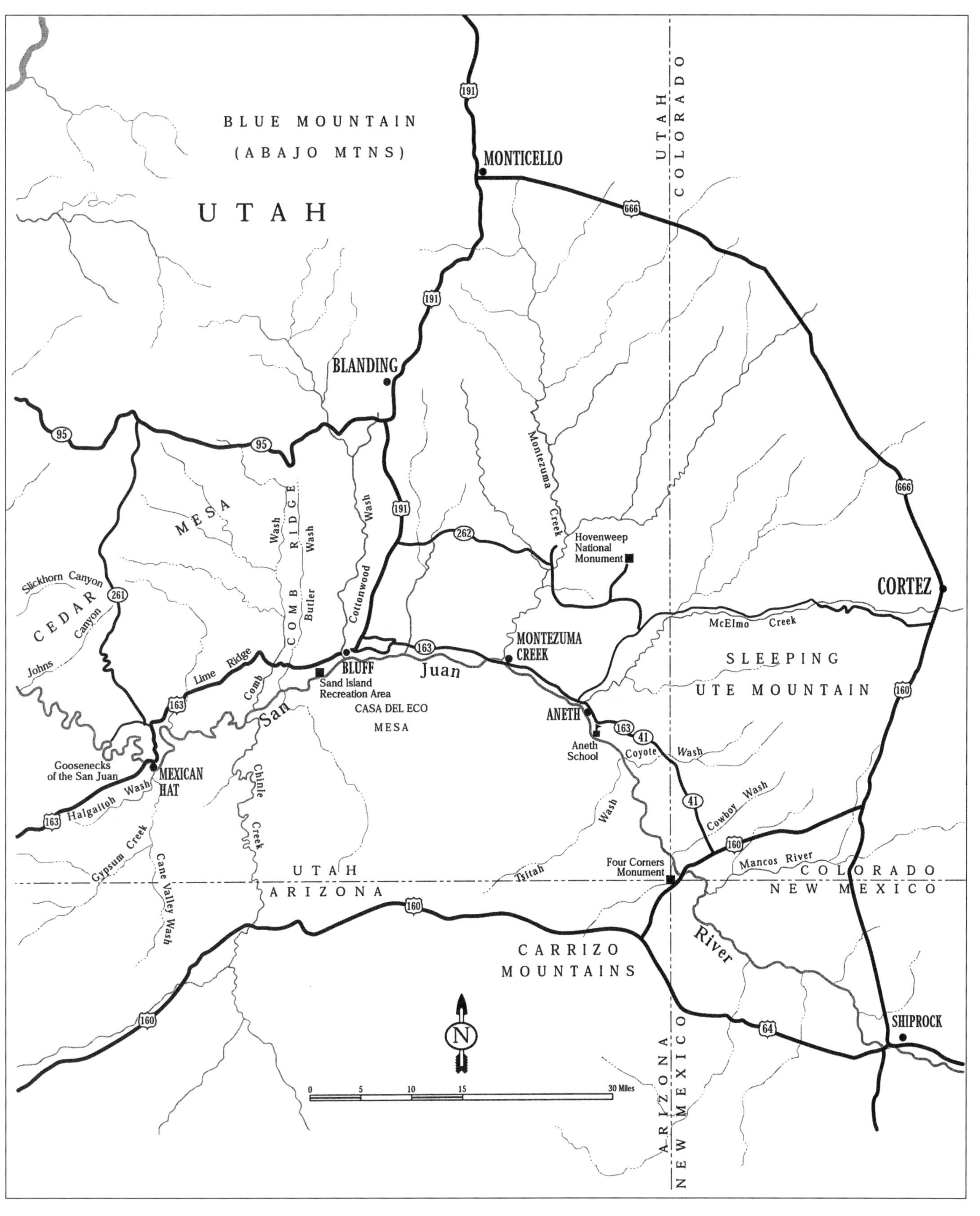

BLUE MOUNTAIN
(ABAJO MTNS)
UTAH
MONTICELLO
BLANDING
UTAH
COLORADO
Montezuma Creek
Hovenweep National Monument
CORTEZ
MESA
Wash
COMB RIDGE
Butler Wash
Cottonwood Wash
Slickhorn Canyon
CEDAR
Johns Canyon
Lime Ridge
Comb
BLUFF
Sand Island Recreation Area
San Juan
CASA DEL ECO MESA
MONTEZUMA CREEK
McElmo Creek
SLEEPING UTE MOUNTAIN
ANETH
Aneth School
Coyote Wash
Cowboy Wash
Gooseneck of the San Juan
MEXICAN HAT
Halgaitoh Wash
Chinle Creek
Gypsum Creek
Cane Valley Wash
Wash
Tsitah
Four Corners Monument
Mancos River
UTAH
ARIZONA
COLORADO
NEW MEXICO
CARRIZO MOUNTAINS
River
SHIPROCK
ARIZONA
NEW MEXICO
N
0 5 10 15 30 Miles
191
666
95
262
261
163
41
160
64

Introduction Twelve Millennia on the San Juan

When the famous explorer John Wesley Powell passed the mouth of the San Juan River on 31 July 1869, he barely acknowledged it. During the next decade, when his geologists and archaeologists fanned out to explore, map, and generally reconnoiter the Colorado Plateau, the last blank spot on the United States map, they ignored the waterway the Utes call River Flowing from the Sunrise. For Major Powell, as for most nineteenth-century Americans, the San Juan River country remained a terra incognita. There were simply few pressing reasons—geological, agricultural, or cultural—for most Americans to know more about it. For the federal government, Powell was the main spokesman on western land affairs in the post–Civil War period, and for most Euro-Americans, the San Juan was a backwater.

Well into the twentieth century, even for Indians like the Utes and Navajos, the Lower San Juan functioned as a kind of refuge beyond the reach of Indian agencies at Shiprock, New Mexico, and Towaoc, Colorado. The San Juan's exclusion from Rinehart's *Rivers of America* book series in the 1940s likewise indicated its relative obscurity. Writing about the Colorado River for that series, Frank Waters noted that the San Juan is "the largest river in New Mexico. Its annual discharge of 2,500,000 acre-feet is over twice that of the noted Rio Grande. Yet it remains one of the least known rivers in America."[1] Past judgments aside, it should be better known—for both local and national reasons.

Today Utah's San Juan River, like nearly all waterways in the West, is a river in demand both regionally and nationally. Its water is becoming ever more valuable in this always-arid landscape. Various Indian tribes are claiming their water rights as granted by the Supreme Court's 1908 decision known as the Winters Doctrine;[2] federal water engineers are controlling the river's flow with two large dams, one near the Colorado-New Mexico border and one past the river's end near the Utah-Arizona border; federal land agencies, obligated by the Endangered Species Act, are trying to save animals like the Colorado pikeminnow (née squawfish), the peregrine falcon, and the willow flycatcher; private and commercial river runners are demanding an equal say in the river's use for their sport and businesses; farmers are trying to maintain their traditional water allotments; towns along the river are clamoring for their share of the water; and, amid all the arguing, Indians and Anglos alike are reasserting the spiritual significance of the river. The San Juan River today stands at a crucial juncture in its twelve-thousand-year history of human occupation and use.

While demands on the river are increasing each year, compared with many rivers draining into the Pacific, the San Juan is sparsely settled and has been intellectually neglected. Because of the area's ruggedness and aridity, especially along the Utah section, relatively few people have settled the river's sandy banks. Although the human population in the region has increased significantly over the past century or so, the San Juan below Four Corners remains an area where the human touch is not always obvious. Despite the increased use of the river and the two dams controlling it, it is still possible to talk about managing it in a "naturalized" way. Parts of the San Juan today, especially in its canyons, strongly resemble the river of hundreds, even thousands, of years ago. Still it is both a natural and social space. Historian Richard White's description of the Columbia

These boys show off a Colorado pikeminnow they caught in the Green River in the early twentieth century. Pikeminnows this size also swam in the San Juan until dams and pollution nearly killed them off. They are the subject of a massive recovery effort as mandated by the Endangered Species Act. (Upper Colorado River Endangered Fish Recovery Program, U.S. Fish and Wildlife Service)

River applies as well to the San Juan: an "organic machine . . . at once our own creation," yet retaining "a life of its own beyond our control."[3]

Planning along the San Juan and litigation over its waters are also relatively recent, compared with other western rivers like the Colorado, the Gila, and the Columbia. National environmental laws and the significant amount of public land along the river intensify the need for coordination among numerous federal agencies, local governments, Indian tribes, and citizen groups. This kind of cooperation, as seen in the recent San Juan River Basin Recovery Implementation Program (SJRIP), is new. With local interest in and demands on the river increasing, this seems a propitious time to narrate the story of the San Juan and the people who have wrested a living from it.

The San Juan's story, however, resonates beyond the Four Corners area. It is now one of the premier river-running destinations in the United States, attracting more than thirteen thousand boaters a year. This is just a few thousand shy of the number who float the Colorado through Grand Canyon. While most come from the Four Corners region, the San Juan attracts recreationists from every state in the Union as well as foreign countries. Given its prominence in the burgeoning river-running industry, its history becomes more important simply because more people are now paying attention to it.

The San Juan is also a neglected component of one of the most studied phases of western history: water development in the Colorado Basin. The flood of books on the topic has crowded the literary shoreline in recent years. Historians and others writing about the Colorado have correctly called its history crucial to understanding western settlement; the rise of the environmental movement; cultural conflict between Anglos, Indians, and Hispanics; and the rise of federal hegemony in the West. They have tended, however, to overemphasize the Colorado River portion of the basin's story at the expense of the San Juan and other tributaries.[4] True, the Colorado is the main attraction and a symbol for water concerns, but the San Juan's story in some ways tells us more about the way some of these issues have played out, especially settlement and cultural conflict. While the San Juan remains sparsely settled, it has certainly attracted more people to its cottonwood- and willow-lined banks than many portions of the Colorado. Moreover, it is one of the most "Indian rivers" in the United States. If the West, as Patricia Nelson Limerick claims in *The Legacy of Conquest,* is where we all met and where the study of race relations is most revealing, then the San Juan is an excellent place to watch that process unfold.[5] With Navajos, various Ute bands, Paiutes, Jicarilla Apaches, Mormons, non-Mormons, and Mexicans all contending for its waters over time, the San Juan provides a superb case study of the way cultures deal with their environment and each other in a cauldron of cooperation, coexistence, and conflict. Few rivers' histories open so many different windows onto race relations and the environment.

Finally, the San Juan's story is important because it typifies much of the rural West today, caught between the resource-extraction era, with its depleted ecologies, and the New West, with its emphasis on environmental protection, tourism, and sustainability. All of these values currently compete for attention, both locally and nationally.

The San Juan is unique in another way. Despite the area's relative obscurity, many of those who have traveled or settled there have recorded their impressions, either orally or in writing. From historic as well as contemporary Native Americans to explorers to various kinds of scientists to Mormon settlers to government agents, the material on the San Juan is rich and offers the researcher a specificity not often found elsewhere. This book's scope is somewhat narrow—the two-hundred-mile stretch of Utah's San Juan—but its coverage is deeply layered, like the eons of limestone deposits along parts of the river. The authors hope what is presented here will stimulate future studies of people and their interaction with western rivers.[6]

How does the Lower San Juan compare to other western rivers? Stacked against those in the Intermountain West—the Gila, Colorado, Little Colorado, Green, and Rio Grande—the San Juan's history holds much in common. These rivers are all significant water sources in arid lands, giving credence to what historian Charles S. Peterson wrote about the Little Colorado: "The River itself organized the people. It dictated the numbers who came and in a large degree molded their experience."[7] All these rivers are controlled to some extent by federal agencies, with large dams on the main stem river and/or tributaries. The Rio Grande has the fewest. The Colorado and Green, because they have the deepest canyons, have the largest: Glen Canyon and Boulder Dams and Flaming Gorge Dam, respectively. All these dams provide flood and sediment control, while some generate power. Unintentionally, they have also exacerbated the spread of tamarisk while negatively affecting habitat for native fish.

In cultural terms, perhaps only the Rio Grande in New Mexico is more Indian and multicultural than the San Juan. The Lower San Juan and parts of the Little Colorado, however, share the distinction of having Mormon settlements. For combinations of Mormons and Indians, the San Juan is unique. The trading posts along the San Juan also developed differently than elsewhere. The Gila and Rio Grande have larger population centers than the Lower San Juan, although in New Mexico the river has some decent-sized towns. It has also seen more oil development along its banks but is still best known for its recreation. Like the Green and Colorado in their canyon sections, the Lower San Juan has seen dramatic numbers of river runners arrive since the recreation boom following World War II. That is why many Americans think of the Utah canyons of the San Juan, having experienced them through river running.[8]

To really understand the San Juan, one must know a little about its recent geological history. Between twenty and ten million years ago,

More than ten million years ago, this Honaker Trail section of the San Juan was a meandering stream flowing over a flat desert. When the country began to uplift—the Monument Upwarp shown in this 1910 photo—the San Juan kept cutting and incising. (E. G. Woodruff photo, #168, U.S. Geological Survey)

The broad alluvial plains between Four Corners and Chinle Wash, seen in this 1929 photo near Aneth, provided the base soil for agriculture and town building from 1500 B.C. to the present. (Herbert E. Gregory photo, #580, U.S. Geological Survey)

The Goosenecks below Mexican Hat are the classic example of the geological principle of an entrenched meander. The snakelike course of the river predated the country's rise and the river's cutting and incising. It takes the river five miles to advance just one. (Tad Nichols photo, Manuscripts Division, Marriott Library, University of Utah)

the river established itself as a flat meandering stream which flowed out of the San Juan Mountains of southwest Colorado and snaked its way across the desert toward the Colorado River.[9] About that time, the country below present-day Bluff began to uplift into what is now known as the Monument Upwarp, a ninety-mile long, thirty-five-mile wide series of north-south–running anticlines and synclines between Comb Wash and Clay Hills Crossing. An uplift associated with Navajo Mountain, the Slick-Rock section, influenced canyon building between Clay Hills and the confluence with the Colorado.

An entrenched meander, the San Juan sliced into these upwarps at a rate comparable to the country's rise, ultimately creating spectacular, thousand-foot canyon walls. In places like the world-famous Goosenecks, the deeply incised river loops back on itself like a folded ribbon. By five to six million years ago, the San Juan had definitively cut through softer, more easily eroded materials and was incising itself into its present course. Upstream from the Monument Upwarp in the Blanding Basin, the river continued its snaking pattern, shifting this way and that across the broad valleys that barely contained it. All the while, it was hauling down quarries worth of sediment from the San Juan Mountains and tributaries north and south.

The greatest effect on San Juan River geomorphology followed four major periods of glaciation during the last one-and-a-half-million years, part of the epoch known as the Pleistocene. Wetter and cooler, the period averaged about twenty inches of rain per year, as opposed to eight now. Consequently, it saw massive flows through the San Juan corridor, probably close to one million cfs (cubic feet per second). Compared to the highest flow of the Holocene (8000 B.C. to the present) of around one hundred thousand cfs, the Ice-Age San Juan was an awesome erosional and depositional force. The river at Bluff during a Pleistocene flood, for example, would have stretched from cliff to cliff—over a mile wide.

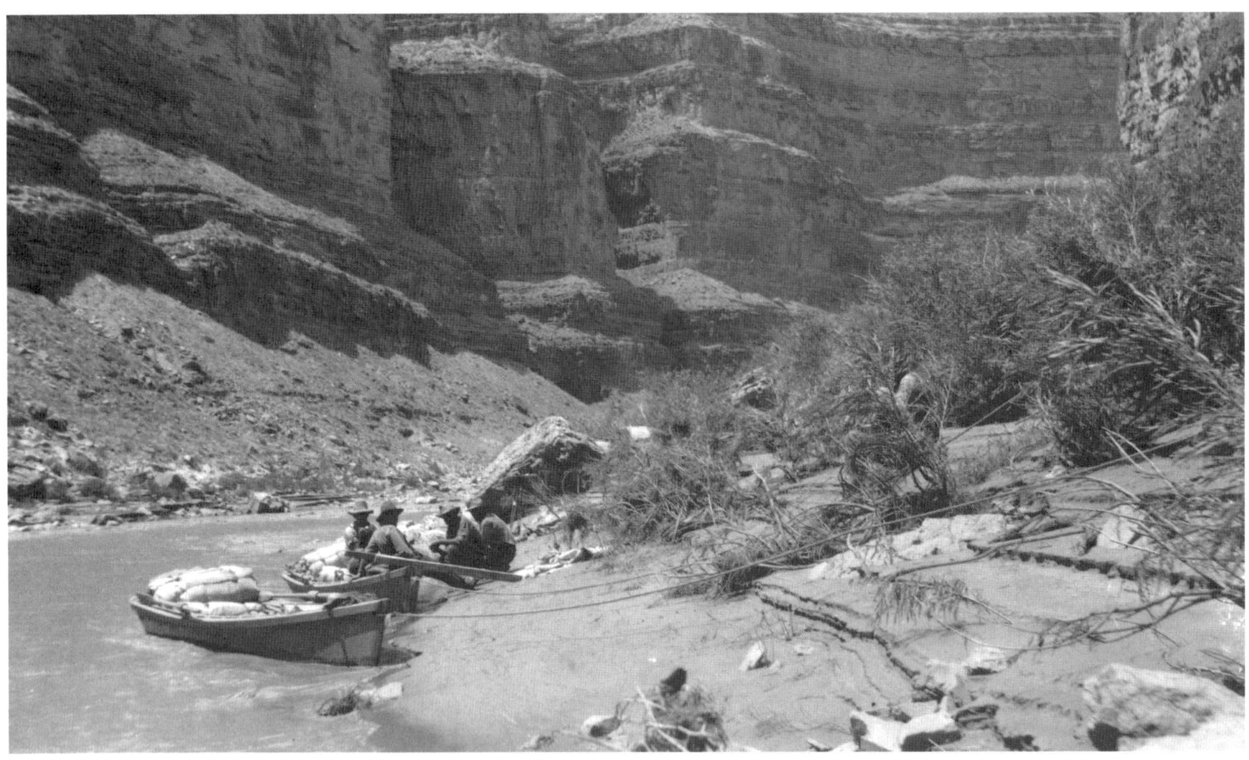

The 1921 Trimble Expedition takes a lunch break between Honaker Trail and John's Canyon. Evidence of recent floods appears in the mashed-down vegetation on the right. Those floods have been cut in half by Navajo Dam. (Hugh D. Miser photo, #434, U.S. Geological Survey)

The results of those floods appear in the form of high alluvial terraces, cobble fields, and dunes between Four Corners and Chinle Wash. Those great sediment deposits of the Blanding Basin provide the base soil on which all plant and animal life along the river has sustained itself. That in turn attracted human beings to the area about twelve thousand years ago. Later those fertile terraces made farming possible along the San Juan, from the Basketmaker Anasazi period, circa 1500 B.C., to the present.[10]

The river still originates in the San Juan Mountains of southwestern Colorado and flows for more than one hundred miles through northern New Mexico before entering Utah near Four Corners. In each of the three states it traverses, it exhibits different characteristics. The southwestern Colorado section is a somewhat-clear, free-flowing mountain river, bordered by big pines, pinyon-juniper forests, and dense vegetation and hemmed in largely by the igneous and metamorphic rocks of the San Juan Mountains. Just before it leaves Colorado, three small rivers join it: the Piedra, Rio Blanco, and Navajo. Not far into New Mexico, at the crease between the Rocky Mountain and Colorado Plateau geomorphic provinces, it suddenly drops to a desert plain, meandering through flatter, drier terrain. Here it begins absorbing great loads of sediment from tributary rivers and washes and assumes its characteristic brown color.

Since 1962, Navajo Dam near the Colorado-New Mexico border has controlled much of the San Juan's flow through New Mexico and Utah. Impoundment, however, has not greatly changed sediment loads. In much of the area above the dam, the river runs over crystalline rocks and is well vegetated. Consequently, the Colorado section contains far less sediment per water unit above the dam than below it, where sedimentary rocks such as sandstone, siltstone, and shale underlie the river and its tributaries. Siltstone and shale are especially erodible and significantly increase the sediment load. Moreover, those areas in New Mexico and Utah are more arid and less vegetated. This likewise contributes to sediment buildup.[11] The dam, however, has cut probably by half the huge floods that formerly raced out of the San Juan Mountains and Nacimiento Uplift on the Jicarilla Apache Reservation.[12]

While the New Mexico section resembles the Utah part more than the Colorado section, there are important reasons why this study focuses on the river from Four Corners to Lake Powell; the division is not merely artificial. Many of the physiographic factors have ultimately influenced the cultural history of the area. Geologists, for example, divide the river below the dam into five distinct geologic sections, three of which fall in Utah.

East to west along the river from Four Corners, the Blanding Basin comprises the first physiographic unit. An area of low mesas, buttes, and shallow drainages, the basin's western boundary is Comb Ridge. From there, a broad anticlinal fold called the Monument Upwarp provides the setting for the incised meanders of the San Juan called the Goosenecks. Its western flank dips down at the Clay Hills Crossing-Paiute Farms area. Here begins the Slick-Rock section, a rugged area of mesas, canyons, and promontories associated in part with the uplift of Navajo Mountain southeast of the confluence of the San Juan and Colorado. Currently, Lake Powell backs up to the east into this section all the way past Clay Hills.[13] The Utah sections are known collectively as the Lower San Juan, an area characterized by uplift and river incising.

Recent, more-comprehensive studies of the riparian corridor by SJRIP scientists have confirmed and refined the importance of geological divisions for all aspects of life along the river. SJRIP researchers divided the river into eight "reaches." They used criteria such as river-valley geometry, riparian vegetation, channel gradient and patterns, tributary influence, human influence, and aquatic habitat to define each reach. The Utah sections comprise the first four reaches according to these scientists, who point out that these areas differ significantly from the Upper San Juan or upper four reaches.[14]

In general the Lower San Juan experienced significantly less human influence than the Upper San Juan. For example, in the Upper San Juan in New Mexico, numerous diversion dams block the river's flow, while in the lower part, the river surges freely. In the Utah sections, irrigation and agriculture are less prominent than in New Mexico, restricted mostly to the area between Four Corners and Chinle Wash. Below Chinle deep canyons largely prohibit farming along the river. Only the small-scale horticulture of Anasazi and later Paiute and Navajo Indians could take advantage of small plots of land along tributary streams.

In addition to affecting human occupation and land use, these divisions tell something about native fish. For example, Colorado pikeminnows appear more prevalent in the lower half of the river. This may have something to do with the concentration of their traditional spawning grounds in the Four Corners area and/or the impediment to upstream migration imposed by diversion dams at Shiprock and elsewhere.

Besides looking at the river's immediate corridor, we will sometimes wander up various side drainages to see what happened there. Rivers are connected to other ecosystems and especially influenced by what occurs along their tributaries. Chinle Wash, Montezuma Creek, Cottonwood Wash, and the canyons cutting Cedar Mesa have exercised an enormous influence on the San Juan. Cottonwood Wash, for example, can dump huge amounts of sediment into the river, often creating havoc for Bluff settlers over the years. If this approach occasionally appears far ranging or inconsistent, we beg the reader's tolerance and hope, in the end, that our geographical boundaries make sense.

The nature of the landscape directly influenced both the prehistory and history of the Lower San Juan. Anasazi, Utes, Navajos, and Jicarilla Apaches found that the upper river in New Mexico provided better camping and farming sites. Small groups of Basketmaker and Pueblo Anasazi lived along the Lower San Juan, but no significant population centers existed there like the Upper San Juan sites of Aztec, Salmon Ruin, Mesa Verde, or Chaco Canyon. Nearby Cedar Mesa, however, was heavily populated at different times during the Pueblo Anasazi period. Historic Indian use has followed that same pattern. Small populations of Paiutes have lived for hundreds of years at Navajo Mountain and along San Juan tributaries like Paiute Farms and Montezuma Creek.[15] During the late-nineteenth century, however, the more populous and mobile Utes and Navajos found refuge on the Lower San Juan from federal troops and the influence of Indian agents at places like Shiprock (for Navajos) and Towaoc (for Utes.)

Navajo (above) and Glen Canyon (facing page) Dams have had the most profound effect on San Juan riparian ecology. They came on-line in 1962 and 1963, respectively. (Bureau of Reclamation, Upper Colorado Region)

Ute and Navajo activities along the Lower San Juan mirrored those in the upper, New Mexico section—hunting, gathering, farming, and grazing—but they took on a different personality. The Weeminuche Utes, in particular, found fewer hunting opportunities on the Lower San Juan. Despite the region's ruggedness, Indians were drawn to it because of the river. It thus became a kind of expansionist frontier for Utes and Navajos as their populations increased, as members of both tribes sought to hunt and gather resources, and as Navajos, in particular, needed more land for their sheep.[16]

If Ute and Navajo use of the area was hesitant to develop, Euro-American hegemony was not much different. The Spanish influence, so prominent in New Mexico, affected the Lower San Juan only indirectly. Utes and Navajos adopted horses, sheep, farming methods, and tools from the Spanish. Except for a few explorers, military expeditions, and slave traders, Spain and then Mexico ignored the Lower San Juan. It lacked obvious agricultural, mineral, and trading potential and posed a prominent geographical barrier to trade with California. Moreover, Spain guarded its topographical information jealously. Although hard to document, the advent of Anglo fur trappers in the early nineteenth century may have wreaked environmental havoc by nearly eliminating beaver along the San Juan and its tributaries. Beaver dams control erosion and provide a rich environment for smaller birds and other animals. Despite its slow beginnings, the entrance of Europeans and Americans into the San Juan, starting in 1765, heralded a change. The technologies and values of the West, with its industrial production and secular view of nature, have continued to exert a profound effect on the San Juan landscape to this day.

By the early 1880s, the process of change had speeded up considerably. Texas cattlemen

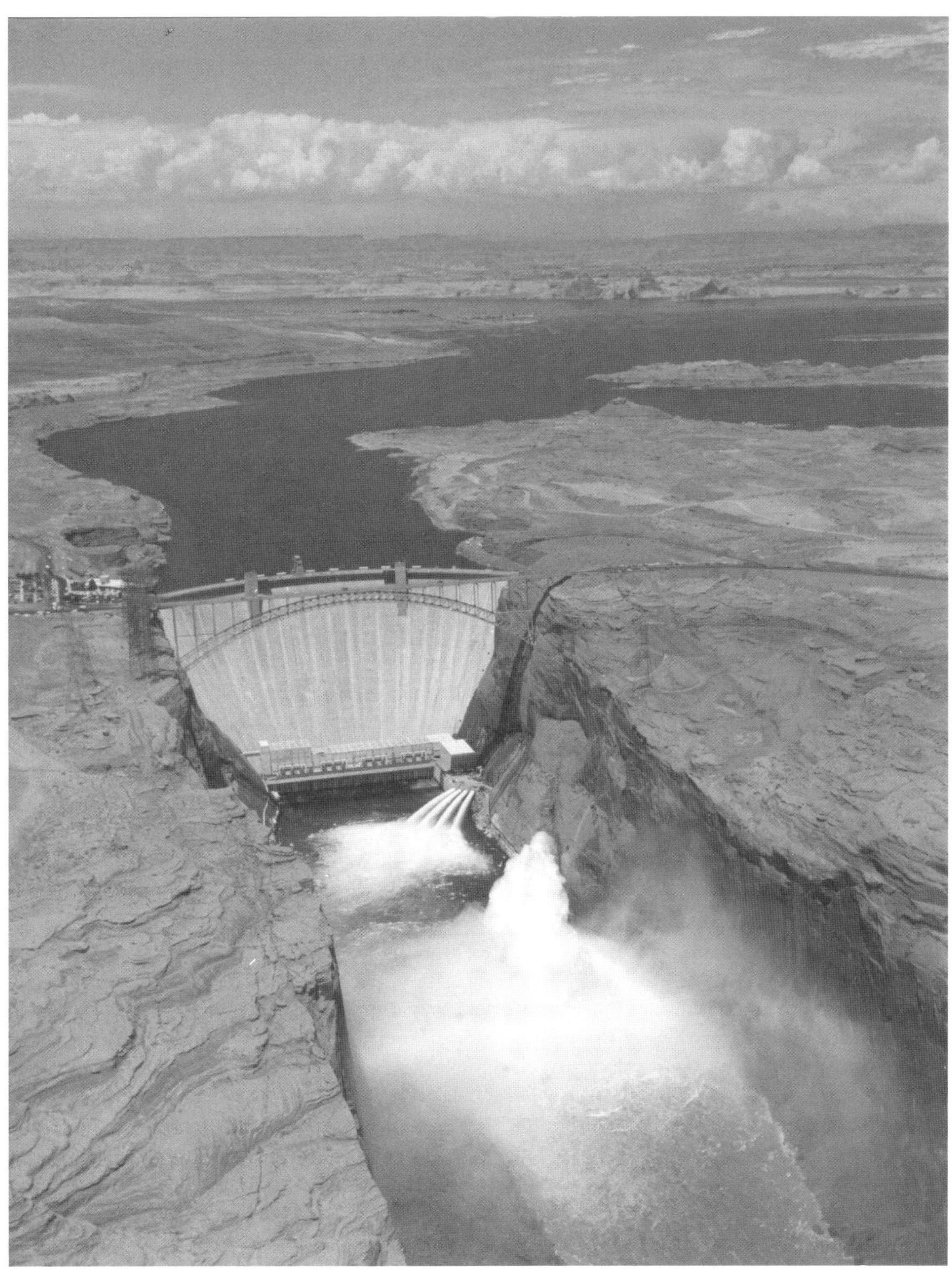

rode into the area, lured by its remoteness from government authorities and the availability of free land. The Texans' reputed lawlessness was one reason the Church of Jesus Christ of Latter-day Saints (also known as the LDS Church or the Mormons) sent a colonizing mission to the San Juan country in 1879–80. The Mormons also wanted to control the entire Utah Territory, sought a warmer climate than the Salt Lake Valley for their converts from the South, and desired better relations with the Indians living in Utah's most remote region.[17] Trading posts, operated by both Mormons and non-Mormons beginning in the late nineteenth and early twentieth centuries along the Lower San Juan, also shared some different characteristics. Isolated as they were, these Utah posts functioned not only as communal gathering places for Indians who were naturally drawn to the river but also reflected Mormon policy and practices.

Mormon relations with Utes, Paiutes, and Navajos differed from those of other Anglos in the Upper San Juan. A distinct theological cast colored Mormon paternalism. Their theology encouraged conversion of Indians rather than eradication or expulsion. They failed to convert many of the area's Indians but enjoyed more peaceful relations than their neighbors. Mormons protested, nevertheless, when southwestern Coloradans tried to remove some Utes to San Juan County following the discovery of gold and silver in the San Juan Mountains and the so-called Meeker Massacre on the White River.[18]

When the Mormons arrived in 1880, their fumbling efforts to irrigate also set in motion a riparian-altering process unprecedented in the history of human interaction with the river. There were two significant results. First, farming, grazing, and, to a lesser extent, mineral extraction on the San Juan have been part of a worldwide phenomenon that has hastened more erosion than a Pleistocene flood.[19] The second result is what appears to be ultimate control. Eighty years after Euro-American farmers planted their first crops, two dams, Navajo and Glen Canyon, came on line within a year of each other, in 1962 and 1963. These dams restrict a major part of the San Juan's flow.

The challenge of water control in the Colorado Basin in turn occasioned the rise of the biggest government agency in world history, the Bureau of Reclamation. The specter of that agency's power and the resulting dams in the Colorado Basin, however, also gave birth and focus to the modern environmental movement and its renewed set of values regarding nature.[20] Those politics and values manifested themselves in a set of national environmental laws in the 1960s and '70s (the Wilderness Act, the National Environmental Policy Act, and the Endangered Species Act, to name but a few), as well as new missions for federal agencies (National Park Service, Bureau of Land Management, U.S. Fish and Wildlife Service, and Bureau of Indian Affairs) to enforce them.

In weaving the story of the riparian landscape together with that of Mormons, Indians, trappers, government agents, and recreationists, this narrative adopts a three-tiered approach to environmental history.[21] In this model, the natural history of the landscape, with both organic and inorganic components—plants, animals, geologic processes, and weather, forms the basis of the analysis of the Lower San Juan.

Next come the technologies people use to control their environment, ranging from a Clovis hunting point to the adoption of corn and dam construction. Related to these technologies are the institutions formed to apply them—a hunting-gathering band, a Mormon colonizing mission, or a government agency like the Bureau of Reclamation.

Finally, one must account for the mythic and ideological levels on which a society functions. Artistic expression, like a petroglyph, a poem, or a photograph, speaks volumes about how people value their landscape and why they apply their tools and institutions to the environment the way they do. For example, consider the comments of two writers seventy-five years apart, speaking about the same San Juan wilderness. In 1875 Hayden Survey topographer George B. Chittenden wrote, "This whole portion of the country is now and must ever remain utterly worthless."[22] He spoke for the federal government and most Americans in valuing land according to its exploitable resources. This point of view underlay the decisions of government builders as they fundamentally changed every aspect of the river's ecological makeup by constructing dams at either end. Novelist Wallace Stegner viewed that same empty space positively in 1949, saying, "This is the way things were when the world was young; we had better enjoy them while we can."[23] Stegner placed recreational and aesthetic values above utilitarian ones and presaged the post–World War II environmental movement that was just beginning to find its voice. That attitude led to the enactment of important environmental laws and irrevocably changed the way people interacted with the river corridor. These two observations say much about

"The beginning of the Monument Upwarp: Lime Ridge at Chinle Wash, 1914." (Herbert E. Gregory photo, #244, U.S. Geological Survey)

the way nineteenth-century frontier attitudes toward the San Juan had evolved by the mid-twentieth century.

Even though these three approaches sometimes receive separate treatment, as historian Donald Worster says, "in fact they constitute a single dynamic inquiry in which nature, social and economic organization, thought and desire are treated as one whole. And this whole changes as nature changes, as people change, forming a dialectic that runs through all of the past down to the present."[24] The history of salt cedar, or tamarisk, in the Southwest, discussed extensively in chapter 8, illustrates the interaction of all three levels of inquiry. This hardy, water-loving tree originated in ancient Mesopotamia (modern-day Iraq), but American seed companies imported it in the early nineteenth century to control erosion. It has now grown out of control in the West, its spread greatly abetted by man-made dams like Navajo and Glen Canyon. Reactions to its unexpected dominance range widely: valued for soil stabilization and erosion control; criminalized as a water thief and beach-invading, insect-harboring weed; accepted as part of the consequence of dam building.

As with many other aspects of the river's history, speaking of the long-term viability of native vegetation or consequences of introduced plants necessitates throwing in a big dash of relative time—geologic and human. San Juan human history, with all its vicissitudes, is little more than an interesting, if perhaps tragic, interlude in the processes that have shaped the river. Recent geologic events, however, such as the deposition of massive alluvial banks, specifically set the stage for the human drama played out in this arid and dramatic river landscape.

This book covers all phases of the Lower San Juan's environmental history but concentrates mainly on the late-nineteenth and twentieth centuries, when the most profound environmental changes have occurred. This is not to say that the San Juan was an untouched paradise before Euro-Americans came on the

In 1875 Hayden topographer George B. Chittenden deemed the San Juan country worthless. Survey photographer William H. Jackson, however, clearly saw it in the more aesthetic terms that characterized mid-nineteenth-century nature appreciation. This is Lime Ridge, the eastern flank of the Monument Upwarp. (William H. Jackson photo, #1157, U.S. Geological Survey)

scene. All the peoples who have lived in the San Juan corridor have sought to shape their environment and wrest a living from it. Negative impacts on plants and animals have not been the sole province of white people. The first Americans, the Clovis hunters, may have applied both a technology and mythology to a landscape they did not entirely understand and ultimately reaped unforeseen consequences. It is to their story that we now turn.

1 PREHISTORY: *From Clovis Hunters to Corn Farmers*

Humans have hunted and herded animals, gathered and cultivated plants, and generally made a living in the San Juan River area for at least the last twelve thousand years. Although always a marginal area, the river valley's population reached a high point during the Anasazi occupation between 1500 B.C. and A.D. 1300.[1] During this prehistoric period, the San Juan landscape was certainly no untouched Eden. To be sure, since Euro-Americans entered the San Juan country and applied the technology of the Industrial Revolution, they have changed the landscape more dramatically than both prehistoric and historic Indians. Yet, before one accounts for that massive environmental change, it is crucial to understand the roughly twelve thousand years preceding it.

Although pre-Columbian Indians in the San Juan basin manipulated their environment, the influence of climatic variation cannot be ignored. During the prehistoric period, the San Juan changed from an Ice-Age climate with cooler temperatures and much more precipitation to the drier, warmer weather it now experiences. The first recognized and established entrants into the San Juan, the Clovis hunters, and their successors, the Folsom hunters, lived during the five-hundred-to-thousand-year transition from the cool, wet Pleistocene to the warm, drier Holocene. Moreover, all the prehistoric groups that archaeologists distinguish—Clovis, Folsom, Plano, Archaic, and Anasazi—had to cope with climatic changes during their tenure on the San Juan. They all made land-use decisions based on the environmental deck nature dealt them, on the skills and tools they had to play the game, and on the imaginative and cultural ideas they brought to the table. Often they hedged their bets wisely, but other times they overplayed their hands. None of these groups lived in perfect harmony with the San Juan landscape, although the Archaic lifeway persisted longer than any other.

Interest in San Juan prehistory has focused largely on the Anasazi from roughly 1500 B.C. to A.D. 1300. The Anasazi fired the imagination of the American public in large part because, in contrast to Indian groups before and after, they built magnificent structures. More than other Native American groups in the area, they reflected a Euro-American definition of civilization. The often-neglected groups of prehistoric Indians in the San Juan area, however, deserve equal consideration. It is crucial to understand how the hunting-gathering Clovis, Folsom, and Archaic Indians manipulated the San Juan environment and changed themselves in the process.

In the late Pleistocene, sometime around 10,000 to 9000 B.C., the Clovis hunters walked into the San Juan area. This is what they found: Weather conditions were cooler and wetter, but today's temperature extremes did not exist. Rather than four seasons, two split the climatic year: a mild, cool summer and a wet, cold winter. The growing season extended longer, and plant species varied considerably, unlike the relatively less diverse environment of the Holocene, 8000 B.C. to the present.[2]

A twentieth-century visitor to the late-Pleistocene San Juan River would be shocked to see what luxuriant vegetation grew in the bottoms as well as how massive the river flows were. That time traveler would find plants flourishing now commonly found on Navajo Mountain, Elk Ridge, or in the Abajo Mountains. A few would be barely recognizable. Tall Douglas firs, white

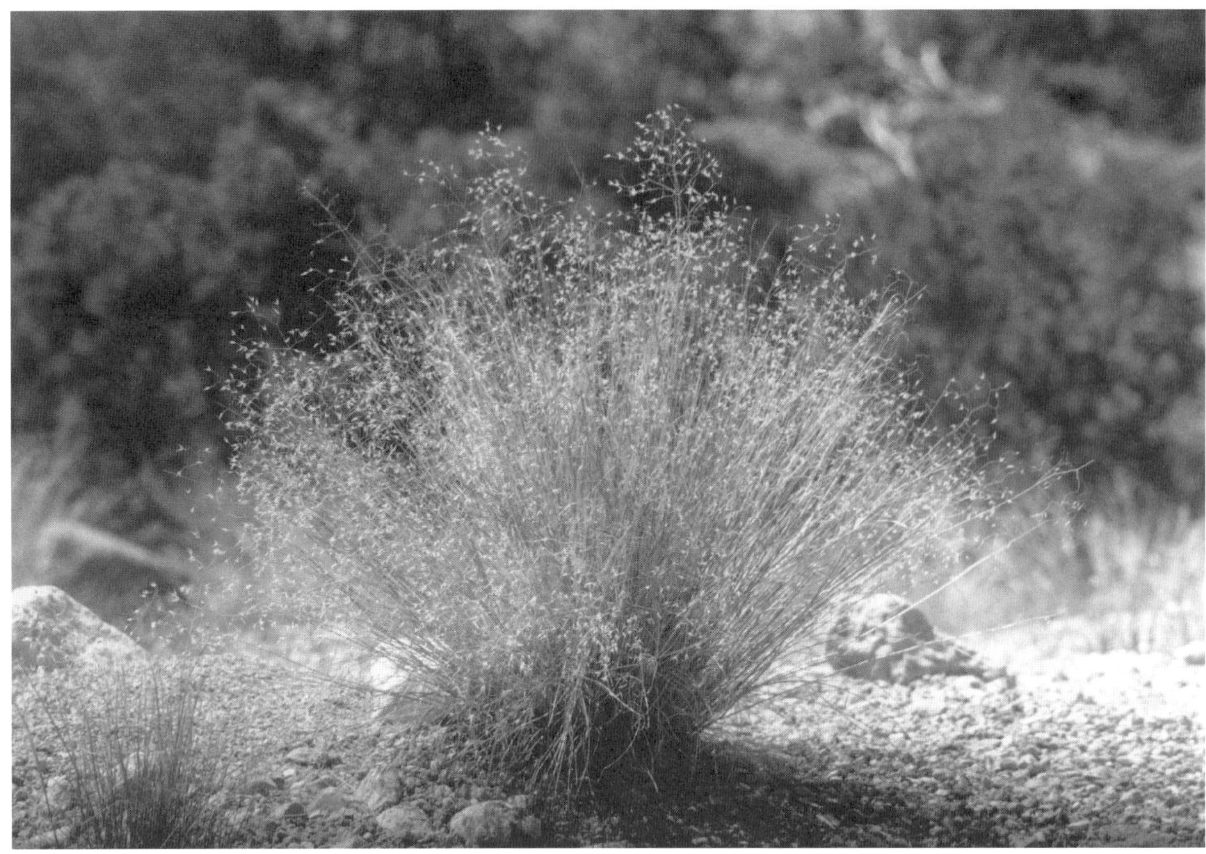

Indian ricegrass has been a staple of southwestern Indian diets since the first Clovis hunters. It was ground into a meal and also made into a drink. (James M. Aton photo)

birch, limber pines, and blue spruce lined the banks of the river and its tributaries. Also common were red osier dogwood, alderleaf mountain mahogany, wild rose, and Rocky Mountain and common juniper. The more recognizable plants would have been Mormon tea, prickly pear cactus, narrowleaf yucca, cattails, big sage, and Indian ricegrass.[3] This green, rich environment was just the kind of place that attracted Columbian mammoths, Shasta ground sloths, Yesterday's camel, and other giant animals of the late Pleistocene. For the Clovis hunters, it probably was "a veritable Garden of Eden."[4]

Although it is unclear exactly who were the first Americans and when they arrived, the Clovis hunters (named after Clovis, New Mexico, where their artifacts were first discovered and identified) remain the first verifiable group of humans in the New World. While a few possible pre-Clovis sites have been excavated by archaeologists at places like Monte Verde in Chile and Meadowcroft Rock Shelter near Pittsburgh, none of them has passed all the criteria established by archaeologists.[5] This situation is changing rapidly, and many archaeologists privately think a pre-Clovis presence will soon be accepted. Clovis points, however, have turned up in every state in the U.S. The majority of these sites lie on the Great Plains, but at least a score of them are on the Colorado Plateau.[6] One sits on Lime Ridge, overlooking Comb Wash and the San Juan River.[7]

What brought these hunters to the San Juan area apparently was the presence of Columbian mammoths and an occasional mastodon. Clovis hunters probably traveled in groups of forty or fewer, including both sexes and all ages. Although they appear to have specialized in these two large animals, they also hunted other large herbivores, such as camels, ground sloths, long-horned bison, giant short-faced bears, horses, and musk oxen. When time and opportunity presented themselves, they also probably caught rabbits, wild turkey, and other smaller animals. Wild

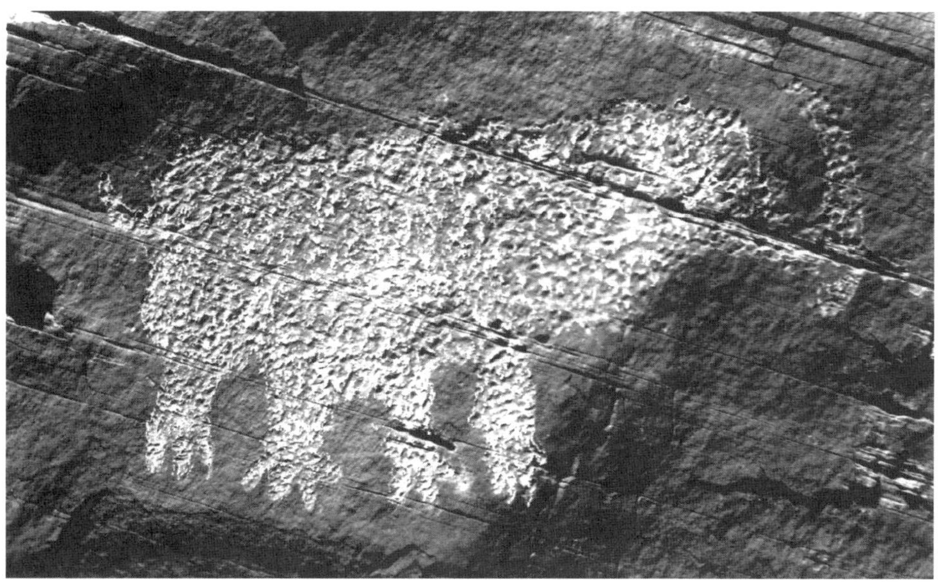

The Moab mastodon—real or fake? This petroglyph was found near Moab, Utah and then "enhanced" by its finder. Archaeologists debate its authenticity, but mastodons and mammoths did roam the Colorado Plateau until about eleven thousand years ago. (San Juan Historical Commission)

vegetables no doubt formed part of their diet during the warm season.[8] Like any hunters, the Clovis people were opportunists, but they probably preferred mammoths. Within five hundred years or less, however, mammoths were extinct. Clovis hunters may have been the culprits.

The extinction question has drawn much attention precisely because one interpretation of it is an archetypal story of the Fall. Subsequent Native American groups might come and go, like the Navajos whose sheep overgrazed the hills north of Bluff, but somehow those environmental trespasses seem less portentous. This creation story says that when people entered the Garden, they destroyed a vital, even totemic, part of that paradise: those magnificent mammoths which waded along the lush bottoms of Comb and Butler Washes. These people—the Clovis hunters—might have committed the Original Sin of the Americas.

We explore this extinction possibility in depth because it reveals crucial information about the changing San Juan environment. It shows what kinds of plants and animals inhabited the area. It demonstrates the way climatic change affected aspects of the landscape. And it throws in the human element: the application of technology to manipulate an environment, along with the cultural and ethical values that accompanied it. Whatever the exact source of the Pleistocene extinctions, this creation narrative frames an important question for the rest of this book. The complete story of the San Juan River demands that we ask not only what the river landscape looked like, but what people found and did there. One can view the mammoth-extinction story as the beginning script of a San Juan River palimpsest.

Standing twelve to fourteen feet tall and weighing upward of twenty thousand pounds, Columbian mammoths appeared in North America nearly two million years before the Clovis people. They grazed on grasses and shrubs, their flat teeth especially suited for grinding.[9] These giant creatures ate prickly pear, gambel oak, grass flowers, sedge, birch leaves, rose, saltbush, big sage, and smaller amounts of blue spruce, waffleberry, and dogwood.[10] All of these plants flourished in the moist bottoms of the San Juan and its tributaries like Comb Wash.

By the time the Clovis hunters arrived, possibly because the environment was drying out, mammoths appear to have been congregating near water sources. This seems especially true on the now-arid Colorado Plateau.[11] It may

From this Lime Ridge campsite, Clovis hunters had direct access to Comb Wash and the San Juan River via the drainage below. Archaeologists believe this high point gave the Clovis hunters a view of mammoths along the river drainages. (James M. Aton photo)

account for the Lime Ridge campsite near the San Juan; it was probably a hunting stand from which Clovis Indians stalked mammoths in either Comb Wash near camp or along the San Juan River, a short distance to the south. The Lime Ridge site was perfectly situated to give Clovis hunters a long view of these drainages, all the while staying upwind. It offers a 360-degree view of the surrounding area, and in particular overlooks a side canyon that runs into Comb Wash. This drainage was probably a corridor for animals to move between the Lime Ridge uplands and the lower riparian zone.[12]

The hunters probably ambushed several mammoths from sites like Lime Ridge. Female mammoths and their offspring would have been especially vulnerable to mass killings because elephants behave altruistically. Studies of elephant behavior in Africa reveal that if one is killed, others (especially females around offspring) will rally around, making them easier prey.[13] Other scholars believe that while Clovis hunters did not habitually kill groups of mammoths, they would have if the opportunity presented itself.[14] But kill mammoths they definitely did. The question is to what extent?

At the end of the Pleistocene, both flora and fauna underwent major changes in the Americas. As the climate warmed and dried along the San Juan, for example, plant communities started to crawl up the drainages and slopes toward the ridges and mountains, chasing a cool, wet climate. The blue spruce-limber pine-Douglas fir communities once lining the San Juan ended up on Navajo Mountain, Elk Ridge, and the Abajos. Pinyon-juniper woodland communities from the lower Sonoran and Mojave Deserts, in turn, replaced them. Desert shrub communities, likewise, took over from pinyon-juniper.[15] Plant environments were changing radically, and species of megafauna in the San Juan and elsewhere, like the much-hunted Columbian mammoth, became extinct. Was it because of climate change or due to the Clovis hunters?

For years scientists had assumed that the giant mammals of the Pleistocene died gradually because the weather patterns altered and the ensuing Holocene environment no longer supported them. Many still hold climate to be the culprit. But in 1967, Arizona archaeologist Paul S. Martin first proposed the "overkill thesis": Clovis people had hunted the megafauna to extinction. In his groundbreaking work, Martin showed that some thirty-one genera of large mammals disappeared about ten thousand years ago. He theorized that these animals had evolved without fear of human hunters. When the first hunters arrived in America, "there was insufficient time for the fauna to learn defensive behaviors."[16] The result was a hunting blitzkrieg.

In a mere one thousand years, he postulated, a band of forty Clovis hunters could have spread throughout the Americas and multiplied to over a half-million people, wiping out the vulnerable mammoths and other megafauna as they went. Unaware of what they were doing, the Clovis hunters kept pushing on to new hunting grounds, taking the easy prey; perhaps at times they even wasted much of the mammoth because there were so many. When the large animals disappeared, Martin said, populations crashed, and hunters turned to other animals and food-gathering strategies.[17] Following this massacre, mammoths, mastodons, and other giants no longer lumbered along the lush bottomlands of the San Juan, eating sedge and rice-grass. After two million years in North America, all that remains of the mammoths are piles of bones and desiccated turds. If Paul Martin is correct, these first Americans were responsible for perhaps the most dramatic of many extinctions in North America.

Not all archaeologists and paleontologists, however, accept Martin's thesis, and there is fierce debate. Many believe that the appearance of Clovis hunters and mass extinctions were a coincidence. Climate alone might have delivered the knockout punch. These Ice-Age mammals had coevolved with certain kinds of plant communities, which began to change between 10,000 and 9000 B.C. For many of these megaherbivores (large plant eaters) like the mammoth, a reduction in the kinds of plants they preferred created greater competition with other animals.[18] Moreover, the change from a two-season to a four-season year meant that many plants that mammoths browsed on no longer had a full growing season. Thus, plant diversity declined, and megaherbivores might have found it increasingly difficult to forage for the high-protein diet they needed. They would have been pushed to eat lower-protein plants with higher toxins. As a result, megafauna with conservative digestive systems would have lost out to animals which could adapt.[19] The Clovis hunters might have merely shown up at places like Lime Ridge to witness the sorry spectacle and take advantage of dead or dying animals. Another explanation postulates that the mammoths and other large mammals were on the ropes when the Clovis hunters appeared; these hunters merely delivered the fatal blow.[20]

One factor that must be considered when discussing the slippery eel of Clovis responsibility for mammoth extinction is what religious obligation they may have felt toward the animals they killed. No one will ever know. But if ethnographic comparison and contemporary hunters and gatherers offer a clue, and we can take a giant leap in time, space, and circumstance, the Clovis probably had little concern for conservation. Robert Brightman, in his study of Rock Cree relationships with the animals they hunt and trap, points out that the gods or overspirits provide the animals. The spiritual relationship with the supernatural controllers of the game, not the animals and their reproductive thresholds, determines the availability and scarcity of meat. Similar conclusions have also been reached about historic, traditional Navajo hunting practices.[21] There is no way of knowing what Clovis hunters camping on Lime Ridge thought about mammoths, leaving archaeologists plenty of opportunity to speculate.

With the collapse of the Clovis-megafaunal hunting lifeway, Paleo-Indians retooled and concentrated on hunting the long-horned bison *(Bison antiquus)*. These hunters, known as Folsom after the initial discovery of their artifacts at

Folsom, New Mexico, settled mostly in the Great Plains area, where bison congregated in largest numbers, even up to the last century. Folsom presence on the Colorado Plateau was less pronounced than Clovis. Sites near Green River, Utah, and along other riparian drainages, together with long-horned bison remains in similar places, suggest that these animals followed the lead of other megafauna: They grazed the waterways. Although the Folsom groups apparently did not bump into each other on the Colorado Plateau and the San Juan, it is quite possible they engaged in less hunting alone and more hunting and gathering combined because fewer bison frequented higher areas like the cavernous plateaus around the San Juan.[22]

The dividing line between various Folsom and Plano groups and the succeeding Archaic culture is unclear. As one San Juan archaeologist put it, "The whole Archaic period is blurred and poorly resolved."[23] Nevertheless, many aspects of Archaic lifeways can be described with confidence. Their presence on the Colorado Plateau is well established and extensive. The term *Archaic* describes a general hunting-gathering lifeway that persisted at least intermittently from 6500 B.C. to A.D. 1. This length of time alone indicates the success of this subsistence pattern. It is wrong, moreover, to assume that theirs was a hand-to-mouth existence, scavenging for every available ricegrass plant or rabbit to fend off starvation. Rather, the Archaic appear to have exploited selected animals and plants in different ecological zones.[24]

The earliest Archaic sites in the San Juan area are near Navajo Mountain in Dust Devil and Sand Dune Caves, the so-called Desha Complex Archaic, dated around 6000 B.C. Elsewhere near Glen Canyon—at Bechan Cave, on the northwestern Colorado Plateau, at Cowboy Cave, and at Sudden Shelter—and at Old Man Cave in Comb Wash, Archaic camps date to the seventh millennium B.C.[25] To the east, excavators have also found Archaic sites in the Middle San Juan basin near Chaco.[26]

Certain generalizations about the San Juan Archaic and their environment are possible. Their population waxed and waned according to wet and dry weather cycles, with a general trend toward increasing as the Anasazi period neared. Over the millennia, the Archaic evolved from concentrating on hunting, like their Paleo-Indian forebears, to gathering plants.[27] The reasons are not clear. Did environmental conditions like the altithermal (a long period of higher-than-normal temperatures between 4000 and 2000 B.C.) lead to less game? Did the increasing numbers of Archaic people result in overhunting? Did the Archaic find gathering plants a more efficient way of meeting their nutritional needs? Or was it a combination of all these factors?

The answers are inconclusive, but the questions raise important considerations about the interaction of people with the landscape along the San Juan. In general hunting supplanted by gathering is a more efficient way of supplying food. It is possible that the Archaic, over a few thousand years, unknowingly pushed game—deer, bighorn sheep, and elk—to their limit and were forced to begin gathering wild plants.

Archaic gatherers were opportunists, but they did not wander aimlessly, searching for plants to eat. They moved in a regular pattern and returned to productive areas. They scouted before gathering and possibly communicated with other bands as to prolific plant locations. In general Archaic bands in the San Juan area followed the seasons: In the spring and summer, they might camp and gather plants on dunal grasslands like those above Bluff and Montezuma Creek. Come fall they moved to the pinyon-juniper uplands near Navajo Mountain, Elk Ridge, and Cedar Mesa to hunt game and gather wood for fuel and shelter. After 2000 B.C. when pinyon became common, they also gathered pine nuts at higher elevations. Throughout the year, they probably dropped down to the San Juan and its tributaries, where plants, game, and fuel were readily available.[28]

Recent work in the Chaco River basin suggests that the Archaic employed a "mapping on" strategy. During the spring and summer, when various greens or seed-bearing plants were reaching harvest stage, Archaic bands located near a particular field of, say, goosefoot and picked its leaves. Then they moved to another area, where, for example, dropseed was maturing and picked its seeds. This high mobility, especially during the warm months, was based on knowledge of their home areas and the way weather affected certain plants' growth.[29]

In contrast to the common belief about arid lands, the high deserts and river bottoms of the San Juan country were a grocery store of plant food. Two plants in particular formed the basis of prehistoric Indian diets throughout the Archaic period and strongly supplemented Anasazi crops: chenopods (goosefoot) and amaranths (pigweed), together called cheno-ams. Interestingly both plants "pioneer" disturbed soil, areas that have been trampled by human feet or disrupted by digging. Thus, when Archaic groups winnowed seeds from these two plants, unknowingly they were replanting for the next year.[30]

Goosefoot grows in alkaline soil, making the salty greens especially tasty. Its seeds were parched and eaten dry or made into a meal. Distillation of the stems made a powerful anthelmintic that dispelled parasites. Pigweed also greens up throughout the spring and summer. In late summer, its seeds were parched, popped, and ground into a meal. Sometimes it was stirred into a drink. Pigweed produces more protein per land unit than corn; it is nutritionally superior to true cereals in protein, carbohydrates, and fat. In fact, caches of pigweed have turned up in archaeological sites worldwide.[31]

Besides goosefoot and pigweed, Archaic bands in the San Juan area collected a variety of grass seeds, especially Indian ricegrass and dropseed. Indian ricegrass continues to play a vital role in southwest Indian diets even today. Its seeds were ground into meal after cooking or parching. While this plant is not viewed as a pioneer, it has been found in disturbed sites, especially on south-facing slopes of slide areas. Although lower in starch and sugar than wheat and other cultivated grains, ricegrass yields 120 calories per ounce.[32] Dropseed is another seed-bearing grass that grows in areas shunned by more palatable grasses. Aside from their nutritional value, these grasses provided both fiber for the diet and bedding.

Many other plants produced edible seeds for the Archaic, such as cattails, fiddlenecks, and composites like the sunflower. Archaic gatherers also feasted on a variety of berries and fruits from vegetation near the San Juan, like prickly pear cactus, blackbrush, blackcap, wild rose, creeping hollygrape, honeysuckle, and serviceberry. The Indians picked other greens, peas, seeds, and roots in season. Supplemented by meat from deer, bighorn sheep, rabbits, birds, and other small game, the Archaic diet was well rounded and met all nutritional needs.[33]

Before moving on, it's important to ask how well the Archaic people succeeded. This hunting-gathering way of life lasted by far the longest in the history of human occupation on the Colorado Plateau. Contrasted with the Paleo-Indians (who shared much with them), the Anasazi, or subsequent Indian groups, as well as Euro-Americans in the Southwest, Archaic practices stand as singularly successful. Coupled with its longevity is the relatively benign impact the Archaic lifeway had on the environment. As one archaeological team asserts, the "long tenure of the Archaic in the San Juan Basin testifies to the overall success of this adaptive system."[34]

It is not clear whether the Archaic ultimately adopted farming or horticulturalists moved in from the south. If the Archaic did begin to experiment with farming—and it is clear that the cultigens (corn, squash, and later beans) as well as agricultural techniques came from the south—the question is why. Hunting and gathering, after all, had worked well for a long time. The usual answer is population increase.

Imagine the scene: An Archaic family along the San Juan near Montezuma Creek finds itself more pinched for space every year. One year a new band moves into the Aneth area, where the Montezuma Creek band has always gathered Indian ricegrass. An enterprising neighbor from Aneth shows the Montezuma Creek band that they can plant corn seeds in an alluvial fan. In late summer, if the weather has not been too dry, they can harvest the corn, eat it, and store some for lean winter times. The Montezuma Creek band asks, "How can we lose?"

In fact, it appears that from before 1500 B.C. to A.D. l, horticulture along the San Juan was a hedge against bad years, a little extra money in the bank when hunting and gathering were not paying off as well as usual. Still, using either the migration or gradualist model, this adaptation took time. Unfortunately, the archaeological record cannot tell us about all the individual decisions that bands of people made year in and year out to change to sedentary farming. Hunter-gatherers leave less garbage for archaeologists to sift through than farmers like the

Anasazi. As one archaeologist so aptly puts it, "In contrast to our lithic-based, foggy view of the ephemeral and elusive PaleoIndian and Archaic periods, the Basketmaker [Anasazi] people leap forth from their dry caves fully dressed (by Basketmaker standards), coiffured, painted, and equipped with a wonderful array of skillfully made baskets, bags, tanned hides, feather and fur robes, and tools of all sorts."[35] The leap appears sudden, but it really was not.

Throughout their tenure in the San Juan area, the Anasazi continued to supplement their diet with the wild plants and animals they had relied on during the Archaic period. The more they did so, like the Kayenta Anasazi (south of the San Juan River and west of the Arizona-New Mexico border), the healthier they stayed. Research indicates that, in general, hunter-gatherers enjoyed better health than horticulturalists because their diets were more rounded.

The wild plants eaten by the Basketmaker (1500 B.C. to A.D. 750) and Pueblo (A.D. 750 to 1300) Anasazi along the San Juan were largely the same ones Indians had been eating since Paleo-Indian times. They also added some. Their staples were cheno-ams, ricegrass, dropseed, juniper berries, four-wing saltbush, yucca, sunflower, globemallow, ground cherry, purslane, Mormon tea, pine nuts, plantain, beeplant, wild onion, tansy mustard, parsley, and buffaloberry.[36] The Pueblo groups, however, intensified horticulture and food storage, presumably because of population increase.[37]

One striking feature of Anasazi horticulture is the way building homes and especially planting fields encouraged the growth of many wild, "pioneer" plants they had always eaten. The greater the population increase and accompanying soil disturbance, the more plants like goosefoot, pigweed, sunflower, beeweed, and prickly pear cactus thrived. It was a true symbiotic relationship for the San Juan Anasazi. Moreover, evidence suggests they encouraged these weeds to grow by watering and tending them.

This same sort of symbiosis occurred with hunting. For example, at Basketmaker sites west of Bluff on the dunes above the San Juan, rabbits, deer, Canada geese, sandhill cranes, and prairie dogs, to mention a few, wandered into the fields the Anasazi planted along the river. The Anasazi then hunted and trapped these invaders to augment their diets.[38] It is hard to know if they realized the ways farming increased the production of many wild foods and animals. But given their long tenure in the area, they probably did. In general, however, throughout the whole two-thousand-plus-year Anasazi period, gathering and hunting decreased as horticulture increased. Growing populations led to overhunting and reduced the range for any one band to locate deer, bighorn sheep, and elk.

Besides hunting, gathering, and agriculture, the Anasazi grew cotton (also imported from Mexico) for blankets and clothing. Some articles, like the so-called Telluride Blanket excavated by pothunters in San Juan County in the 1890s, have survived and demonstrate extraordinary craftsmanship.[39] The Anasazi also raised turkeys to incorporate the feathers into their fur robes and use in ceremonies. Toward the end of the Pueblo III period and approaching abandonment, however, they began to eat their turkeys. This practice indicates a period of pronounced economic, cultural, and environmental stress. As one archaeologist put it, "It is like us eating our dogs."[40]

Like the Archaic, the San Juan Anasazi built homes near their crops. Basketmaker pithouses were especially wood intensive, using perhaps hundreds of pinyons, junipers, cottonwoods, or ponderosa pines for just one large dwelling. One distinguishing feature of the Pueblo period is the introduction of wattle and daub or stone into building techniques. Masonry obviously created a more permanent structure, while pithouses only lasted about ten years before termites and rot undermined them. It is possible that depleted resources hastened the change from wood to rock.

The south-facing, passive-solar position of many Anasazi masonry structures like River House (or Snake) Ruin on the river is well known, thanks in part to the budding solar-energy movement of the 1970s. These structures provided excellent solar heating during the winter when the sun was low on the horizon. Conversely, in summer the overhanging cave roofs cooled residences when outside temperatures were reaching one hundred degrees. Their use of solar energy and some apparent

River House (or Snake) Ruin along the San Juan River receives full sun at the winter solstice. At the summer solstice, it is in full shade. The Pueblo II Anasazi (A.D. 900–1100) who built this and many other houses in Four Corners country understood solar gain. (James M. Aton photo)

solar petroglyph calendars in places like Chaco Canyon and Hovenweep prove the Anasazi watched the sun closely and knew how to predict astronomical events.[41]

Another interesting feature of Anasazi farming was their sophisticated irrigation systems. Near Navajo Mountain at Beaver Creek in Cha Canyon, the Anasazi constructed intricate, rock-lined ditches to direct water from the creek into their fields. The ditches ranged in length from ten to thirty yards. Many of them featured small, tapered stones which slid in and out of the faces of other notched stones to allow water into a ditch or move it along to the next one. In addition to these complex irrigation channels, the Anasazi farmers constructed stone windscreens on the upwind sides of their fields to prevent sand from blasting their plants and drying out the soil. At nearby Desha Canyon, just east of Cha, the Anasazi also built terraced plots, which tied into their ditches.[42] In all, the Anasazi, like their descendants at Hopi, Zuni, Acoma, and elsewhere, were skilled farmers who utilized a variety of methods to water crops in a high-risk, arid environment.

What may not have been so obvious to them was farming's detrimental effect on their environment. In pinyon-juniper uplands like Cedar Mesa north of the San Juan, the Anasazi likely practiced slash-and-burn horticulture, torching trees and then clearing the stumps. For a few years, the fields produced large crops before depleted soils forced the farmers to clear a new patch. Still, for the first few years of a fallow period, an abandoned field continued to grow garden weeds like amaranths, purslane, and goosefoot, as well as shrubs with edible berries like currants and three-leaf sumac. Nevertheless, in a seventy-five-to-two-hundred-year period on Cedar Mesa, the Pueblo II and Pueblo III Anasazi

The Anasazi used check dams like this reconstructed one at Hovenweep (facing page) to catch precious water in an arid environment. (James M. Aton photo)

effectively destroyed the arable lands they created by slashing and burning. These methods probably shortened the duration of their occupation and hastened abandonment of Cedar Mesa before A.D. 1300.[43] Forest depletion also occurred at Chaco National Monument and probably contributed to the Anasazi's demise there near A.D. 1150.[44] Unlike those on Cedar Mesa, Chaco's surrounding forests never recovered, probably because populations along the Lower San Juan were smaller.

In the Dolores River area, not far from the San Juan, the Anasazi's razing of forests led to the loss of sage grouse, disruption of large-game migration, increased erosion, and sage and wood depletion in general. Likewise, farming at the Coombs Site near Boulder, Utah, markedly reduced pinyon, juniper, and sage during Pueblo occupation. "Environmental degradation," the Coombs Site archaeologist writes, "is an apt description of its severity."[45] In short the Anasazi's intense use of wood for fuel and structures greatly affected the forest ecology and erosion in these areas.

This same sort of environmental impact was felt on the San Juan River. Unfortunately, the Middle and Lower San Juan have not attracted the intense scientific scrutiny of the Chaco or Dolores areas. A Basketmaker III site west of Bluff, however, demonstrates some interesting facts about erosion. Cutting cottonwoods and reeds to construct wood-intensive pithouses probably intensified the bank erosion along the San Juan which followed Basketmaker III times.[46] All told, San Juan Anasazi horticulture probably had a substantial impact on the ecosystem. Still, the Anasazi did not fundamentally reduce the carrying capacity of the land. Historic activities like logging, mining, farming, and grazing have altered the landscape "much more than any prehistoric impact."[47]

One of the most discussed aspects of Anasazi culture, of course, is the general abandonment

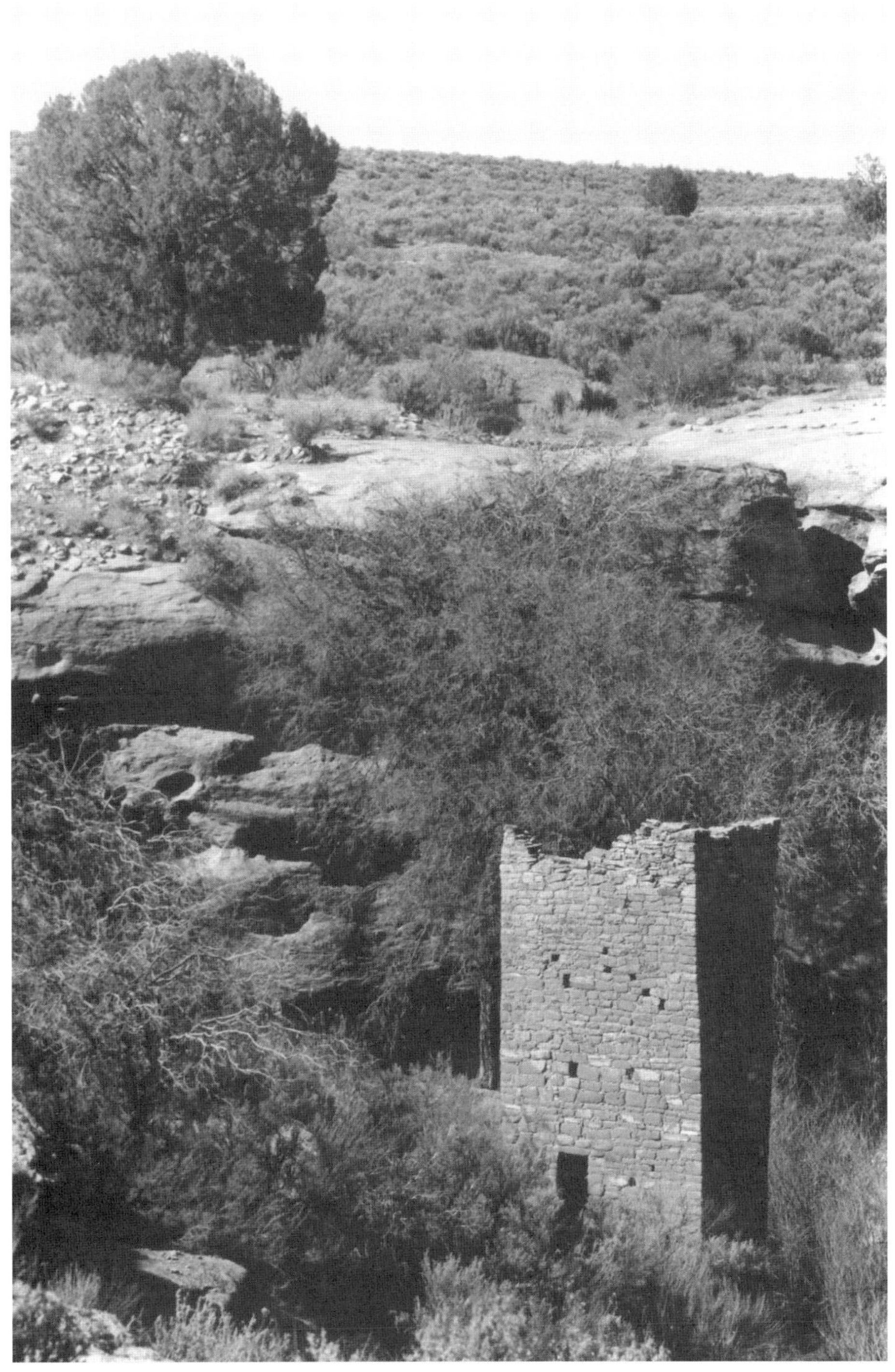

of the San Juan River circa A.D. 1300. It is well known that many Mesa Verde Anasazi (north of the San Juan) and Kayenta Anasazi (south of the San Juan and west of Chaco) migrated south to settle on the Hopi mesas. Other Anasazi groups moved east to live along the Rio Grande. The prevailing question remains why did they leave the area? It does seem, as one archaeologist put it, as if "someone should have stuck around."[48]

Nearly all the hypothetical answers relate in some way to the environment. The discovery of tree-ring analysis by A. E. Douglass in the 1920s gave archaeologists an especially valuable tool to measure rainfall in a particular year. Dendrochronology in the San Juan country

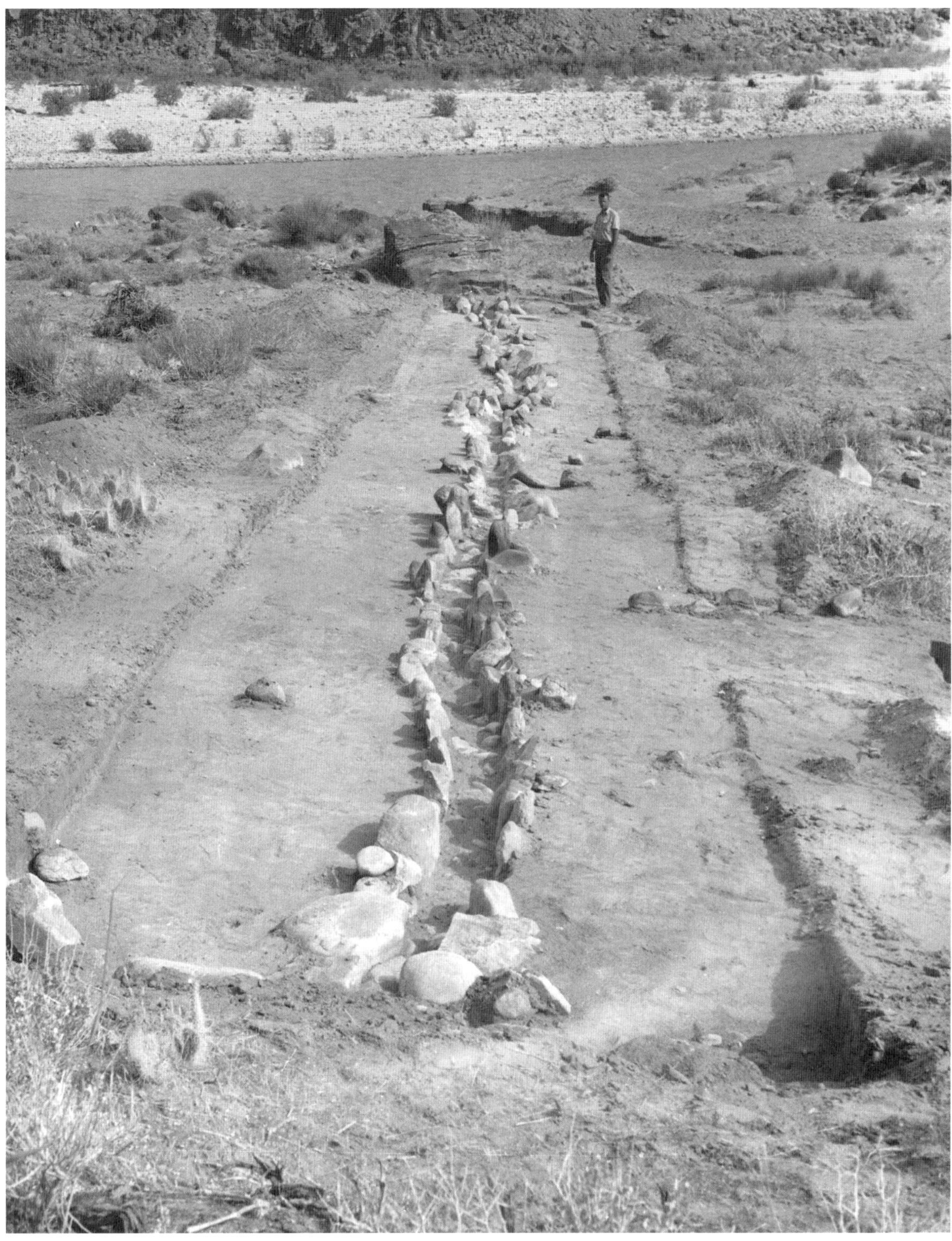

The Pueblo III Beaver Creek Anasazi community (A.D. 1100–1200) used a variety of techniques to irrigate their corn fields, including this rock-lined ditch which took water directly out of the San Juan. (Museum of Northern Arizona Photo Archives, NA 7175)

shows that a great drought persisted for at least fifty years through 1276+, apparently pushing Anasazi farmers out of marginal areas. Besides the great drought, another theory advanced is arroyo cutting due to environmental degradation. As already noted, deforestation at Chaco would have contributed to arroyo cutting, leaving the fields high and dry. The extent of arroyo cutting elsewhere in the San Juan drainage, however, is less clear.

Early scientists like A. V. Kidder have suggested that warfare was the deciding factor in abandonment. Pueblo III structures seem to have defensive postures. Moreover, there is increasing evidence that the Numic-speaking Paiute probably occupied the Lower San Juan by at least A.D. 1300. But even if there was no conflict between the Anasazi and these Paiutes, there is evidence that warfare among Pueblo groups increased during the Pueblo III period. That warfare, says Steven A. Leblanc in his provocative book, *Prehistoric Warfare in the American Southwest,* directly resulted from climate change in the thirteenth century: "in the 1200s . . . the climate deteriorated significantly and warfare became virulent. . . . Indeed, there is some evidence that this Late Period of intense warfare was not just a pan-Southwestern phenomenon but a pan-North American phenomenon as well."[49]

Other theories which archaeologists discuss but find difficult or impossible to document are "the bright lights theory" and the "religious revolution theory." The former postulates that when crops continued to fail, people tended to congregate where there were more potential marriage partners, more social activities, and more crop surpluses. In other words, they moved to places like the Hopi mesas and along the Rio Grande.

The second theory has been articulated by, among others, the Navajos, who later moved into the territory the Anasazi abandoned and whose name for their predecessors ("enemy ancestors") was adopted by archaeologists. They believe that the Anasazi were a brilliant culture which went astray. As one contemporary Navajo elder put it, they "shriveled and died because the people transgressed the laws of the holy beings and of nature as they sought ease through the power which they abused. . . . A holy way gone bad."[50] Some archaeologists concur, speculating that the religious life of the Anasazi might have grown too extreme, too abstract, too involved in something that had nothing to do with the land. It might have become a system too rigorous to contend with problems that occur with agriculture in a marginal area. In some ways, this theory meshes with the Zuni idea that the San Juan Anasazi moved because they were looking for a center place where they could regain spiritual balance.[51]

All the environmental stress factors—drought and arroyo cutting—could easily have been part of that cultural-religious transformation. Unfortunately, western science does not have very good tools for measuring prehistoric social and religious change. It seems obvious that environmental factors alone could not have caused such complete abandonment; a change in religious systems may be the only way to account for it. But scientific methods do not help to interpret such a change.[52] No doubt the factors were complex and interrelated.

The problem with studying abandonment of the San Juan area is that we have only physical evidence to map the actions of these highly religious people. The Clovis and Folsom people who first loomed large on the landscapes of North America left very little indication behind of a spiritual life, yet they must have had one. The earliest rock art in the Southwest that is firmly dated is the so-called Archaic abstract style. Close to the San Juan River, this Archaic rock art was first found around the base of Navajo Mountain and is now under the waters of Lake Powell. These panels date from between 2000 and 6000 B.C.[53] The figures suggest they were largely the work of men because of the subject matter: hunting (sheep), religion (kachinalike figures), weaving (design motifs), and farming (maize and sunflowers).[54] The especially high number of sheep represented indicates a hunting shamanism similar to the split-twig figurine complex in the Grand Canyon.[55]

First discovered in the Grand Canyon in the late 1930s and then elsewhere on the Colorado Plateau, the split-twig figurines have generated a flood of commentary and speculation. They provide one key to understanding the psychic relationship between the late Archaic and their landscape. They may even tell us what these people thought and felt about their prey. These figurines have turned up in California and Nevada as well as the Colorado Basin.

The Grand Canyon figurines, however, which have been dated to 2000 B.C., raise the most puzzling questions because of small, pointed sticks piercing the bodies of the animals. A number of factors, especially their location in isolated caves not used for habitation, suggest that the figurines represent deer or bighorn sheep which the Archaic ritually killed prior to the hunt. As one scholar put it, "If a miniature figure of the animal to be hunted were ritually killed, success would be more certain in the actual quest."[56] Much of this is speculation based on analogy with hunting cultures worldwide. It is possible the figurines had more prosaic functions, but the spearlike sticks certainly indicate they were something more than toys or dolls.

Besides the split-twig figurines, rock art from that period suggests the same kind of hunting magic was being pecked on sandstone walls. In the Lower San Juan-Glen Canyon region, petroglyphs of bighorn sheep and deer were probably part of the same hunting and ceremonial traditions as the split-twig figurines.[57] Sheep, deer, and other animals have continued to be depicted in southwestern Indian art to the present. While the Archaic cultures turned increasingly toward plant gathering and the Anasazi toward horticulture, it is clear that hunting did not diminish in psychic importance. Stalking, killing, and eating animals loomed large in the religious lives of most native cultures in North America.

Perhaps the power of Anasazi rock art reveals itself most dramatically near the confluence of Butler Wash and the San Juan River on the so-called Kachina Panel. Because it has so many different kinds of figures, the panel is an outstanding example of the variety of Anasazi rock art and what it says about its makers' relationship with the environment. The huge, trapezoidal human figures seem to be shamanic. Some are phallic, suggesting an association with sexual potency; others contain small, humanlike figures and are probably female. The spectacular headdresses also hint at shamanic flight.

Rock-art scholar Polly Schaafsma believes that the anthropomorphs "not only had ceremonial impact" but "they were probably representations either of supernatural beings themselves or of shamans. Images such as these may have been thought to contain the soul force of the beings they represent. The many hand prints around or in the torso area . . . support this possibility; they . . . identify the supplicant who had offered prayers to, or through, the beings portrayed."[58] In other words, the hand prints said to the spiritual powers, "I made this offering. Please recognize it." The bighorn sheep and yucca plants on this panel emphasize the sacredness of the twin subsistence activities of the late Archaic-early Basketmakers: hunting and gathering.

Much like the rest of Anasazi culture, Pueblo rock art seems literally to have exploded around the San Juan and Colorado Plateau. This may have had something to do with increased population and sedentarism. As populations grew and consolidated, multiclan villages developed. Some of the rock art may have helped different clans maintain their separate identities during a time of increasing social complexity. Besides the actual clan identity, certain symbols apparently documented who "owned" which fields, check dams, and so on.[59]

Even with the shift to sedentary horticulture and decreasing numbers of game, both the depiction of sheep on rock walls and studies of contemporary Pueblo Indians reveal that a lot of social and ceremonial organization still went into hunting animals for food and other uses. Horned sheep have always had supernatural significance for San Juan Indians. Horns not only suggest shamanic power but are also associated with one of the most widely known kachina figures, Kokopelli, the humpbacked flute player.[60] Known by his Hopi name, this figure first appeared on rock walls around A.D. 1000, during Pueblo II times, with his flute, humpback, and phallus. Yet earlier flute players appear in rock art from Basketmaker III times. This figure may have been significant even in late Archaic times.[61]

In Hopi mythology, Kokopelli is a kachina figure associated with increased rain, crops, and fertility. He plays his flute over springs to attract rain clouds. Additionally, he is a hunting magician and often appears with sheep and deer. Sometimes he carries a bow and arrow rather than a flute. His hump may contain babies,

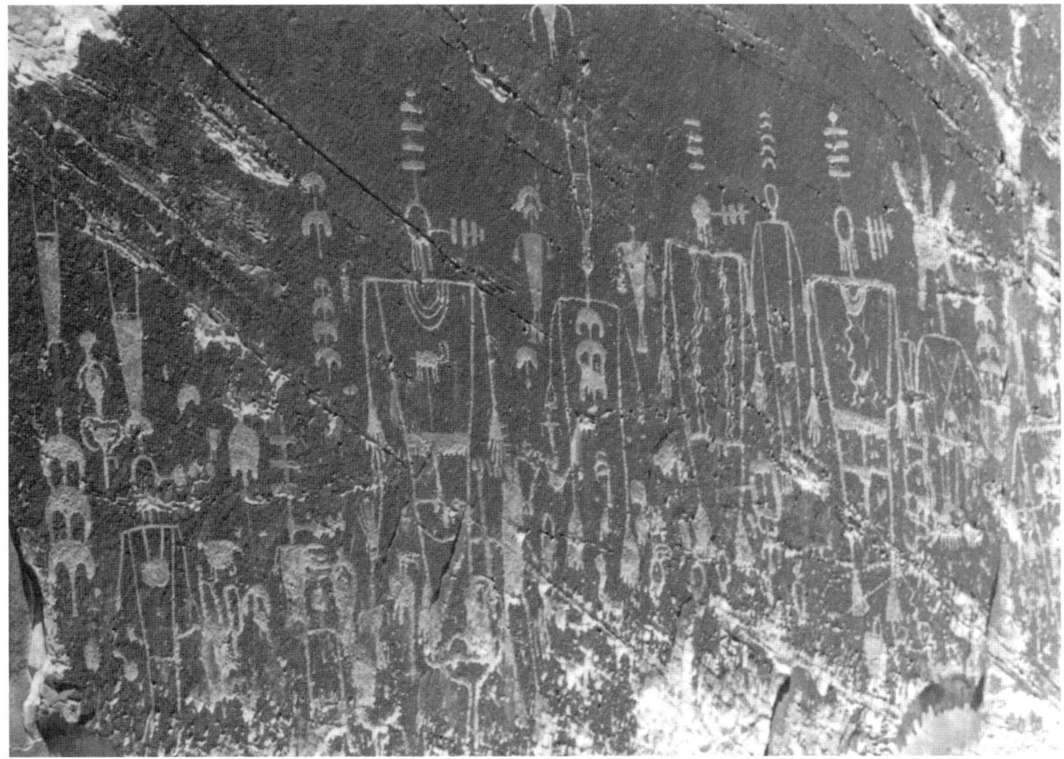

The Kachina Panel at the confluence of Butler Wash and the San Juan displays some of the most spectacular rock-art anthropomorphs in the Southwest. They date to between A.D. 50 and 500, the Anasazi Basketmaker II period. (James M. Aton photo)

blankets, belts, or seeds. These he gives to the women he seduces. Thus, both his humpback and phallus are associated with fertility and procreative powers. In many ways, Kokopelli may be compared to the trickster archetype, who, in spite of unrestrained sexuality, changes from an unprincipled, amoral force to a creator who brings order and security, in the form of meat and corn, into the world.

Less prominently depicted than the male Kokopelli is a female Kokopelli Mana figure. While not Kokopelli's wife, she shares his spirit in terms of sexuality and fertility.[62] Kokopelli and Kokopelli Mana appear to have had major ritualistic and religious significance for the Anasazi. This society believed success in hunting, raising crops, and producing offspring—all of vital ecological importance—depended on these figures' sacred help.

Most other natural features of Anasazi life were also depicted on rock walls along the San Juan: corn, badgers, bear tracks, dogs, stars, crows, suns, frogs and lizards, mountain lions, rabbits, turkeys, water skates, snakes, and ducks. The water skate, known as Tekeowati or "the mother of animals," was seen in visions by Hopi who were thinking of game. Snakes were symbols of water and, hence, prosperity and abundance—similar to their associations in planting cultures worldwide. They also help hold the world together because they are magically associated with gravity. Ducks have long been connected with shamanism. In the Rio Grande Pueblo world, they serve as seed bearers and messengers to the rain clouds of the four sacred directions and the gods. Also, inversely, holy beings may assume the form of a duck.[63]

In conclusion, rock art was intricately tied to the way the Anasazi made a living from hunting, gathering, and farming. In ways we will never know, the Anasazi depicted their sacred relationship with plants and animals on the walls, the "canvases," where they lived. The rock art of the San Juan River and elsewhere in the Anasazi world shows not only some of the changes in their way of life but the manner in which they attempted to cope spiritually with ecological and environmental changes. Since the San Juan Anasazi have persisted for over three thousand years, including their modern-day counterparts at Hopi and elsewhere, it is clear that religious figures on rock walls have been a factor in their survival.

Ultimately, the story of prehistoric Indians' relationship to the San Juan landscape is very complex. As we learn more about these peoples' interaction with their environment, the story will become both clearer and more complex. Some elements of the story are known, however, and probably will not change. For at least twelve thousand years, Indians along the San Juan River gathered the same plants for food and other uses. They continually hunted virtually every animal available, from megafauna like the mammoth to very small game like the rabbit. Some, like the mammoth, they may have driven to extinction. Other populations, like deer, they may have altered. In general, hunting remained a strong element in the spiritual lives of Anasazi even as corn, beans, and squash filled their stomachs.

While farmers now do not have the luxury of moving so easily to another region when weather patterns change, prehistoric Indians did so for nine to ten thousand years. Their flexibility with climate serves as a valuable lesson about adaptability. Prehistoric Indians along the San Juan sometimes made unwise land-use decisions, but they generally had a relatively benign impact. In most cases, land and animal populations can recover from neglect and abuse—the pinyon and juniper on Cedar Mesa, for example. The environmental history of prehistoric Indians shows just how hard it is to live well in a landscape as arid as the San Juan. Even when a culture has intimate knowledge of a region's ecological processes and components, and a mythology to match, it still struggles to endure.

2 NAVAJOS, PAIUTES, AND UTES: *Views of a Sacred Land*

Close to the time (roughly A.D. 1300) when the Anasazi abandoned their alcove dwellings and floodplain farms for lands south of the San Juan River, the tribes that would be present at the start of the historic period arrived to take their place. Fortunately, because of written records and a healthy oral tradition, there is a much better understanding of the importance of the river in the lives of these Native American groups: the Utes, Paiutes, and Navajos. All three tribes took a physical, pragmatic stance toward the river, encouraging use of the riparian ecology in a high-desert environment. They also, however, held strong beliefs about its spiritual powers, based upon mythological teachings. What follows is an overview of traditional Native American perspectives that reflects a mundane, yet sacred, relationship between the land and its people.

Let's begin with a brief sketch of these peoples' prehistory and early history. The Numic-speaking Paiutes and Utes were the first to arrive on the brown waters of the San Juan. Anthropologists argue about when the ancestors of these people set foot in the Four Corners area. Some believe there were two different migrations of Numic speakers, one around A.D. 1 and the second around A.D. 1150. The latter movement generally coincides with Anasazi abandonment of the San Juan basin, but evidence of turmoil between the two groups is sketchy. Other anthropologists believe the Southern Utes came much later; most agree that by the 1500s, both groups were well established in the region.[1]

By historic times, the Southern Utes comprised three bands: the eastern-most group was the Muache, who lived in the Denver area; the Capote ranged through the Sangre de Cristo Mountains of Colorado and south to Taos, New Mexico; and the Weeminuche hunted and gathered on lands bounded by the Dolores River in western Colorado, and in Utah, the Colorado River to the north and west, and the San Juan River to the south. All these groups were highly mobile and journeyed far into the Great Basin, throughout the Colorado Plateau, and onto the plains. The Weeminuche Utes dominated southeastern Utah, playing the most critical role along the San Juan River.

The Paiutes shared a cloudy prehistoric past with their linguistic brothers, the Utes. At the time of early white contact, sixteen identifiable bands comprised the Paiute tribe, with the San Juan being the only group to occupy lands south and east of the Colorado River. Perhaps this is why their name has been translated as "people being over on the opposite side" or the "San Juan River people."[2] In southeastern Utah, the San Juan Paiutes lived close to the Weeminuche. While Southern Paiute territory centered in southwestern Utah and Nevada, its most eastern members, the San Juan Paiutes, pushed into Monument Valley on the Utah-Arizona border. So it is not surprising that the historical record tells of groups of these people living at the base of Navajo Mountain, in Monument Valley, at Douglas Mesa, in Allen Canyon to the north, and around the Bears Ears and Elk Ridge, whose canyons drain into the San Juan. Intermarriage between Utes and Paiutes creates even greater confusion in separating the two groups. Southeastern Utah was truly a mixing pot, in every sense of the word.[3]

The major distinction between the Utes and Paiutes in this area was a cultural, not a linguistic,

This Ute petroglyph along the San Juan River was etched during historic times and emphasizes mobility, an essential characteristic of Ute lifestyle. (James M. Aton photo)

one, brought about by the environment and the technology related to it. In white documents and correspondence, the Utes and Paiutes of southeastern Utah are often described simply as Paiutes. From a more scholarly point of view, the Paiutes operated in family groups, and only infrequently, when resources allowed, came together as bands. They hunted and gathered in an austere desert land and had no centralized chieftain, collective religious practices, or common goal (other than survival) to unite them.

The Weeminuche Utes shared many of these characteristics but were generally able, because of a richer environment and access to the horse, to operate in larger, more-cohesive groups. The farther east one traveled, the more the Ute culture took on a Plains Indian look. The Utes in the Lower San Juan area used brush wickiups (characteristic of the Paiute culture) in the summer and elk and deerskin tepees (identified with the Plains Indians) in the winter, suggesting this cultural mix. To the white settlers, there was little or no distinction between Utes and Paiutes on the Lower San Juan. For ease of identification, this book will simply refer to the two groups in this region as Utes.[4]

These peoples' interaction with the land spoke of deep cultural ties. Though not as well documented as some historic groups, the Utes named places and endowed the land and its creatures with significance. They also had a descriptive classification system that helped locate a spring, canyon, or resource.[5] Thus, names for the San Juan River included Water Canyon, River Flowing from the Sunrise, and Lower River (compared to the Colorado River, known roughly as Cedar Trees and Canyon Runs through It).

Canyons that join the river and places around it had similar names. For example, there were Greasewood or Sagebrush Canyon (Montezuma Canyon), Slick Rock Mound (Comb Ridge), Red Wash (Cottonwood Wash—the water runs red when it rains), Down by the River (Bluff), Two Rocks Canyon (Cow Canyon), Water Runs Every Day through There (Recapture Canyon), and Bitter Root or Many Yucca Mountain (Sleeping Ute Mountain).[6]

The life of a nineteenth-century Ute, before intensive white contact forced drastic changes, was tied closely to the rhythms of nature. The People followed a seasonal pattern of migration that was carefully bound to the plants and animals ready for harvest. Not surprisingly, water and grass played a dominant role. The People selected campsites based upon the availability of springs, streams, and rivers for drinking water, grass for livestock, and firewood and trees for shelter and preferred lower elevations to avoid the deep snows of winter. As the deer moved down from higher elevations in the late fall, the People followed the same pattern,

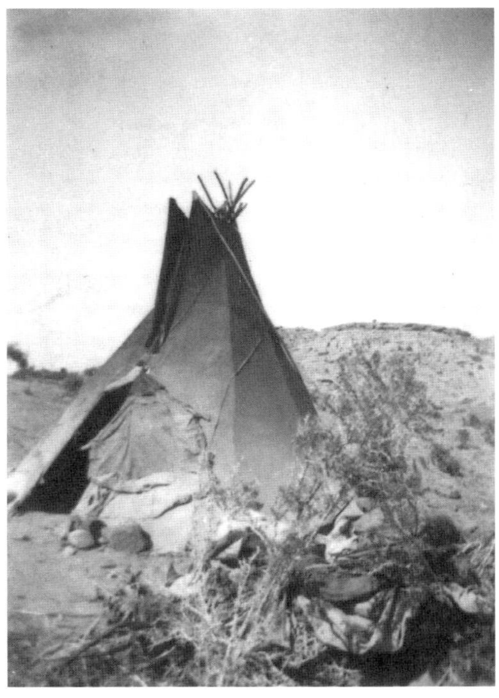

Southern Ute tepees, made of deer and elk skins, served as winter quarters for the people as they sought lower elevations and protection from winter storms. (San Juan Historical Commission)

descending to valley or canyon floors where shelter and food were available.

This natural cycle was incorporated into the descriptive names for the seasons. For example, fall was called "leaves turning yellow," winter "heavy snow" or "hard-times month," spring "snow melting," and summer "leaves coming out" or "much warmer for growing things." The three spring months had specific titles: March—"warm days beginning," April—"green grass appearing," and May—"mother of the two preceding months." The People started to move back to the mountains "when the doves sound soft."[7]

The Utes established their winter camps in locations such as Montezuma Canyon, with its neighboring Cross, Squaw, and Benow Canyons; Dry Valley, Harts Draw, Beef Basin, Westwater and Cottonwood Canyons, Butler Wash, White and Douglas Mesas, and along the San Juan River—especially near Bluff, Recapture Canyon, and Sand Island. The east side of Comb Ridge, where the winter sun warmed the rocks, was a favorite camping area that extended down Butler Wash all the way to the San Juan River. As the weather became milder and grasses appeared, streams like La Sal, Deer, Coyote, Two Mile, Hop, Geyser, Taylor, and Beaver flowed out of the La Sal Mountains, and Spring, North and South Montezuma, Cottonwood, Recapture and Indian Creeks poured off of Abajo or Blue Mountain, as it is known locally. Numerous springs such as Dodge, Piute, and Peters also invited the Utes to scatter and camp as they searched for food.[8]

Favorite areas to plant small garden plots in corn, beans, squash, and melons were Montezuma and Allen Canyons, Indian Creek, Paiute Farms, and Paiute Canyon. Favorite hunting places for deer and other large animals were the La Sal, Blue, Navajo, and Sleeping Ute Mountains and Elk Ridge, while antelope were stalked in the Dry Valley area. Elk, desert bighorn and mountain sheep, wild turkeys, rabbits, badgers, beavers, bears, and fish enriched the diet. The women provided many of the edible wild plants, including pine nuts, chokecherries, yucca fruit, Indian ricegrass, wild onions and potatoes, sunflower seeds, bulrushes, serviceberries, and raspberries.[9]

The technology to work this environment evolved with time. Ute homes reflected the mobility of a hunting and gathering society. Deerskin and elk-hide tepees were later replaced with canvas tents, with an average diameter of fourteen feet. Brush wickiups, in a four-pole pattern or with poles leaned against a tree, provided shelter in the summertime.[10]

Information on fishing techniques is sketchy. Accounts indicate fishing was a male activity, but fish were part of the general diet, with certain restrictions. Northern Utes, who are

Comb Wash was an excellent place for harvesting Indian ricegrass; the other side of Comb Ridge (background) was a favored winter camping spot because of its exposure to the sun and warmer temperatures. (O. C. Hansen Collection, Utah State Historical Society)

closely allied in beliefs and practices to their southern relatives, used weirs of willow screens to direct fish into shallow waters to be speared or shot with barbed arrows. Fishing lines of braided horsehair with a bone, wood, or later a metal hook, as well as squawbush nets, provided the angler with other tools for capturing his prey. If not eaten immediately, the catch was dried, placed in deer or elkskin sacks, and stored underground in a dry place for future use.[11]

While fish were not a mainstay in the Southern Utes' diet like deer and other animals, they were important enough to be incorporated into some taboos. For example, for thirty days following childbirth, a mother could not eat meat or fish without spoiling her husband's chances of obtaining game. Likewise, if a woman ate fish during menstruation, she permanently damaged her male relatives' hunting ability.[12] In an animistic universe, rules prescribed acceptable interaction with nature.

Herbs and plants that grew along the river and in its surrounding canyons were also an important part of the Ute lifestyle. Today Ute informants bemoan the loss of knowledge about plant use for both food and healing. One gets the impression that all the world was once a combined pharmacopoeia and storehouse. Comb Wash was a favorite place for harvesting Indian ricegrass; the Bears Ears supplied pine nuts; in the washes and along the San Juan River, the inner layer between the bark and wood of the cottonwood tree provided a sugary sweet and food extender, and serviceberries made a tart condiment or mush. When sickness struck, Ute patients drank tea brewed from sagebrush leaves; a sore throat was cured by boiling pinyon sap with grease, then applying it to the neck; the roots and flowers of sandpuff remedied stomach and bowel problems; spearmint leaves cured an upset stomach; and gumplant served as a cough syrup.[13] Nursing mothers who wanted to wean their children rubbed masticated sagebrush leaves on their nipples.

The People, as part of the larger ecosystem, often struggled for survival as life went through cycles of feast and famine. Family groups remained relatively small, joining together for hunting and gathering in the late spring and hunting in the fall. Each of these bands had a leader, selected because he made wise decisions about where to obtain food and how to keep the group out of trouble. The size of these groups varied from one to ten families, but as the People lost more and more land to white encroachment, they were forced into larger concentrations, primarily in Montezuma and Allen Canyons.

Often each band also had a spiritual leader, who understood the supernatural powers associated with the land and how best to appeal to them. He went to these "power points" during the appropriate season, and, on behalf of his group, prayed, left an offering, and asked for help. The individual members of the band also prayed, but not at the power point used by the medicine man.[14] Different types of spirits lived in caves, rocks, springs, rivers, and mountains and helped or harmed depending upon the way they were treated.

The world was much more than just a physical realm to sustain life. It was a gift from the Creator of All Life, Sinawav, imbued with spiritual powers. Myths and tales tell of supernatural, mystical experiences, filling the Four Corners region with a power and sense of divine meaning predating contemporary times. One story tells that Sinawav became lonely and so formed fish of different sizes and shapes from the small end of his staff, then gave them the breath of life. Next he took leaves from various trees and tossed them in the air, creating different types of birds. From the center of his staff came animals like deer, rabbits, coyotes, desert bighorn sheep, and other creatures. He believed that he had done well, but as he watched the strong prey upon the weak, he decided to create one more animal—the bear—from the large end of his staff. To this animal fell the responsibility of maintaining peace and interpreting and teaching the rules of harmony to the other animals.[15]

Another story tells about the origin of the large canyons of the Four Corners area. During the time when animals and gods talked, Hawk and Sinawav went hunting together. Sinawav caught many more rabbits, making Hawk jealous, so a conflict ensued. Hawk let out a piercing scream that shook the earth, cracked its crust, and fragmented it into the canyon system that exists today.[16]

One of the most interesting mythological beliefs that ties directly to the San Juan River is about Pa' ah a pache (Water Boy), sometimes called a "water baby" or Roams along in the River. Descriptions of what this creature looks like vary. Some people say it resembles a fish with long black hair and a flowing mustache. Others say it has legs like a man instead of a tail. There are stories of both male and female water babies, one version telling what happens when a young man sleeps by a river. He may wake to find a beautiful woman in a green dress lying next to him. After he sleeps with her, she may lure him into the water to remain with her people.[17]

There are many other accounts of this creature's activity in the rivers and lakes of Utah. One tells that a woman left her baby strapped in a cradleboard by the river, then went off to do her work. While she was gone, a water baby removed the infant and climbed into the cradle. The mother did not realize what had happened, nursed the water baby, and was swallowed by the creature. Another story relates that two female water babies pulled a man into the river and took him to their home beneath the water. They wanted to marry him, but he thought they were ugly and eventually escaped. Water babies cry like humans and are often heard near the river. They supposedly accept tobacco for smoking and haunt a person's dreams when they are mistreated, but they can also be playful, especially with older people. They also have the power to raise the level of the water temporarily.[18]

Water babies exist in the San Juan River. Local tales claim they can walk on land as well as swim in water. A human baby should never be left by the river, or it may be lost; when people camp by the water, they hear the water baby crying, but when someone goes to investigate, it slips away undetected; and people should avoid going down to the river at night. The Utes took frequent baths but only in the shallows. Even though they considered themselves good swimmers, they often held onto the mane or tail of their horse when crossing deep rivers. Swimming unassisted in deep water was considered dangerous because a water baby could pull a person down in a whirling funnel of water to drown.[19] Recent sightings of water babies have occurred in the Colorado River.

A more powerful people called the Diné or Navajo eventually joined the Utes and Paiutes along the banks of the San Juan River. Scholars still debate when they entered the Southwest. Some argue that by the fourteenth century, the Diné, or "the People," were migrating into the Four Corners region as the Anasazi departed. Navajo lore is replete with stories of interaction between the two groups. Most anthropologists agree that by the end of the 1500s, the Diné were spread throughout northern New Mexico, a portion of southern Utah, and part of northern Arizona. They also concur that the Navajos migrated from northern Canada with other Apachean peoples, who are linguistically related as Athapaskan speakers. Studies suggest northern groups separated from those migrating south around A.D. 1000 and that the division between Apaches and Navajos occurred about three to four hundred years ago. However, these are only rough estimates.

Navajos reject these theories, claiming there is nothing about a land bridge across the Bering Straits and subsequent descent from the north in their oral tradition. Instead, their religion teaches that they traveled through three or four worlds beneath this one and emerged in the La Plata Mountains of southwestern Colorado or the Navajo Dam area of northwestern New Mexico. The gods created the four sacred mountains—Blanca and Hesperus Peaks in Colorado, Mount Taylor in New Mexico, and the San Francisco Peaks in Arizona—intending them as supernatural boundaries within which all was safe and protected. In addition, the gods also established four sacred rivers—the Rio Grande, Colorado, Little Colorado, and San Juan—to be defensive guardians.

In addition to its religious importance, the San Juan River also acted as a line of demarcation between Navajo and Ute territories, although there were exceptions. Groups of Paiutes lived in the Monument Valley and Navajo Mountain areas which were south of the river, while small bands of Navajos hunted, gathered, grazed sheep, and lived north of it. The historic record indicates that generally, however, the San Juan River was a territorial boundary during aboriginal times.

Like the Utes, the Navajos were interested in the rich resources of a riparian environment. But unlike the Utes, who often traded for agricultural products because they practiced horticulture only on a minor scale, the Navajos depended heavily on corn, beans, and squash. The fact that the waters of the San Juan were being used in many different ways even in aboriginal times is important in understanding the later cultural and ecological history of the river.

Beyond agriculture, natural plants flourishing along the river's banks and in tributary canyons were also intensively used. Navajo informants provide excellent information about Native American use of river plants. Wild onions, turnips, squawbush, Indian ricegrass, Rocky Mountain beeplant, and goosefoot offered a supplement to their diet of corn, beans, squash, and mutton. Cottonwoods lined the river and were used for cradleboards, fire drills, and summer cooking because their wood gives light but not much heat. Rabbitbrush steeped in water alleviated coughs, colds, headaches, and menstrual cramps and made a yellow dye for wool, while sagebrush rid sufferers of indigestion, the pain of childbirth, cold swellings, and tuberculosis.[20]

Many older Navajos remember that the banks were "thick with squawbush and cottonwoods," that "there were plenty of plants used as medicine herbs," and that, "in the spring, one could see the vegetation's rippling waves across the meadows every time the breeze blew."[21] Another person described that

> the main wash from [the mouth of] Montezuma Creek all the way to Hatch [approximately twenty miles] was filled with cottonwoods. Up on top [of the mesas], the greasewood bushes were big with huge stems. They grew higher than the hogan in some places. The horse trails went under and through this tangled top brush; it was that thick and high. But it is not like that now.[22]

This abandoned hogan, photographed in 1921 at the mouth of Chinle Creek, testifies to Navajo occupation of this important crossing site of the San Juan. (Hugh D. Miser photo, #565, U. S. Geological Survey)

Now a lot of the natural vegetation along the river and in tributary canyons has either been washed away, removed by people, or choked out by the newly imported tamarisk, leaving only a few large cottonwoods dotting the sides of the river.

This is also true of some of the fauna. Attracted by the large cottonwood stands, beavers built their homes along the banks. They were said to be plentiful until the Navajos killed them so that "medicine men could use the skin in their medicine bags" and as material for clothing worn in the Yeii'bichai ceremony.[23] The scent from the beaver's castor invoked the holy beings' power during prayers. Raccoons, said to be doctors, also inhabited the thick vegetation along the river, while prairie dogs and rabbits preferred more open spaces and provided Navajos and Utes with meat, as did the antelopes on the plains and the deer in the Sleeping Ute, Blue, La Sal, and Carrizo Mountains.[24]

Until recently, Navajos did not eat fish from the San Juan River. This may be attributed, in part, to a story in which the Navajos fought their cruel taskmasters, the Anasazi. The Navajos drove their enemy into a big bend of the river, but to avoid capture, the Anasazi leaped into the water and were transformed into humpback fish.[25] Eating fish was taboo and was definitely not allowed after a person had a No-toah (Waterway) ceremony performed. The holy beings and creatures associated with water would be offended.

In addition to its resources, the river also supplied both a thoroughfare and a barrier. Although the river was an easily recognized boundary for Navajos and Utes during the 1860s, some Navajos still ventured beyond it and settled in Ute country around the Aneth-Montezuma Creek region, the Bears Ears, and Navajo Mountain. Dry summers facilitated travel across the river because it shrank so that people could walk or ride to the other side. During the high-water stages, Navajos tried to avoid fording the river, but if it was necessary, they crossed holding onto their horses. Oral testimony indicates that boats were rarely used by Navajos, and when they were, it was only for crossing, never for traveling any distance on the river.[26]

Once across the river, a traveler faced a network of trails that crisscrossed the high-desert country. This trail system fed into locations near Montezuma Creek, Aneth, and the Four Corners Monument, partly because the way was easy and partly because of the existence of a series of canyons, comprising McCracken, Montezuma, Allen, and McElmo to the north, and Desert Creek, Lone Mountain, and Tsitah to the south. Farther downriver, where canyon walls made access increasingly restricted, there were firm-bottomed crossing sites at Sand Island, Butler Wash, Comb Wash, Mule Ears (Chinle Creek), Goodridge (Mexican Hat), Clay Hills, Paiute Farms, Copper Canyon, and Trail Canyon/Wilson Creek. Minor paths connected the major network of trails that laced the barren stretches of high country to mountainous or other well-watered sites.[27] On rare occasions, such as in 1918 when the thermometer dipped to thirty-two degrees below zero, the river became a frozen road and shortened the distance between trading posts. One trader remembers the wagons almost pushing the horses along as they skidded over the ice between Aneth and Montezuma Creek.[28]

The Navajos, like the Utes, gave place names to thousands of geographical features throughout the Four Corners region. Often one place had two or three different titles, not all of which were generally known. Names could be derived from mythological events, personal experience, the type of resource available, a historic occurrence, the shape of a land feature, or where certain people lived.

Take the Aneth region, for instance. This area played a key role in the history of Navajos living along the Lower San Juan. Because McElmo Creek empties into the river near a wide floodplain suitable for planting crops and travel, it was natural for people to congregate here to plant crops. T'aa biich'iidii is its most popular name, derived from the government farmer, Herbert Redshaw, who lived there in the early 1900s. He walked slowly, deliberately, in an almost-robotic fashion; hence, one explanation of his name is that it means "just his devil or ghost within." Another is that he used to cuss and tell people to "go to the devil," while a third asserts that he was as "slow as the devil." Whatever the reason, the name stuck and has become the official title of the Aneth Chapter. Another place name for Aneth is Big Ears or Wiggling Ears, a description of a trader with a prominent physical feature. Still other names tantalize with the stories they imply, such as Barely Enough Pep to Make It and A Good Place to Stay Away From. Aneth is also known as Black Mountain [Sleeping Ute] Wash [McElmo Creek] Joins In.[29]

Montezuma Creek is called Where the Sagebrush Wash Drains into the River but also has other epithets such as Black Hat, alluding to Bill Young, who established a post there; Mosi or Cat, after an earlier trader called Old Cat; Flew Back Out, and Large Eyes. Some place names in the Aneth-Montezuma Creek area are associated with economic activity, such as Among the Prairie Dogs, because Navajos transplanted these animals to add to their food resources. Other spots are called Clay (used in ceremonies), Spring in the Sour Berry [Squaw] Bush, Gather Yucca, and Corn Bush.

Place names between the Four Corners Monument and Montezuma Creek also mark events, such as Soldiers' Crossing, given during the 1906 Bai-a-lil-le disturbance; Reclaiming the Horses, in remembrance of a woman who caught some Utes stealing her horses so she whipped and scolded them; and To Look at One Another, bestowed on a trail on a hill that was narrow enough to make passersby acknowledge each other.[30]

As Navajos settled this area, geography also helped establish limits for land use. One Navajo tells that her two relatives, Woman from Blanding and Old Gray, came over a hill above Montezuma Creek and outlined the boundaries of their new home. Woman from Blanding declared, "From that juniper-covered hill to White Point, down the gray ridge to Stair Formation Rock, and across to Fallen House—this is how big our land will be."[31]

In one case, the action of the river even suggested a name. According to Cyrus Begay, a Navajo elder who has lived in this region for close to a century,

> [the San Juan] would rise, causing some erosion of the banks and washing the trees and vegetation away by their roots. This vegetation accumulated in certain parts of the river, creating dams higher than this hogan and causing the river to take an alternate path. Before too long, the riverbed had widened. Just this side

Travel is an underlying motif in the Navajo worldview. These men, captured in this turn-of-the-century photo, had come into town to trade at the Bluff Co-op (background). (San Juan Historical Commission)

> [eastern end] of Montezuma Creek is a place called Revived Vegetation. This spot was formed in two years after the river switched to the other side, giving it a chance to thicken with assorted green vegetation. It was beautiful. But after a few years of occasional flooding, the area washed away. This is its [section of the river's] history.[32]

Yet beyond the physical resources and dynamic shifts of the river, there lies a fascinating body of lore, based on mythology and spirituality, that is deeply rooted in Navajo thought. Since everything is connected within the Navajo universe, to speak of the river as a single, separate entity does violence to prevailing viewpoints. On the other hand, references to San Juan River appear in many of the myths, which provide the basis for Navajo interaction with the river. Here is a summary of pertinent aspects of these beliefs.

Navajo tradition tells that the People lived in either three or four worlds (depending on the version of the myth) beneath this one. In the preceding worlds, everything was created spiritually before it was conceived physically, including the San Juan River. Indeed, the four rivers that bound Navajo lands today were all in place in the world beneath this glittering world. When the holy beings entered this sphere, they brought the knowledge and materials to recreate a physical replica of the world they had left and imbue it with animistic forces.

Water was the force that caused the Navajos to abandon the previous world. One account of this story central to Navajo beliefs tells that Coyote, the traditional trickster, stole Water Monster's two babies. Water Monster (Teehooltsodii—One Who Grabs in Deep Water) controlled all the waters in the earth as well as those on the surface and, when he recognized the theft, flew into a rage. He opened all the gates that held back the waters and successfully flooded the entire fourth world. Coyote, along with the other inhabitants, fled before the wall of water.

Eventually, through trial, error, and sacrifice, the People found a way into this world.

They then discovered that Coyote had concealed the water babies in his coat and the flooding waters were sent as revenge for his action. The Diné returned the babies to their parent and offered *ntł'iz*—a ceremonial gift of precious stones and shell—to appease Water Monster.[33] Implicit in this story is the suggestion that these creatures are associated with rain. Sacred offerings of *ntł'iz* at springs and rivers may summon desired moisture which Water Monster controls, as do other holy forces in nature.[34]

This abbreviated account of the creation story is important because it introduces Water Monster, a creature whose offspring inhabit rivers, lakes, and oceans. In this world, the main Water Monster resides in the ocean to the east (Atlantic) and is chief of the Water People there. The mythological First Woman is said to have recognized some types of fish, clams, crabs, seals, and other forms of water life as her neighbors in the world below this, so Navajos today do not eat them because they could be friends from an earlier time.[35]

Water Monster lives in a home within the depths of a body of water. Spinning, funnel-shaped whirls are entrances into his chambers, where he drags his victims. Outside his home is Water Monster's pet, a water horse *(teeh lii*—'deep water pet' [horse]), which is a guardian. Water Monsters have fine fur like an otter and horns like a buffalo, while their young may be spotted with various colors. Some people say they look more like a buffalo or hippopotamus. Water animals such as beavers, otters, muskrats, fish, frogs, and turtles, as well as waterfowls, live within the domain of Water Monster and are not eaten, though otter and beaver skins may be used for clothing and rattles. A turtle shell with pebbles also makes a good rattle. Even a sheep, an animal free from most restrictive taboos, cannot be eaten if it has drowned in the river.[36]

The Navajos have a deep respect for the power of water, lightning, and other natural forces. One story relates that a mythological hero, Monster Slayer, visited the home of Water Monster and demanded back all the people who had been drowned, struck by lightning, or lost in quicksand or marshes. Water Monster had no desire to let them go, so Monster Slayer set the water on fire and forced their release. The people were ecstatic over their newfound freedom; Water Monster only grumbled that he would "take some of your people once in a while," thus explaining what happens to those struck by lightning or drowned today.[37]

The Waterway ceremony removes the effects of a damaging experience an individual has had with drowning, near drowning, or dreams of drowning. The mythological basis for the ceremony explains that a man visited Water Monster to beg release of a drowned grandson. The captive, as well as the rescuer, was covered with green slime, but both were finally released. Frog, Turtle, Otter, Beaver, and the Thunder People performed a bathing ceremony that eventually cleansed the captives from the limiting effects of the slime.[38] This ceremony is still performed today.

Another story tells of a mythological character named He Who Teaches Himself, who journeys down the San Juan River inside a hollow log fashioned by the holy beings and protected by clouds, rainbows, and other supernatural aids. After a series of adventures, the hero is brought to Water Monster's home, freed only after the gods intervene, and returned to his normal state by Frog, who shows him how to prepare a special cigarette. It is painted black for Water Monster, blue for the water horse, yellow for otters and beavers, and white for frogs and great fish. When a person nearly drowns, he or she smokes this specially prepared cigarette to alleviate the water sickness.[39] Not everyone is fortunate enough to escape the effects of the river and water creatures. The San Juan River has claimed its fair share of lives. In 1993 a Navajo teenager was swimming and drowned in the river at Mexican Hat. Law enforcement officials, river rangers, and community members made numerous attempts to recover the body but failed. Religious leaders in the area believed the drowning represented Water Monster taking one of the People home to his kingdom as a sign that the Navajos must return to traditional ways. This view is part of the teachings concerning life on the river.[40]

The San Juan River is not only destructive, however; it is also portrayed as a helpful, protective power. For instance, it is designated as one of the four sacred rivers and marks the northern boundary of Navajo lands. Known as Old Age River, Male Water, One with a Long

Body, and One with a Wide Body, the San Juan has been described as an old man with hair of white foam, a snake wriggling through the desert, a flash of lightning, and a black club of protection to keep invaders from Navajo lands.[41]

The river has a spirit of its own that can be asked for help. Many older people today stop their cars and offer corn pollen as they cross the water. Charlie Blueeyes, a longtime resident near the river, explains,

> This water can hear you. You offer it corn pollen when you are going for something, such as buying a horse, as a shield against harsh words said to you, when going to play cards at Towaoc (Utes), on a hunting expedition, or just traveling around. When you are on foot, you say, "I am going over you, my grandmother." You do not tell it I am going into you. You put the corn pollen on the edge of the river. The river is holy.[42]

Other people tell that they "plead with the river's holy being," that "the holy people right there are listening," and that "all nationalities—white, Mexican, and other Indians—would not discriminate" against the traveler, whose wishes will be answered.[43] Another says, "It is our boundary or shield. Corn pollen is given to it to bring good health to the mind and body and your transportation, whether it is a horse or an automobile. . . . You sprinkle the pollen with the flow of the river. When you are coming back, you use the other hand to sprinkle the corn pollen because it is like you are traveling."[44]

This concept of the river and other topographical features serving as a shield is a common motif that runs throughout Navajo thinking.[45] Nowhere is this more dramatically revealed than in the teachings about Navajo Mountain and Rainbow Bridge. Karl Luckert, a specialist in Native American religions, published a series of interviews with Navajo elders and medicine men in *Navajo Mountain and Rainbow Bridge Religion*. A brief overview of some of these teachings illustrates the intensity of religious thought concerning the river and environs.

The San Juan is considered a male river, the Colorado a female, and where they join (Water Comes Together) near Rainbow Bridge is the place where clouds and moisture were physically created. Prayers and offerings of corn pollen and *ntł'iz* prompted the holy beings to bless the land with water and provide protection from non-Navajo enemies. Thus, Protectionway ceremonies focus on this area because of the mythological teachings linking the mountain, the arch, and the river.[46]

The sacredness of this area and the canyons bordering the San Juan nearby is attested to by both Navajo and white observers. Ernest Nelson, a prominent medicine man from the Shonto area, commented,

> The Black Club [San Juan River] was laid down in the north so that people other than the Navajo people would be prevented from wandering about in this sacred area. And even we [the Navajo people] are not to wander into those sacred places without a purpose. And if we do [go there, we should do it] only in a prescribed manner, by placing offerings and by speaking ceremonial prayers at places which were put there in those times by the holy people.[47]

Historical testimony indicates these beliefs were practiced. Walter Mendenhall, a miner on the Lower San Juan in the 1890s, noted that it was very difficult to induce Navajos into the canyons bordering the river. He explained, "We never could get an Indian to go down with us into a canyon. They hear the rocks rolling down there and say it is the Great Spirit. They attribute the noise from rolling rocks to a supernatural cause and seem to believe that the canyons are inhabited by spirits."[48]

Today the situation has totally reversed. Because of the dammed waters of Lake Powell, the junction of the San Juan and Colorado can only be guessed by medicine men, so no creation of new water can occur. The rise and fall of the lake create concern about the erosion of the base of Rainbow Bridge and the possibility of collapse. And most importantly, the high waters of the lake and a new boating dock near the bridge have made access by tourists an easy, pleasurable adventure but a frustrating experience for traditional Navajos, who in the past worshiped here.

Indeed the canyons of the Glen Canyon Recreation Area, instead of serving as a shield against foreign elements, act like a magnet to draw crowds of vacationers to this sun-soaked, redrock country. For example, a marina in

Traditional Navajo thought teaches that Rainbow Bridge was formed by the gods and held cloud and moisture-producing powers. (San Juan Historical Commission)

Forbidding Canyon near Rainbow Bridge became "the largest waterside gas station west of the Mississippi River and the single most profitable Chevron station anywhere."[49] Because of congestion, it was moved in 1984 to the less-restricted Dangling Rope Canyon. Still, this has not deterred the growing swarm of visitors to the bridge. In 1997 there were approximately 180,000; in 1998, 196,000; as of September of 1999, 210,000, or roughly a 10 percent increase each year.[50] The future portends more of the same.

Court decisions in 1974 and again in 1980 gave no help to the Navajos trying to protect the bridge. Business and the waters from the Glen Canyon Dam held sway over the ruling, which said that Rainbow Bridge would remain accessible to the public. In 1995 a small group of medicine men, youthful Navajo advocates, and sympathetic whites closed entry to the bridge for four days to renew this sacred site for worship through blessing.[51] But generally, it is no longer desirable for ceremonial use. The holy beings have fled, and in their place, or at least accompanying them, is a growing politicization of Native American religious rights.[52]

Although as many as one thousand boaters a day visit Rainbow Bridge, small groups of Navajos, San Juan and Southern Paiutes, and White Mesa Utes voice increasing opposition to this abuse of a sacred site. Park Service signs and rangers can request respect but cannot prevent tourists from wandering beyond boundaries and off established paths, littering, and in other ways showing disregard for traditional land ethics. Even the shin-high wall built in 1995 to keep people contained does not stop those determined to do what they want.[53]

What answer is agreeable to both sides of the issue? Court cases attempting to enforce protective elements of the American Indian Religious Freedom Act (1988) in other parts of the country have been generally unsuccessful. Even the short-term closure of the bridge to tourists in 1995 drew strong opposition from a number of groups. Perhaps education grounded in mutual respect will prove the most effective means to change things. As the public becomes more sensitive to Native American beliefs, a greater tolerance for practices will follow. It is all a matter of perspective.

In summarizing the traditional attitudes toward the San Juan River by Utes and Navajos, a spiritual, religious view emerges as strongly as a pragmatic use of riparian resources. For both groups, the two approaches were not separate. The gods were as much a part of the physical realm as water, minerals, and the dynamic forces of nature. Just as human beings are composed of spiritual and physical sides, so, too, is the river. That is why a resident from Navajo Mountain described the river in one breath as a male body of water loaded with spiritual significance, and in the next, told of its physical wealth. He concluded by saying, "Similarly that is how our life is, and life is progressing. Birth and growth: This is what the river represents. This is how it is told. It is not just a river that flows."[54]

3 Exploration and Science: *Defining Terra Incognita*

Navajo, Ute, and Paiute sacred views of the San Juan River and its environs were about to meet their greatest challenge when the Spaniards arrived in the eighteenth century. The ways in which the Indians eventually adopted European ways of life, however, were slow and selective. In fact, the process was indirect at first because these Spanish and later American explorers never settled in the San Juan area. Nevertheless, the exploration of the San Juan basin by Spaniards and Americans from 1765 to the mid-twentieth century forms an important precursory chapter in the story of Anglo exploitation of resources that began in the late nineteenth century.

European and American exploration of the San Juan occurred during what historian William H. Goetzmann calls "the Second Great Age of Discovery."[1] An outgrowth of the European Enlightenment, this age marked the emergence of science, whose prime objective was no less than a complete empirical rendering of the planet and its peoples. Material progress was equally important. The exploration of the San Juan by geologists and archaeologists in particular contributed significantly to unraveling the great scientific issue of the later nineteenth century—time. In that sense, those scientists thrust the San Juan onto an international stage. In the mid-twentieth century, scientists with the Glen Canyon Survey put it there again. Their work established benchmarks for ecological studies and archaeological salvage operations.

A full hundred years or more before that first group of scientists and even Anglo settlers came to wrestle with the San Juan, however, the Spanish ventured up from Santa Fe in search of silver, slaves, and converts. Spaniards and Mexicans crossed and skirted the river and commented about the area, but generally stayed away from it. During the early nineteenth century, American trappers operating out of Taos and Santa Fe penetrated the area in search of beaver. They spent considerable time along the river—more in the Upper San Juan—and probably hastened erosion by overtrapping. Unfortunately, their comings and goings are poorly documented. Soon after, the first in a long series of United States military and scientific expeditions set out to explore, map, and catalogue the resources of the San Juan region. The legacy of those largely government-sponsored expeditions continues today. Their progeny—various federal agencies—still have jurisdiction over the area. Intermixed and sometimes connected with government expeditions have been numerous archaeological explorations that have helped publicize the area not only to the rest of the nation but to the larger world beyond. In all, scientists have mattered most in San Juan exploration.

Besides knowledge and its practical applications, the ideas informing these pursuits need to be considered as well. At the same time that these explorers were traversing and studying, formally or informally, one of the most difficult-to-penetrate landscapes in North America, they were evaluating it, commenting upon its features, and passing on that information to prospective settlers. In short, the exploration of the San Juan area between 1765 and the mid-twentieth century established the contradictory terms by which we still measure the basin today—wasteland, treasure trove of resources, adventureland, home, and sacred space.

Although these explorers did not remain very long, their stories are important vignettes in the narrative sequence of the San Juan environment. The Spanish, for example, had a very direct effect on the Utes' impact on the landscape because they introduced horses and perhaps guns. The horse greatly expanded the Utes' range and pushed them from a hunting-gathering lifeway, which emphasized the latter, to one which relied more on the former. It increased the Utes military strength, as it did their southern neighbors and frequent enemies, the Navajos. Guns compounded that strength. To a lesser extent, other kinds of trade goods undoubtedly changed some Ute, Navajo, and Paiute subsistence patterns. The acquisition of cloth, metal goods, and even foodstuffs altered the way these Indians interacted with their environment. Agriculture and herding profoundly affected the San Juan landscape.

Spanish exploration of the Southwest in general and the San Juan in particular appears more important today than it actually was. Now that the region, especially the canyon country, has become scenically and scientifically famous, Spanish exploits have captured some of the attention. But as historian Stephen J. Pyne has written, the Spanish did not make the canyon country famous. Instead, they surveyed and explored the edges of the region, then turned their backs to it. Moreover, what they knew and wrote about it stayed locked away until twentieth-century historians began combing archives in Mexico and Spain. Spanish exploration in the New World was marked by the outlook of the Catholic Reformation and hence was conservative and suspicious by nature. Of all the European nations affected by the Enlightenment and science, Spain, Pyne claims, was the "most retarded in its capacity to absorb its discoveries within the context of the new ideas and new sensibilities that raged across the rest of Europe."[2] What maps, diaries, reports, illustrations and other scientific discoveries Spanish explorers and scientists produced largely ended up lost, unpublished, or secreted away in royal archives. The Spanish Enlightenment imploded because of prevailing attitudes that generally hid geographical information from enemies and because of the influence of the French Revolution. Outside of Spain, few contemporaries in the European scientific world read the considerable data amassed by conquistadors.[3]

The first known Spanish entrada into the San Juan corridor took place in the summer of 1765, when Juan Maria Antonio de Rivera explored the area north and west of New Mexico in two separate expeditions. The first took place in June, the second in October. His reports commented on the landscape near the river and the surrounding area but gave the general impression of inaccessibility and unsuitability for settlement. Rivera had obtained an official license from Governor Tomas Velez Capuchin. Royal order prohibited trade with the Utes, probably because Spain was interested in converting Indians, and traders often reflected some of the worst aspects of Catholicism. In reality traders had preceded Rivera into the area because he obtained guides from one group which had already contacted the Utes. These guides were probably part of a covert group of contraband traders operating out of northern New Mexico.[4]

Governor Capuchin ordered Rivera to search for the source of a silver ingot that a Ute Indian had brought into Abiquiu. Word had come from cash-strapped Madrid to the Royal Corps of Engineers in the New World: Locating mineral sources was a growing priority. Rivera's mission formed part of that effort to replenish the royal coffers with gold and silver. The expedition also had the equally important but hidden goal of military reconnaissance: locate the great river (the Colorado), find a way across it, scout for settlement opportunities, and establish relations with the Indians who lived on the far side.[5] According to historian Iris H. W. Engstrand, Spanish exploration and science in the New World always had a very pragmatic goal: to improve everyday life.[6] In this way, it prefigured much of the work of American reclamation scientists more than one hundred years later.

Rivera's 1765 trips took him into southwestern Colorado, followed by southeastern Utah. During the first foray, a detachment led by Gregorio de Sandoval possibly worked south from the Hovenweep area to the San Juan River near present-day Aneth and Bluff. There they were greeted by a group of Weeminuche Utes, whose three encampments stood on the south side of the river. When the Spaniards appeared, "one of them [Indians] dove into the river to see who our people were. At the same time one

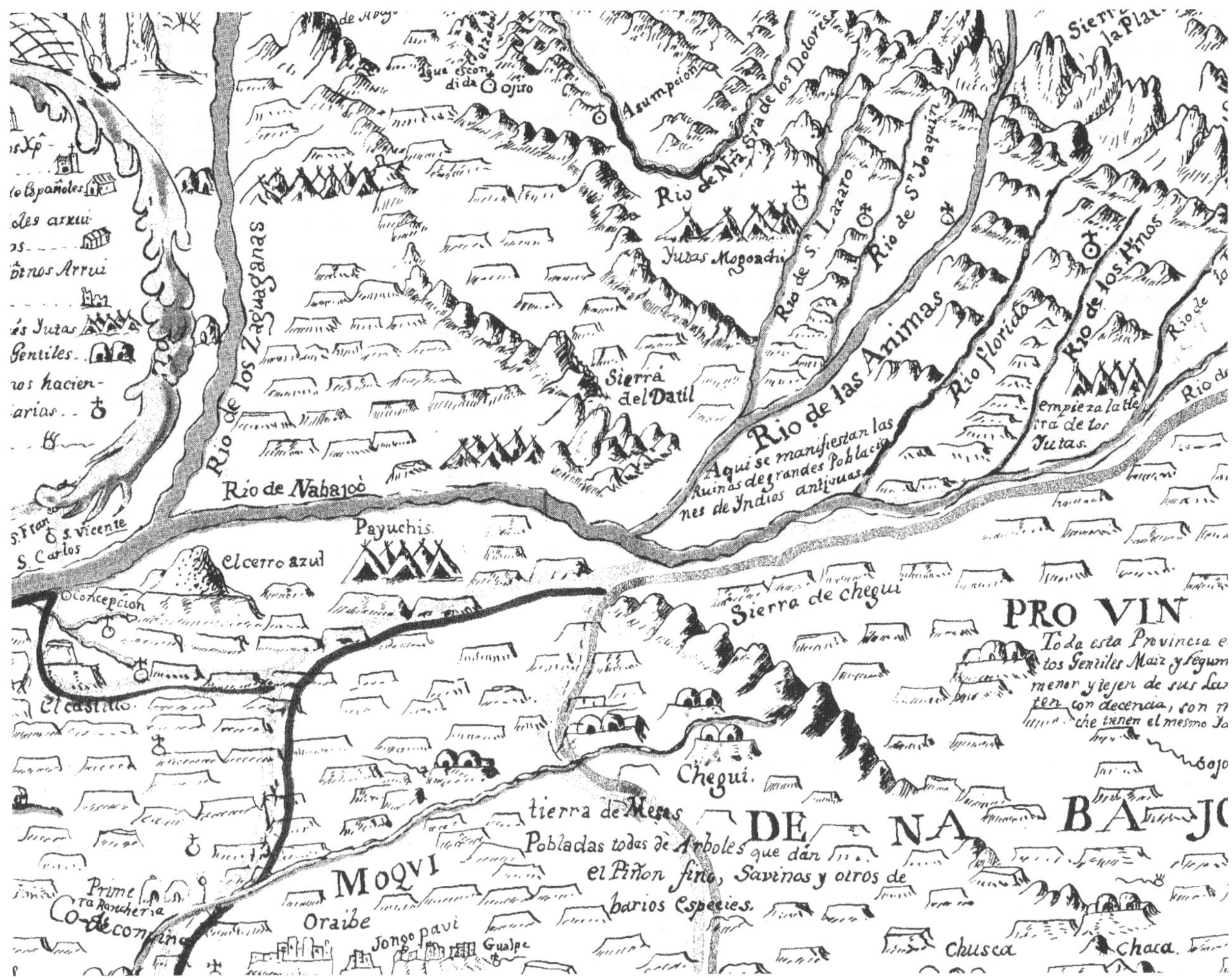

The most important scientific product of the Domínguez-Escalante Expedition, this section of the map by Captain Bernardo de Miera y Pacheco accurately shows the relationship of the San Juan (confused with the Rio de Nabajoo) to topographical features and Indian tribal areas. (Utah State Historical Society)

of ours dove in and they met in the middle of the river . . . ours persuaded him . . . to cross to our side and converse."[7] Their eagerness to talk to their visitors suggests lack of fear; the Utes probably wanted to trade, reinforcing that other traders had preceded Rivera into the area.

On the second expedition, Rivera penetrated the canyon country all the way to the Colorado River at the present site of Moab. He found no gold and silver, but his report provided valuable topographical and anthropological information to his superiors. Moreover, these reports shed light on the way eighteenth-century Spain viewed the Four Corners landscape: They hoped as always to locate mineral wealth and make converts to Christianity; they probably had a geopolitical interest in finding another route to the West Coast (even though Junipero Serra would not travel to California for another four years); but they did not view the area as a potential settlement. Rivera's October-November route eventually became part of the Old Spanish Trail, developed during the next century. That trail brought many Mexicans into the San Juan area, increasing knowledge of the river basin and its inhabitants. But none of Rivera's discoveries seeped outside the Spanish world.

The immediate benefactors of Rivera's topographical information were insiders, fathers Francisco Atanasio Domínguez and Silvestre Vélez de Escalante, who skirted the San Juan area in 1776. They initially crossed the river near the present New Mexico-Colorado border. The natural tendency of explorers like the fathers was to follow a river, but they kept on their northward course, apparently aware of the impassibility of the San Juan canyons. On their return to Santa Fe, they passed within forty miles of the confluence of the San Juan and Colorado when they forded the latter at Padre Creek. Now under Lake Powell, this famous spot was known as the Crossing of the Fathers for years. The expedition struggled through this slickrock area but eventually limped back to Santa Fe via the Hopi villages in Arizona.

The expedition's most important scientific accomplishment was a set of fairly accurate maps of the Colorado Plateau by Captain Bernardo de Miera y Pacheco, a retired military engineer who accompanied the padres. Although de Miera mistakenly identified the San Juan as a tributary of the Navajo River (Rio de Nabajoo) rather than the other way around, he showed that it ran east-west out of the San Juan Mountains. He indicated how the river related topographically to various mountain ranges like the Abajos (Sierra de Abajo) and tributary rivers like the Los Pinos and Animas. He also correctly located the tribal areas of the Payuchis (Paiutes), Yutas (Utes), Nabajoos (Navajos), and Moquis (Hopis) in relation to the river. Finally, he clearly showed where the San Juan emptied into the Colorado. It was a wonderful piece of work which actually had some influence outside Spain. The great German geographer, Alexander Humboldt, apparently saw a copy of a de Miera map in Mexico City and included some of its features in his *Political Essay on the Kingdom of Mexico* (1810).[8] This was a rare instance of Spanish science crawling out from behind its rock into European light.

De Miera's maps included the names Rivera gave to many of the area's rivers—Animas, Dolores, and, most importantly, the San Juan. Also significant from a contemporary point of view were these Spaniards' comments on the landscape, which mirrored the shift in thinking about nature, especially wild nature, that was taking place in Europe and the New World during the late eighteenth century.[9]

Since the Renaissance, western thinking had presumed that nature was made for human exploitation. This attitude still largely prevailed with Rivera and Domínguez-Escalante. Rivera was looking for trading routes and silver, while the padres were searching for trading routes and souls. But as for landscape aesthetics, a subtle but revolutionary change was occurring in the West, one reflected in these Spaniards' writings. In the seventeenth and eighteenth centuries, the most beautiful natural scene for Europeans was a humanly modified one—the neatly plowed field, the symmetrical hedgerows, the grazed pasture. This kind of human order demonstrated nature's usefulness.

Both the padres and Rivera reflected this idea when they described the beauty and utility of certain natural scenes around the Colorado Plateau. For example, the priests commented extensively on the attractiveness of Cedar Valley, Utah (San Jose), because it was well watered and thus farmable. But they also appreciated the wild, redrock formations in Paria Canyon, downstream from the San Juan, as "a pleasingly jumbled scene."[10]

A decade earlier, Rivera had continually remarked about the beauty of flowing water, lush meadows, and striking vistas. That he did so at least seventeen times on his first expedition is especially striking given the sparseness of his journal. In his first encounter with the San Juan near Pagosa Springs, Rivera enthusiastically described the valley as "a river very much larger than the last one, much wider, very lovely and fast flowing, which we called the *San Juan.* It has many meadowlands, well-provided with grasses." On his second expedition, after he and his men crossed through a mountain pass near Placerville, Colorado, Rivera waxed romantically rhapsodic, saying, "There we stopped and viewed the vastness of its beautiful valley, its meadows with various springs that flow directly west."[11]

Both the fathers and Rivera, then, in their descriptions of the landscape represented their times in appreciating both the ordered agricultural landscape and the wild, unordered scene. At the beginning of the eighteenth century, Europeans had valued the regularity of nature, seen in man's orderly imprint on the land. But by the end of the century, wild nature had assumed greater prominence as a place to experience the most sublime and intense emotions. Whether they realized it or not, both Rivera in his spare diary and Domínguez-Escalante in their expansive journal reflected evolving ideas about nature as they described the landscape on their respective journeys: utility and pure aesthetic delight. In many ways, they embodied the competing opinions that persist about the San Juan country.

Economics and politics, not aesthetics, however, dominated affairs in New Mexico. After Mexico achieved independence from Spain in 1821, official policy toward the San Juan/Ute area changed. Northern New Mexicans were eager to trade with both the Indian tribes to the north and Americans. Royal decrees had previously forced them to buy goods from Chihuahuan traders at inflated prices. While there had definitely been contraband trade in goods and slaves between 1765 and 1821, overland trade now became legal. Traders were already exchanging horses, guns, and other manufactured goods with the Utes for deer, antelope, and bear pelts as well as slaves.[12] The Utes, who had military supremacy because of the horse, raided Paiute villages, stole children, and sold them to traders, even though Spain and later Mexico officially outlawed the practice. Scant records about trade exist, but at least two other official expeditions passed near the San Juan country: Vizcarra in 1823 and Armijo in 1829. The former was a punitive raid against Navajos who had been stealing livestock, and the latter a trading trip to southern California. Reports from both excursions emphasized the aridity of the San Juan area and thus encouraged avoidance.[13]

So it appears that both Spaniards and Mexicans flirted with the San Juan region up to the early nineteenth century, contacting and sometimes trading with various Indian groups. Some official and a lot more unofficial knowledge of the area spread throughout northern New Mexico, but little of it seeped outside the Hispanic world. Clearly people knew enough of the San Juan area to realize that for mineral and settlement purposes, it was best left alone. Spaniards and Mexicans recognized a few agricultural possibilities and admired some of the scenery. Mainly, however, because of the defensiveness of Spanish culture and politics, the San Juan canyons remained terra incognita.

Even though New Mexicans traded with Utes and Comanches for deer, antelope, and buffalo pelts, they did not generally trap animals. The San Juan became a little better known after it was exploited by American fur trappers, who wandered into Taos and Santa Fe after 1821. The first trapper known to have ventured into the San Juan country was William Wolfskill. The twenty-four-year-old Kentuckian arrived in Santa Fe with William Becknell in 1822 on the latter's second expedition to New Mexico. Two years later he outfitted a party that trapped first in southwest Colorado, then split up and moved, with Ewing Young and Isaac Slover, down the San Juan. It is difficult to know how far downstream they traveled, but it is doubtful they penetrated the canyons below Chinle Wash.

In June, though, they returned to Taos with a whopping ten-thousand-dollars worth of furs and the distinction of being the first known trappers to venture to the west. According to historian David Weber, they also motivated an exodus to the area, probably on many of the trails blazed by Spanish traders.[14] Other trappers who entered the San Juan area shortly after Wolfskill's group were Thomas L. (Peg-Leg) Smith and Antoine Leroux.[15] In 1825 an alarmed but exaggerated report to the Mexican government in Santa Fe claimed that Americans had built a fort on the San Juan, probably above Four Corners. In all likelihood, it was merely a trapper's encampment.

Another mountain man who may have trapped up the San Juan from the lower end was James Ohio Pattie, author of the self-aggrandizing, often-inaccurate, but nevertheless-important *The Personal Narrative of James Ohio Pattie of Kentucky*. Working along the Gila and lower Colorado in 1826, Pattie joined Ewing Young's group and apparently pushed up the Colorado, across the Arizona Strip north of the Grand Canyon, then down to the mouth of the San Juan in Glen Canyon. Anthropologist A. L. Kroeber thinks he trapped up the San Juan a few days and then continued east to Navajo country. If Kroeber is right, Pattie and his colleagues were the first white men to see the lower San Juan and trap its beaver.[16] Their hasty departure suggests that trapping was poor. Biologists in the 1950s concluded, however, that the area contained as many beavers as any place in Utah.[17]

Fur trappers, then, explored more of the San Juan River than their Spanish and Mexican predecessors. If Wolfskill's 1824 haul is any indication, their impact on beaver populations may

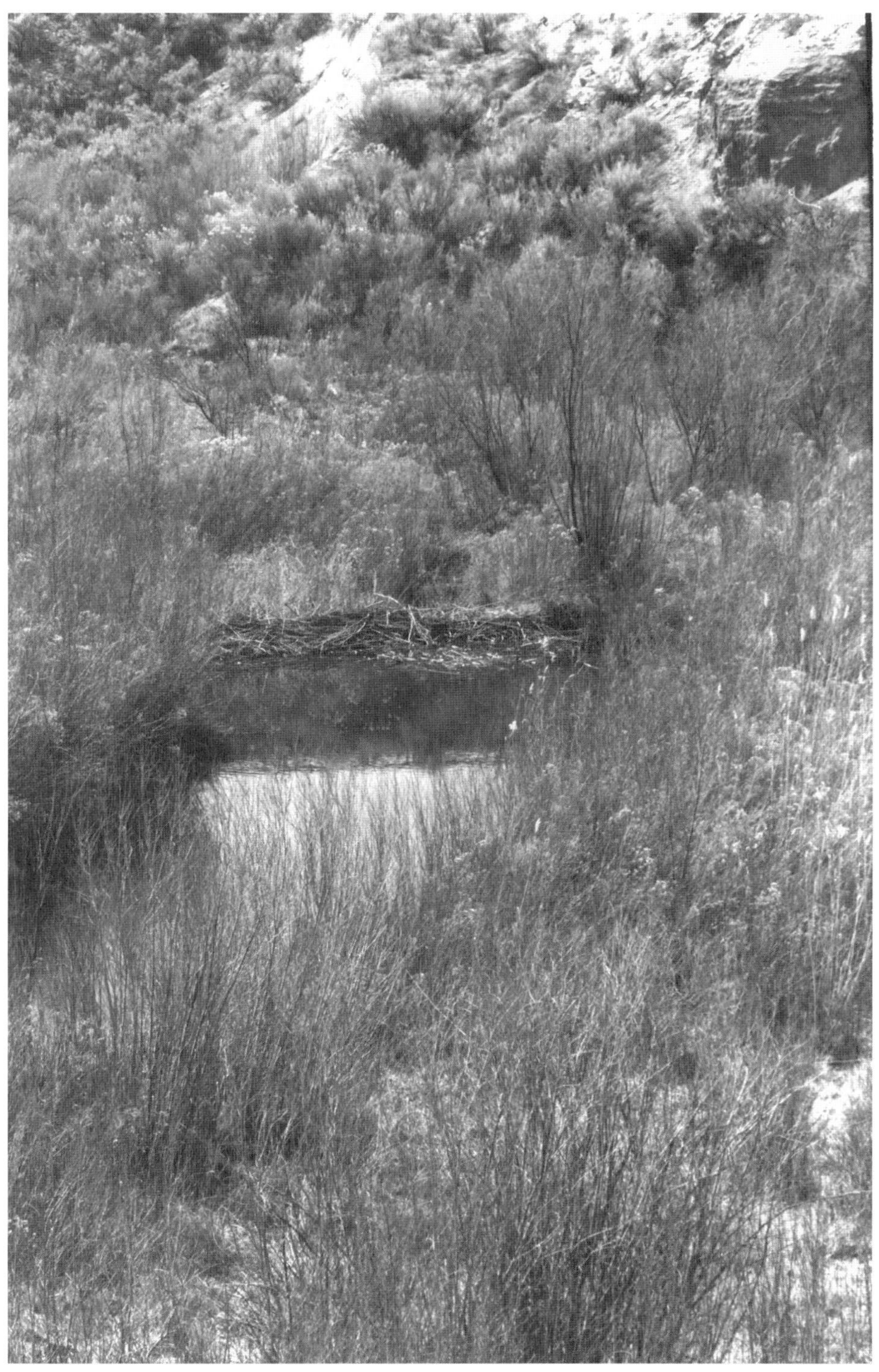

Beaver are coming back in the San Juan drainage. Dams like this one at Butler Wash prevent erosion and provide a rich habitat for birds and other wildlife. Locals say that because of its beaver dams, Butler Wash, a large side canyon of the San Juan, never flash floods. (James M. Aton photo)

have been significant. Historian David J. Wishart believes that overtrapping hastened the Rocky Mountain fur trade's demise. As exemplified by Wolfskill, trappers approached their trade with an attitude "that emphasized short-term exploitation rather than long-term sustained yield."[18] By 1838 Mexican officials had recognized this fact and declared a six-year moratorium on trapping along the Rio Grande. If the Rio Grande was overtrapped, the same situation probably existed on the San Juan and elsewhere in the Southwest. Unfortunately, officials did not patrol the northern Mexican frontier, although it probably would not have made much difference. Historian William deBuys believes that the ruling came too late to help beaver populations and was unenforceable anyway.[19] Nevertheless, the law indicated the severity of the problem on southwestern rivers.

Fortunately for western beaver, European fashions changed in the late 1830s and the fur trade diminished. Close to extinction, the beaver rebounded somewhat, as they have along the San Juan and tributaries like Butler Wash. Beaver populations will probably never reach their pre-trapping high. Man-made dams, continued trapping, and a host of other environmental factors have conspired to keep their number down along the San Juan. Recent anecdotal evidence suggests, however, that these aquatic rodents may be making a comeback near the river.[20]

The decline of beaver in the early nineteenth century did not just affect the animals. Whole watersheds suffered. Most beavers construct their dams on tributaries of major rivers like the San Juan. Their ponds function as silt traps and hence form "a second line of defense against significant erosion." Moreover, these ponds create a moist environment that protects water-loving plants. These in turn attract dense, riparian bird life. The mountain men, who "trapped every beaver they could locate, with no thought for the morrow," did not consider the long-term effect of exterminating the beaver on rivers and creeks: bigger and muddier floods when abandoned dams broke upstream. This may have led to greater erosion along the San Juan and other rivers. Moreover, the riparian life around beaver ponds would also have suffered serious impact. There is no way of knowing exactly what the Taos trappers did to the San Juan, but the riverine environment encountered by the first settlers in the late 1870s was probably vastly different from what it was a mere fifty years earlier.[21]

Fur trappers can hardly be called scientists, but the nature of their work required them to be keen observers and gatherers of information about topography, ecology, and native cultures. Robert M. Utley and William H. Goetzmann have shown that most of the "scientific information" trappers collected passed through an informal communication network, which ended in government map rooms, ethnographic society meetings, and laboratories.[22] Indirectly, then, the fur trappers were unofficial, advance "scientists," whose information about the San Juan was noted and classified. The trappers did not, of course, essentially disagree with their Spanish predecessors: The San Juan did not promise much in terms of resources, settlement, or travel routes.

Travel to and through San Juan country did not end when the fur trade declined. The section of the Old Spanish Trail through present-day San Juan County remained an especially active trading route from 1829 to 1848. We know little or nothing, however, about the traders' side trips. In contrast, a well-documented Mormon expedition entered the country from the opposite direction shortly thereafter in 1854. Having arrived in Utah seven years before and settled the Great Basin, the LDS church sent W. D. Huntington and his men to explore the San Juan area. Brigham Young directed Huntington to survey the territory and establish relations with the Utes and Navajos in preparation for settlement. Huntington "discovered" the ruins around Hovenweep and commented extensively on them in a report published in the *Deseret News*.[23]

In May of the next year, the Mormons established the Elk Mountain Mission at Moab. By August they had sent an exploratory-trading-proselytizing expedition, led by Alfred N. Billings, south down Comb Wash to the San Juan and forty miles up Chinle Wash. In his journal, Billings described the landscape where Comb and Chinle Washes enter the San Juan: "The most Sandy Barron [*sic*] Country I ever Saw the soil is A fine red sand . . . the setlement [*sic*] is on Cottonwood Creek [Chinle Wash] from the cottonwood that grows on its Banks from the

Dr. John S. Newberry was the first geologist to study the canyon country—the Grand Canyon in 1857–8 with the Ives Expedition and the San Juan in 1859 with the Macomb Expedition. He named the geological province the Colorado Plateau, and his insights into the power of erosion gave the region visibility in the world science community. (Manuscripts Division, Marriott Library, University of Utah)

St Johns River [San Juan]." These observations are echoed in the journal of Ethan Pettit, a member of the trip.[24] The Elk Mountain Mission fizzled out after conflict with the Utes, but the Mormons had explored the area, noted its natural resources and native people, and prepared themselves, perhaps unknowingly, for settlement two-and-a-half decades later. Unlike the discoveries of the trappers and the Spaniards, Mormon geographical knowledge stayed inside the confines of Zion and did not benefit American science. The Civil War, however, soon made this country interesting to the U.S. government and thrust the San Juan into the consciousness of the international scientific community.

Shortly after the Elk Mountain Mission members scurried back to Salt Lake, the Mormons entered into conflict with the federal government, the so-called Utah War of 1857–58. Although the war, such as it was, ended quickly, it forced the U.S. military to realize how little they knew of supply routes into Utah. Since Santa Fe was the nearest supply center, the army dispatched Captain John N. Macomb of the Topographical Corps in 1859 to explore the area north and west of Santa Fe, the San Juan country. Macomb was the first of a century and a half of systematic, scientific explorers, backed by government or educational funding, who surveyed and studied the San Juan drainage.[25] Applying modern methods of mapping, topography, and geology, the Macomb report, which appeared in 1876 because of Civil War delays, represented a benchmark in scientific knowledge of the San Juan. It also contained the first published aesthetic appreciation of the river area.[26] Like many army surveys, it primarily compiled topographic information for troop and supply movements. But in common with other government surveys, it ultimately stimulated commercial activity and facilitated settlement.[27]

Macomb was fortunate to acquire the services of geologist John S. Newberry for the expedition. A year before, he had served under Joseph Christmas Ives in his upriver exploration of the lower Colorado River and Grand Canyon. Like Ives, Captain Macomb found the canyon country "a worthless and impracticable region."[28] But geologist Newberry, influenced by Romantic landscape aesthetics and the geologic wonders before him, disagreed with his boss. In his "Geological Report," Newberry waxed eloquent about the "grand view" of the San Juan flowing through Comb Ridge: "The features presented by this remarkable gate-way are among the most striking and impressive of any included in the scenery of the Colorado country."[29] He could barely contain himself as he rhapsodized about the beauty he saw around him. For example, he said, "Illuminated by the setting sun, the outlines of these singular objects came out sharp and distinct, with such exact similitude of art, and contrast with nature as usually displayed, that we could hardly resist the conviction that we beheld the walls and towers of some Cyclopean city hitherto undiscovered in this far-off region."[30] Newberry expressed the first true appreciation of the landscape in a language not very different from the hordes of twentieth-century nature lovers who currently flock to the San Juan. Although a renowned geologist, Newberry was also the first nature-loving tourist to visit the region and report on its scenic wonders.

If Romantic aesthetics inspired Newberry's love of the canyon country, erosion brought out the true geologist in him. With his work in the lower Grand Canyon and on the San Juan River, Newberry made a significant contribution on erosion to world geology. Up to that time, most geological authorities had argued that marine activity or structural catastrophes had created eroded regions like the Colorado Plateau (named by Newberry). But Newberry demonstrated clearly in his reports that, as Stephen J. Pyne has written, "rivers shaped the land, not merely the landscape its rivers." This theory is called fluvialism.

After Newberry's two reports, the canyon country immediately became the "textbook case of American Fluvialism."[31] But perhaps more importantly, his arguments for the power of erosion contributed to the larger debate about the earth's age. Fluvialism buttressed Darwin's case in the *Origin of Species* for the antiquity of the earth. Thus, the San Juan and Colorado Rivers, thanks to Newberry, became world famous among geologists as the place to read the geologic book of time, one primarily crafted by erosion. His study of erosion helped geologists push back the age of the earth and rethink geomorphology.

Besides helping rewrite American geology, Newberry, along with his boss Macomb, commented on the extensive Anasazi ruins near the river. These were the first descriptions of these sites by western scientists, and both men theorized about why they had been abandoned. Macomb thought the Anasazi froze to death, while Newberry more accurately speculated that warfare and drought were chief causes.[32]

The Macomb report and especially John Newberry's contributions marked the beginning of an important period for the San Juan River. The region became geologically significant. In addition, the report also revealed the San Juan as the site of "lost civilizations." Indeed, as Newberry commented, "from the time we struck the San Juan we were never out of sight of ruins."[33] American archaeology grew up in the Four Corners region. Anasazi ruins, more than any single factor, brought scientists, pothunters, tourists, and other visitors into the country. Ferdinand V. Hayden was one of the first.

Government surveyor Hayden sent two of his men, W. H. Holmes and W. H. Jackson, to survey and photograph prehistoric ruins in the San Juan drainage. They were part of Hayden's United States Geographical and Geological Survey in 1874 and 1875. Although his work was similar to the other three major surveys of the postwar era—King, Wheeler, and Powell—in its orientation toward resource exploitation and agricultural possibilities, Hayden had a special knack for playing to the expansionist ideas of nineteenth-century America. He cranked out popular scientific reports that became what one historian has described as an "annual geological Cook's Tour of the territories."[34] He also knew how to use Jackson's photos to interest the general populace in his work, seizing upon the appeal of Anasazi ruins along the San Juan.

Jackson's photos of these ruins appeared at the Great Exposition in Philadelphia in 1876, where they astounded audiences. Due to Jackson's and Holmes's written reports, as well as those by journalists like E. A. Barber and F. W. Ingersoll, who accompanied Hayden, San Juan country suddenly became familiar to the eastern public and even Europeans.[35] The four Hayden surveyors who commented on the Anasazi ruins—Holmes, Jackson, G. B. Chittenden, and Hayden himself—compiled the first environmental history of the Anasazi in the San Juan basin. Their discussions of the ways these Indians lived in the landscape and thought about it and why they abandoned it posed pertinent questions and hazarded still-relevant answers about prehistoric lifestyles and attitudes. They also piqued curiosity about southwestern prehistory among preprofessional archaeologists, pothunters, and tourists.

Archaeological ruins and environmental history notwithstanding, the Hayden Survey also gave more precise descriptions than Macomb and Newberry had of the "most excellent" grazing potential of White Mesa, mining opportunities in the nearby Abajo and La Sal Mountains, and general settlement possibilities along the river. On the last item, Hayden and Jackson disagreed with topographer George B. Chittenden. He saw the river bottom as "utterly worthless" farmland, but they believed that the San Juan corridor "will undoubtedly prove a rich agricultural possession at no distant day."[36] Both prehistoric and historic experience has shown that farming on the San Juan lies somewhere between these extremes. For the Anasazi, as we already learned, the San Juan's agricultural possibilities proved a little closer to Hayden's and Jackson's views; for the Mormons, Navajos, and others, they more nearly matched Chittenden's dour predictions. Nonetheless, the Hayden Survey was significant because it described the San Juan environment, popularized the ruins along the river, extolled the agricultural potential of the region, and initiated the study of environmental prehistory. It put the San Juan, literally and figuratively, on the United States map.

Besides Macomb's and Hayden's government-funded expeditions, a number of private institutions financed scientific study in the region. Many of these were archaeological expeditions, along the lines of Holmes's and Jackson's surveys. But at least one of the privately funded scientists came to study plants. Her name was Alice Eastwood, and her explorations and collections of San Juan flora in 1892 and 1895 constitute another important chapter in the development of San Juan environmental history.

This Canadian native grew up in Denver, where she taught high school. During the summers, she collected plants all over Colorado, eventually meeting the Wetherill clan of Mancos in 1889. By 1892 she and Al Wetherill had arranged to horse-pack from Thompson Springs, Utah, south through Moab and

Alice Eastwood made two significant botanizing expeditions to the San Juan country in 1892 and 1895. She collected 475 specimens representing 162 species, many of which were new. Eastwood was one of the most important scientists to work in the San Juan country. (California Academy of Sciences).

Monticello, then down Montezuma Creek, and up the San Juan. Three years later, at age thirty-five, Eastwood again met Wetherill, and they rode down the San Juan past Bluff and Butler Wash, over Comb Ridge, through Mexican Hat, under the Muley Point Overlook, and up into John's Canyon. By that time, Eastwood had shifted jobs and was working for the California Academy of Sciences in San Francisco. Her colleagues there included Hayden Survey botanist T. S. Brandegee and his wife, Kate, also a botanist.[37]

Eastwood was a fearless and tireless collector of plants. A feminist, Sierra Club member, and flouter of social conventions, she and her journeys are memorable not only because she was the first woman botanist in the Four Corners region but also because her collection provided a baseline study of San Juan flora. In her many published reports and memoirs, Eastwood painted an excellent picture of the area's biota. Her general comments, for example, noted the abundance of tall grass, box elders, greasewood, cottonwoods, and willows along San Juan bottomlands between Four Corners and Comb Ridge. She also complained that the combined odor of beeplant and jimsonweed (sacred datura) "made the atmosphere almost unbearable."[38]

Eastwood observed the Bluff settlement's continuing struggle with its irrigation ditch as well as the many Anasazi ruins along the river. Although she had little time to explore the ruins, she nevertheless intelligently discussed the way the "cliff dwellers" had farmed corn, beans, and squash and used yucca.[39] Had she diverted her attention longer from plant collecting, Eastwood might have pioneered the field of Anasazi ethnobotany.

First and foremost, however, this woman was a botanist. During her two trips, she collected 475 specimens, representing 162 species and varieties. Nineteen species were completely new, and almost all were rare.[40] In addition to her important contributions to San Juan flora identification, Eastwood brought something to her work that was uncommon for scientists of the time: an almost religious passion for the sacredness of life. She shared with fellow Sierra Club member John Muir a sense of the uniqueness of all life. Her collecting was not just a dry exercise in taxonomy but belonged to the larger effort of preservation. An incident from her 1892 trip illustrates her fervor.

She and Wetherill camped in a small cave to escape inclement weather while traveling from Moab to Monticello. After they started a fire to dry off and warm up, Eastwood suddenly looked up and saw their fire was suffocating cliff swallows, which had built their nests in the cave's roof. Writing about this many years later, she said, "I am distressed even now when I think of the destruction of the little birds."[41] This sympathy for nature was unusual for nineteenth-century science.

If the Hayden Survey reports, photographs, and subsequent photo display at the Great Exposition in Philadelphia in 1876 splashed the San Juan in front of the American public, the discovery of Mesa Verde in 1888 and the international Columbian Exposition in Chicago in 1893 made the region world famous. The San Juan basin became known as an archaeological wonderland. And just like today, many of the people who traveled to the region in the 1890s and early 1900s came because of the "cliff dwellers' ruins." This gold rush of a different sort attracted looters, relic collectors, museum-directed excavators, tourists, and budding archaeologists. The distinctions between these enthusiasts for Anasazi ruins, however, were much vaguer in 1890. The discoveries of amateur archaeologists like Heinrich Schliemann and Austen Layard at Troy and Mesopotamia had excited Europeans and Americans about the wonder of "lost civilizations." But archaeology, as a scientific discipline, was in its infancy.

The period following Mesa Verde's discovery by Charlie Mason and Al and Richard Wetherill in 1888 started a stampede to the San Juan country. When reports blew east of a lost American civilization in the Four Corners area, the public and especially eastern museums jumped at the chance to collect and exhibit an American counterpart to relics excavated by Europeans in the Near East. In fact, part of the motivation behind these ventures—both European and American—was nationalistic. As the self-perceived preservers of civilization, European and American museums had no qualms in appropriating any treasures from lost cultures that their scientists unearthed. Not surprisingly, most artifacts dug up in the San Juan between 1890 and 1910 ended up in eastern museums or the private hands of looters from Colorado, Utah, Arizona, and New Mexico.[42]

The first excavation in Utah was probably conducted by Charles Cary Graham and Charles McLoyd in Grand Gulch in the winter of 1890–91. Both were friends of the Wetherills and had helped them excavate Mesa Verde. Depending on which of their contemporaries one believes, McLoyd and Graham either looted sites in Grand Gulch and left a mess or excavated as scientifically as their limited backgrounds allowed.[43] But the specter of amateurs looting important cultural sites quickly prompted Frederick Putnam of Harvard's Peabody Museum to organize and sponsor an expedition, headed by Warren K. Moorhead, to the San Juan area in 1892.

Supported by the Peabody, the Smithsonian, the American Museum of Natural History, corporate sponsors like Armour, and the *Illustrated American Magazine*, Moorhead's group spent April to August of 1892 photographing, mapping, measuring, and, in a few cases, excavating along the San Juan valley. As science, the expedition failed, even though the series of articles that appeared helped publicize the area. And as explorers, group members were inept. For example, their ill-fated attempt to boat the Animas and then the San Juan to Noland's Trading Post at Four Corners ended in near disaster; they bailed out at Farmington, right above the confluence of the two rivers. Moorhead later wrote this puffery of the Animas trip: "The most dangerous feat of river navigation attempted since Major Powell and his party floated down the Colorado River has been accomplished by the *Illustrated American* Exploring Expedition."[44] River runners today would laugh at this incredible boast.

Moorhead also described the famous San Juan sand waves, though he could not account for their cause. Apparently, however, the group camped next to the San Juan at flood stage because the river inundated them. To add to their misfortunes, they found the landscape threatening. The red sandstone wonderland that had so moved a geologist like Newberry more than thirty years earlier hit Moorhead with a dull thud. He wrote, "You cast your eyes about to something of beauty, but you see nothing save great frowning sandstone cliffs, an occasional cow, a coyote, or a sand crane. You sigh for the green fields and shady woods of the East."[45]

Moorhead and his men possessed an arrogance about their scientific credentials and experience which ultimately torpedoed their

Warren Moorhead described the San Juan's famous sand waves, which can happen anyplace in the river when the sediment load is high. A series of waves moves upstream, sometimes reaching a height of ten feet. They can flip an unwary boatman. (Hugh D. Miser photo, #423, U.S. Geological Survey)

efforts. The strangeness of the landscape and its aridity, the haste of their travels, and their general unfamiliarity with Anasazi ruins combined to make the expedition a study in ineptitude. To put it bluntly, they did not know where they were in every sense of the phrase.

Two scientists who accomplished something of lasting value, in good part because they took the time to get to know the region, were T. Mitchell Prudden and Byron Cummings. Prudden was a New York physician who spent many summers between 1892 and 1915 exploring the San Juan watershed. Like many archaeologists of the time, Prudden taught himself stratigraphic excavation. He was the first to describe in print the Basketmaker culture which Richard Wetherill had discovered in Cottonwood Wash in 1893. His 1897 article in *Harper's New Monthly Magazine* was followed by the first overall description and mapping of San Juan ruins, "The Prehistoric Ruins of the San Juan Watershed" (1903). His 1907 memoir, *On the Great American Plateau,* also helped publicize the area and professionalize excavation.[46] While Prudden talked mostly about archaeology and ethnology, his description of the San Juan sounded like much turn-of-the-century nature writing: "The San Juan, muddy and treacherous, rolls sullenly westward through hot reaches of desert, and then rushing along deep gorges, merges at last into the Colorado."[47]

Cummings, a classics professor and dean of arts and sciences at the University of Utah, was another self-taught scientist. He began excavating up Montezuma Creek at Alkali Ridge in 1908 under the guidance of Edgar L. Hewett, director of the School of American Research in Santa Fe. Two of his student excavators, Neil Judd of Utah and Alfred V. Kidder of Harvard, later earned distinction in the field of southwestern archaeology. Besides helping professionalize San Juan archaeology and train future archaeologists, Cummings, along with W. B. Douglass, became famous as the discoverer of Rainbow Bridge in 1909. This spectacular arch on the west side of Navajo Mountain near the confluence of the San Juan and Colorado continues to attract many tourists. Some of them, like Theodore Roosevelt and Zane

Professor Byron Cummings (center, front) of the University of Utah led an expedition to discover Rainbow Bridge. The famous canyon country guide, John Wetherill, is seated to his right. W. B. Douglass, Cummings's rival, sits to his left. (Stuart Malcolm Young Collection, Cline Library, Northern Arizona University, NAU.PH.643.1.130)

Grey, wrote about the area in national magazines and increased the tourist traffic.[48]

Others besides Cummings and Prudden also excavated the San Juan area at the time. Some of their stories, like those of Richard Wetherill and his brothers, have been told before in many places.[49] Others, like Mormon patriarch Platte D. Lyman of Bluff, have had their tales related in different contexts. But most who dug possessed the same get-rich mentality as their fellow prospectors of the 1892–93 San Juan gold rush and later oil exploration. Locals like McLoyd, Wetherill, Graham, and Lyman did not publish their findings. Eastern, foreign, or scientific visitors like Moorhead, Cummings, Prudden, and Frederick Chapin, however, did. They not only promoted the archaeological wonders of the area to a national and worldwide audience but also extolled the beauty of the landscape where the ruins sat.

While most who wrote about the San Juan appreciated its beauty, not all professed Alice Eastwood's love of wildlife. *Illustrated American* leader Warren Moorhead offers a striking contrast to Eastwood's sympathy for animals. Describing hunting down a rattlesnake in the bushes, he wrote, "With great pleasure you put a bullet through its head."[50] Nevertheless, one of the many attractions of the San Juan between 1890 and 1910 was the setting of its magnificent ruins in a stark, redrock landscape. Little has changed except that visitors one hundred years later have these archaeologists' writings to guide their own explorations.

The most important scientist who traversed and wrote about the San Juan country was Yale University and U.S. Geological Survey (USGS) geologist Herbert E. Gregory. During many summers between 1909 and 1929, he explored the country south and north of the river. Although he never actually floated the San Juan, he crossed it, camped near it, and studied it and its tributaries. As a result, he produced a superb series of scientific and historical articles

Herbert E. Gregory, a Yale geology professor, spent numerous summer seasons between 1909 and 1929 in the San Juan country, geologizing, photographing, studying Indians, and recording history. His reports and images form a baseline for any study of the region's environmental history. (Manuscripts Division, Marriott Library, University of Utah)

and professional monographs.[51] As one historian put it, he was "the preeminent field geologist of the Colorado Plateau whose reports have been revised and supplemented but never superseded."[52] If previous explorers added bits to the geographic, geologic, hydrographic, historical, topographic, archaeological, and biological knowledge of San Juan country, Gregory surpassed them by doing it all and doing it better. Not only did he write groundbreaking geologic reports—his main field of study—but Gregory also completely covered the territory. He was essentially writing environmental history well over a half century before it became common among late twentieth-century historians. Nearly every scholar who studies the area—no matter the discipline—begins with Herbert E. Gregory.

Gregory first ventured into what he called Navajo Country, south of the San Juan and east of the Colorado, in 1909. The USGS and Office of Indian Affairs sent him to survey the region's water resources in the hope of developing them for the Paiutes, Hopis, and Navajos. A progressive-era conservationist, Gregory saw his scientific work in the paternalistic, culturally biased terms of the times: "I believe also that the sanest missionary effort includes an endeavor to assist the uncivilized man in his adjustment to natural laws. . . . To improve the condition of this long-neglected but capable race . . . by applying scientific knowledge, gives pleasure in no degree less than that obtained by the study of the interesting geologic problems which this country affords."[53] Like his hero, John Wesley Powell, Gregory hoped his scientific knowledge would hasten the "natural" cultural evolution of Native Americans toward civilization. If his values seem a bit dated, his research and writing are still extraordinarily fresh and full of crucial information for contemporary scholars.

Besides completing an exhaustive study of water sources in this arid region and suggesting ways to develop them, Gregory made crucial studies of San Juan flora and fauna. While he drew in part on the work of earlier botanists such as T. S. Brandegee and Alice Eastwood, his field notes and reports showed he went far beyond just copying them. He not only used his own powers of observation but also interviewed locals, both Anglo and Indian, to understand the area's animals, plants, and environmental conditions. For example, from local government trapper Seth Shumway, he learned about a vigorous campaign in the 1880s that exterminated wolves and bears and nearly eliminated mountain lions from the San Juan. Local Paiutes and Navajos told him that mountain sheep and antelope populations had crashed after Anglo stockmen took over the ranges; the Indians also informed Gregory that overgrazing had intensified a twenty- year period of severe arroyo cutting and encouraged the proliferation of nonnative weeds.[54] Gregory was a Gifford Pinchot-Theodore Roosevelt conservationist who frowned on the mismanagement of natural resources. While no tree hugger like Alice Eastwood or John Muir, Gregory would have felt

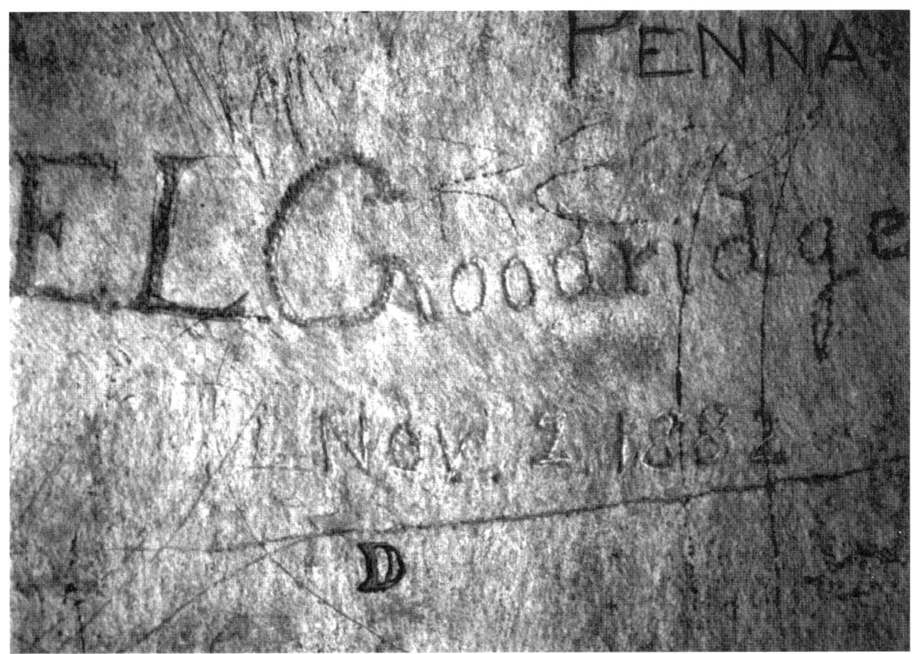

Emery L. Goodridge floated the river in 1882 (and possibly earlier in 1879) and left his inscription near Mexican Hat on November 2. He came from southwestern Colorado and ended his journey at Lees Ferry. He was prospecting for gold but found oil near Mexican Hat, Slickhorn, and elsewhere along the river. He returned to Mexican Hat in 1908 and drilled the first oil well. (James Knipmeyer photo, San Juan Historical Commission)

at home with the "land of many uses" philosophy of the later Bureau of Land Management or Forest Service.

In addition to compiling the most complete natural history of the San Juan to date, Gregory possessed a great interest in both prehistory and history. Large sections of his texts summarized the most up-to-date information about the Anasazi. He did the same for the Spanish and Mormon history of the area, consulting both published texts and living participants. For example, he interviewed Mormon pioneer Kumen Jones of the Hole-in-the-Rock group and E. L. Goodridge, the first man to float the San Juan in 1882 and discoverer of oil at Mexican Hat and Slickhorn Gulch. Gregory's descriptions of Paiute, Navajo, Ute, and Hopi economies are also quite accurate.

Although his love for the stark beauty of the San Juan landscape clearly underlies everything he wrote, he mostly kept his personal feelings in check and stuck to informing the reader. He nevertheless foresaw that tourism would soon be a major part of the San Juan's economy.[55] In fact, one of the first tourist-adventurers to come to the area, a wealthy cotton broker from New York named Charles L. Bernheimer, said he was drawn partly by Zane Grey's novels and partly by Gregory's paper, *The Navajo Country*.

Gregory's guide, John Wetherill, also deserves mention here. The third son of the famous Wetherill clan of Mancos, Hosteen (or Hastiin) John, as the Navajos respectfully called him, figured in many of the important archaeological and geological expeditions of the San Juan. Few white men knew the country between Mesa Verde and Navajo Mountain better than Wetherill. Likewise, few knew the Utes, Paiutes, and Navajos better than Wetherill, who operated a trading post with his wife, Louisa Wade. In addition to guiding scientists like Gregory and Cummings and wanna-be scientists like Bernheimer, he introduced celebrities like Zane Grey and Theodore Roosevelt to Rainbow Bridge and the surrounding country. Wetherill was not a writer, so his considerable knowledge did not get published. But no one, as historian Gary Topping has written, was more at home in the desert.

Indirectly Wetherill contributed greatly to the publication of information about the San Juan. He was an important link in the chain that ended with the work of scientists like Herbert E. Gregory.[56]

Gregory's reports, complete with excellent photographs of landscapes and native peoples, stand as landmark studies of the San Juan and its residents. They appeared just as land use in the area was fundamentally changing from native subsistence to western agriculture and mineral extraction. Gregory admitted he had come to the area to facilitate that change. But he was both scientist and humanist enough to document as fully and clearly as possible the people and nature of the San Juan watershed. As much as anyone, Herbert E. Gregory deserves the title of "chronicler of the San Juan."

If Gregory's initial foray into the country was prompted by the desire to survey water resources, the next explorers, the Trimble Expedition, stand as the first sentence in the climax of the water story. The Bureau of Reclamation and Southern California Edison (SCE) sent the expedition to survey the river for dams. The Federal Water Powers Act of 1920 had made damming the Colorado River system politically and economically feasible because water development became a joint venture between the federal government and private industry. The technology for generating hydroelectric power was also coming of age. In the same year, the Kincaid Act authorized the secretary of the interior to make a geological and topographical survey of the Colorado, Green, and San Juan Rivers. Under the terms of the act, Southern California Edison, a private power company in Los Angeles, agreed to fund part of the survey and provide men and materials.[57]

By July of the next year, the USGS-SCE joint expedition met in Bluff. Under the leadership of Kelly W. Trimble, they spent the next six months mapping and studying the San Juan between Bluff and Lees Ferry on the Colorado. Part of their mission was to suggest potential dam sites along the San Juan, but they also wanted to see how far a reservoir would back upriver from a proposed dam near Lees Ferry. Besides Trimble, the USGS sent geologist Hugh D. Miser. The portly Missourian eventually wrote the final report, took many of the expedition's photos, and, by general consensus, held the group together with his unfailing good humor. Engineer Robert N. Allen represented the power giant's interests, and Bert Loper, later known as the "grand old man of the Colorado," signed on as head boatman. He, in turn, hired young H. E. Blake, who later worked for USGS-SCE expeditions on the Green River and in the Grand Canyon. Two local Mormons, Hugh Hyde and Heber Christensen, rounded out the seven-man, two-boat crew.[58]

Miser's report, now considered *the* classic study of the San Juan, attempted to do for the river what Gregory had done for the surrounding country. In a letter to crew member Heber Christensen four years after the trip, Miser said that he had intended to write a "more or less popular report" of the region, combining the technical aspects of a geologic, topographic, and mineral survey with the day-to-day experiences of expedition members.[59] This trip narrative is one area where Miser's report differs from Gregory's. On the other hand, Miser's sections on history and natural history pale in comparison to Gregory's. He does include, however, valuable historical information on San Juan River travel.

The photographs from the Trimble Survey have proven invaluable for comparing vegetation changes along the river. So have some of Miser's field notes. For example, he noted that they slept at Slickhorn Canyon on a mattress of Russian thistle, which "served well." Thistle's presence indicates that cattle had come down the trail and overgrazed Slickhorn. Today this area has recovered, and Russian thistle is rarely seen.[60] Miser also noted that thistle covered the bottomlands of Paiute Farms. His observations contrasted with boatman Bert Loper's experience during the gold-rush days of the early and mid-1890s. According to Loper, the whole wide bottoms of Clay Hills and Paiute Farms had been covered with cottonwoods. The floods of 1911 and thereafter, exacerbated by watershed destruction on the San Juan and its tributaries, as well as overgrazing by livestock, probably accounted for the altered landscape in 1921. Like Gregory, Miser believed that overgrazing had caused severe arroyo cutting in the more than twenty-five years since whites began running stock around the river.[61]

Hugh D. Miser of the 1921 Trimble Expedition works a plane table below Mexican Hat Rock. Note the rodman across the river. Miser wrote the classic text for the expedition and was the acknowledged "glue" of the trip. (Robert Allen photo, #570, U.S. Geological Survey)

His report also contained anecdotal environmental evidence that healthy populations of flannelmouth suckers swam the San Juan in 1921. At the foot of the Honaker Trail on August 1, a flash flood so loaded the river with silt that hundreds of suckers surfaced in the eddies, trying to get oxygen. The party harvested scores of them to eat. Loper told Miser that he had seen similar floods "last long enough to kill thousands of fish."[62]

The end result of the USGS-SCE surveys on the San Juan, Green, and Colorado Rivers in the early 1920s, however, was dams. And those dams, Navajo and Glen Canyon, have had the most profound effect on life along the San Juan. Although the Trimble Survey did not focus on dam sites along the San Juan, it formed part of a larger effort to alter the flow of every river in the Colorado system and so marked a defining moment in San Juan environmental history.[63] What had begun with expeditions by Rivera and other Spanish explorers to discover and describe the territory, trappers to extract beaver, and archaeologists to uncover lost civilizations ended with the government-funded, scientific surveys of Macomb, Hayden, Gregory, and Trimble. The story of dams on the San Juan, covered in chapter 8, is a big one. The Trimble Survey marks the transition point.

But between government surveys, an unlikely scientific expedition was organized during the 1930s to explore the Navajo country—Monument Valley, Navajo Mountain, and the San Juan River. Organized by National Park Service educator Ansel F. Hall, the Rainbow Bridge–Monument Valley Expedition (RBMVE) was a multidiscipline effort that spanned six summers between 1933 and 1938 and involved more than 250 people. Privately funded, its purpose was twofold: to allow a diverse staff of scientists to explore, map, study, and record one of the last scientifically unexplored areas of the United States (then under consideration as a national park) and to provide young men with the chance to live outdoors and study nature. Most student members paid three to four hundred dollars for the summer adventure-classroom, a considerable sum during the Great Depression. Hall recruited many well-known scientists from universities and museums across the country. Many came from

Hugh Hyde and Robert Allen of the Trimble Expedition at the Honaker Trail after clubbing scores of flannelmouth suckers that had risen to the surface seeking oxygen when the river was at flood stage. (Hugh D. Miser photo, #584, U.S. Geological Survey)

his alma mater in Berkeley and went on to distinguished careers in various scientific disciplines.[64]

The RBMVE produced more than forty technical publications, even though most were mimeographed and poorly circulated. Some scientific work went unreported, but other discoveries became finished pieces of outstanding science. Many valuable archaeological finds came out of the RBMVE; much of this work, especially defining pottery types, was conducted by Lyndon L. Hargrave of the Museum of Northern Arizona. Perhaps even more significant for both southwestern archaeology and San Juan environmental history was the work of

Angus Woodbury was a National Park Service naturalist before becoming a biology professor at the University of Utah. One of the first scientists to call himself an ecologist, his studies of flora and fauna with the Glen Canyon Survey were groundbreaking. (Manuscripts Division, Marriott Library, University of Utah)

Ernst Anteus of the Carnegie Institute and John T. Hack of Harvard. They collected sediments that were "the first in the Southwest to be analyzed for fossil pollen." Hack ultimately concluded that erosion was an important factor in the great abandonment of the San Juan country by the Anasazi. Pollen analysis has since been extremely important in charting human impact on past environments.[65]

Perhaps the most important product of the RBMVE was Angus M. Woodbury's and Henry N. Russell's monograph, "Birds of the Navajo Country."[66] It was a model of ecological science. The study discussed habitats and ecological relationships, in addition to cataloguing all the avian life of the region. Woodbury, from the University of Utah, and Russell, from Harvard, called themselves ecologists long before that was standard practice or even fashionable. Woodbury's participation, as we will see, was an important prelude to his work in the Glen Canyon-San Juan region nearly two decades later. Likewise, the RBMVE, in its own way, began what was later accomplished by the Glen Canyon Survey: a complete study of the ecology and cultural history of the area.

If the Rainbow Bridge–Monument Valley Expedition in the 1930s marked a first attempt to study the lower, south side of the San Juan, the Glen Canyon Survey of the late 1950s was a landmark multidisciplinary, scientific study. When the Colorado River Storage Project (CRSP) passed Congress in 1956, the 1935 Historic Sites Act required that funds be provided for salvage. The contract that the National Park Service signed with the University of Utah and the Museum of Northern Arizona called for salvage and study of the archaeology, biology, geology, paleontology, and recent history of the whole area to be inundated by Glen Canyon Dam. This meant Glen Canyon and the lower San Juan.

Archaeologist Jesse Jennings of the University of Utah formulated and headed the Glen Canyon Survey between 1957 and 1963. His team set a new standard for archaeological and historical salvage operations. (Manuscripts Division, Marriott Library, University of Utah)

Under the general direction of Dr. Jesse D. Jennings of the University of Utah's anthropology department, the different survey groups began reconnaissance in 1957, were in the field by the next year, and worked continuously until 1963, when the backed-up waters of the reservoir stopped them. In the process, the survey helped rewrite the methodology of historic salvage. For the first time, the entire Glen Canyon-San Juan area came under the scrutiny of an organized team of scientists. As director Jennings stated, "The survey's comprehensive multi-discipline approach . . . will surely remain a hallmark in the history of scientific salvage endeavor." Indeed it has. Jennings also noted, with some pride, that by the time the survey completed the work and published it in 1965, its methods had become the norm for salvage operators.[67]

The Museum of Northern Arizona handled archaeology on the lower San Juan, concentrating largely on Anasazi sites. The work was summarized in *Survey and Excavations North and East of Navajo Mountain, Utah, 1959–1962* (1965) by Alexander J. Lindsay, Jr., et al. Historic research for the San Juan canyons fell under the direction of Dr. C. Gregory Crampton of the University of Utah. He published numerous monographs at the time on the Glen Canyon and San Juan areas and four popular books later. His *San Juan Canyon Historical Sites* (1964) was a thorough, mile-by-mile history of the river.[68]

Many archaeologists (and pothunters) had combed the San Juan area, some biologists had collected plants and studied fauna, and others like Gregory and Miser had analyzed geological processes. No one, however, had attempted to synthesize all that information before the Glen Canyon Survey. The concept that drove the survey's scientists and historians was ecology. Thus, researchers examined data,

Professor C. Gregory Crampton (far left) of the University of Utah directed all historical research of the Glen Canyon Survey. This 1962 trip was the last one to document historical sites up the San Juan. Bureau of Reclamation public information officer and river historian W. L. Rusho stands on the far right. (Bureau of Reclamation, Upper Colorado Region)

asking about the relationships between the landscape's resources and its inhabitants and how people made a living?[69]

Although the researchers worked in concert, perhaps the most important studies of animals and plants again came from Angus Woodbury. His Glen Canyon-San Juan reports drew upon his previous study of birds, which he expanded into a complete baseline analysis of flora and fauna along the two rivers.[70] Unfortunately, his untimely death in a 1964 car accident prevented him from completing further studies on the ecology of these river corridors. Nevertheless, Woodbury and his associates' findings were significant in many ways. They gained insight into the way different plants and animals along the rivers occupied certain biological territories. For example, Woodbury defined three distinct plant communities: the narrow streamside or riparian zone, where most vegetation and animal life is concentrated; the terraces; and the sparsely vegetated hillside. He showed how and why these zones' plant communities remained distinct, as well as how invasion and competition occurred. In addition, one of Woodbury's reports contained a valuable history of biological study in the area. Woodbury also broached the question of disease among prehistoric populations. While a common subject in environmental history today, it was new territory in the 1960s. Finally, Woodbury planted valuable seeds for discussion when he introduced the idea that parasites, allergens, and other environmental diseases affected populations.

Like the rest of the Glen Canyon Survey, Woodbury's research focused on relationships. It was not enough, for example, to list the ninety-six kinds of birds in the canyons. He wanted to know how they functioned in communities, which birds lived in which plant zones and what they ate, which predators preyed on what birds, and how climate changes or localized environmental phenomena affected populations. Woodbury envisioned his reports becoming the "standard of comparison with the biological resources of the future reservoir,"[71] and that's ultimately what the Lake Powell Research Bulletins, discussed in chapter 8, became in the 1970s. He set a fine standard for research.

The Glen Canyon Survey stands at the end of two centuries of Spanish and American exploration of the San Juan River. A common goal of all explorers was topographic information. Most viewed that information as part and parcel of exploitation—of silver and gold, scenery in photographs, Anasazi pots, beaver pelts, or water. Many were repulsed by the starkness of the area, but probably more found the sandstone landscape beautiful. These people, along with the artists and writers discussed in chapter 9, helped shape the consciousness of the many tourists who visited the region in the twentieth century. Coupled with an aesthetic appreciation, the salvage work of Woodbury and his colleagues unveiled a new, ecological approach to studying the San Juan. In total the survey ranks with John S. Newberry's work in possessing international importance in two scientific fields: archaeology and ecology. Once again, research in the San Juan rippled outward into the larger world of science.

At the same time, however, the survey's organizers, the Bureau of Reclamation, proposed the grandest scheme of exploitation southwestern canyons had ever seen: damming the San Juan and Colorado Rivers for power, water, and flood control. In a sense, the ecological ideas of Woodbury and the Glen Canyon Survey were too new to have political impact and keep the San Juan flowing free. Theirs was a kind of pre-National Environmental Policy Act (NEPA) environmental impact statement (EIS). But the result, unlike a normal EIS, was already known: The area would be drowned and could no longer be studied, much less appreciated or enjoyed, by humans. Federal law gave scientists the money to study the Glen Canyon-San Juan area, although the patient was scheduled to die.

However, ecological surveys such as the Glen Canyon one ultimately led to laws like NEPA, which required that areas be studied *before* decisions are made about their fate. Before we get to dams, perhaps the most important story in this narrative, we need to look at the uncontrolled river and late nineteenth-century settlement. The civilizing process that occurred simultaneously with livestock, agriculture, city building, and mining continued during the construction of the dams and eventual harnessing of the San Juan River.

4 LIVESTOCK: *Cows, Feed, and Floods*

As the San Juan River has coursed through the Four Corners area, it has both encouraged and denied economic opportunities to Native American and Anglo-American entrepreneurs alike. Its system of canyons and floodplains offers forage for livestock, channels movement, suggests strategic locations for trade, and provides possibilities for agriculture. On the other hand, the river can swell uncontrollably to flood stage, ripping out everything in its path; it has served as a clearly defined legal boundary, restricting access to resources by people on both banks; and, due to the mere presence of its water in a desert environment, has created countless disputes over who should use it.

This chapter and the next focus on the role the river has played in two acts of the human drama staged across its narrow belt of riparian wealth. This chapter discusses the evolution of both the Navajo and Anglo livestock industry, the growth of trading posts that encouraged large herds to depend on the river's resources, and the subsequent development of a road system to move ranching products to market. It is a multifaceted history that extends far beyond the San Juan and throughout the Four Corners region.

The next chapter looks at Navajo farming, especially activities supported by the federal government to move the tribe to economic independence. With both livestock and agriculture, the key to success lay in access to water along the banks of the river. For this reason, the upper portion of the Lower San Juan, where there are broad floodplains and the water flow is less constricted by canyon walls, was the scene of much of this drama.

The earliest reports of Navajo use of the San Juan River date back to the 1820s and 1830s. Military accounts suggest that the lands surrounding the river, especially on the Upper San Juan, were favored planting areas, while in times of trouble, the Lower San Juan provided an escape route for those pursued.[1] Certain bands of Navajos enjoyed friendly relations with Utes living north of the river, while other groups were denied favored status. When intertribal strife reached its peak in the 1860s, the Utes became inveterate enemies of most Navajos. With government encouragement to round up the Navajos and move them to Fort Sumner, the Utes chased their neighbors far south, away from the richer agricultural sites and grasslands bordering the San Juan. Only small groups of Navajos remained, usually in peripheral areas.[2]

The main body of Navajos (around eighty-five hundred), between one-half to two-thirds of the entire population, spent four years (1864–68) in abject poverty and misery at Fort Sumner, New Mexico. This group always hoped to return to their lands, yet feared encountering their Ute neighbors to the north. As early as 1866, Navajos told soldiers at Fort Sumner that "without protection from the Utahs who are our enemies, we would not care to go back."[3] Thus, even though the government released the Navajos from Fort Sumner in 1868, large numbers did not return to the Lower San Juan until the last quarter of the nineteenth century. One of the major forces that encouraged this move north was the demand for more grass to feed growing livestock herds. Ever since the Navajos had first stolen or traded sheep from the Spanish more than two centuries earlier, livestock and grazing had become increasingly important in their economy.

When the Navajos returned from Fort Sumner, they said, "We will go back to our land.

These sheep near Mexican Hat are a small representation of the large Navajo herds that grazed both sides of the San Juan. Scarcity of water and feed coaxed flocks into limited areas. (San Juan Historical Commission)

The people will multiply, the horses and the sheep too, the corn will reproduce itself, plants of all kinds will grow . . . and it will rain."[4] This was not just poetic thinking. To the Navajos, the sheep held supernatural powers that attracted rain and encouraged the growth of plants. The Diné explained their relationship with livestock in the simple but profound belief that "sheep are life."

In the time of myths, when the holy beings created the world, the landscape was predestined to support livestock. The holy beings provided wealth in animals and instructed herders to ask them for supernatural help. The sheep, therefore, became partners with the holy beings to benefit human beings.[5] One of the four sacred mountains, Dibé Ntsaa or Big Sheep Mountain (Hesperus Peak in southwestern Colorado), was "made of sheep—both rams and ewes."[6] The holy beings associated with this mountain poured forth their riches in livestock and were petitioned by herders for supernatural assistance. The holy beings worked through this and other mountains to provide livestock to support the Navajos: "The mountains were put here for our [Navajos'] continuing existence. . . . All of the living creatures, like sheep, horses, cows, etc., said we will help with furthering man's existence."[7]

Medicine men still gather soil, *dziłleezh,* from these mountains and bring it home to protect Navajo land and livestock. One person explained that blessing the animals with prayers through *dziłleezh* brings rain to nurture the land:

> livestock is what life is about, so people ask for this blessing through *dziłleezh.* From the sheep and cattle, life renews itself. Who would give birth in a dry place? This does not happen. You get many lambs and calves from the plants around here. On the tip of these plants are horses, cattle, and sheep. They are made of plants which are sheep.[8]

Thus, Navajo expansion north and the use of natural resources were based in religious faith, not scientific practice. For this reason, Navajos believed the more sheep there were, the more rain and plants would be available to feed them.

This philosophical belief, however, took its toll on the landscape. Information about Navajo activity along the river in this early stage of expansion comes from military reports and citizen correspondence, most of it anecdotal.

When numbers of livestock are given, they obviously are guesstimates, since no one but the Navajo owners were traveling about to count animals.

But no one can deny the unparalleled growth of the herds between 1880 and the beginning of livestock reduction in the 1930s. A quick survey of eyewitnesses reveals the intensity of livestock use by Native Americans. In 1883 Bluff settlers complained that the Indians had been given permission by Henry Mitchell for "absolute possession of every spear of feed on the north side [of the San Juan River] and if it continues it will do us great injury."[9] Later that year the same people reported eighteen Navajo herds of sheep in Recapture, Cottonwood, and Comb Washes, as well as some in Montezuma Canyon. It is not difficult to accept reports of herds as large as twenty thousand animals ranging along the Lower San Juan.[10] Complaints of "thousands of sheep and hundreds of horses north of the river" continued for years.[11]

Navajo testimony confirms these reports. In 1905 agent William T. Shelton estimated one thousand Navajos lived in the vicinity of Aneth. Just one of these people, Mexican Clansmen, said that he owned almost three thousand sheep and a "good bunch" of horses. Another person, named Headman, claimed fifty horses, twenty-five cattle, and seven hundred sheep. Herds of up to three thousand sheep grazed on the Montezuma range, while others just as large munched on grass along Mancos Creek, attracted by feed and water. The result? "They [herds of sheep, goats, and ponies] keep the grass from seeding and destroy the feed for the coming winter."[12] A scholarly estimate places the total number of sheep on the reservation as high as 1,700,000 by 1892.[13]

As Native American herds grew, so, too, did the cattle and sheep industry in Anglo- American settlements. The largest livestock center on the San Juan was the Mormon town of Bluff. When these settlers arrived in 1880, they brought more than one thousand head of cattle and a large herd of horses.[14] Like the Navajos' herds, Anglo livestock holdings did nothing but expand.

In 1887 Francis Hammond estimated that there were fifty thousand head of sheep and eight-to-ten-thousand cattle grazing the ranges of San Juan County. Of that number, six thousand sheep and two thousand cattle belonged to the Mormons and the rest to outsiders. Ten months later, the Mormon sheep herd had doubled.[15] That same year the Bluff Co-op Store sold the wool from the biannual shearing of its eleven thousand sheep for between eleven and fourteen cents a pound in Durango, Colorado.[16]

Add to these figures the activities of settlers from Colorado, and the "invasion" of Mormon ranges on Elk Ridge and in Comb Wash, Recapture Wash, and Montezuma Canyon by non-Mormon cattle companies, and it becomes clear that the problem of overgrazing and erosion skyrocketed. Indeed some of the outfits near Blue Mountain pastured as many as twenty to thirty thousand cattle on county soil, and this number did not include the estimated one hundred thousand cattle brought in from Colorado for winter graze and a lower tax assessment.[17]

Intensifying the stress on the environment caused by overstocking the range was a decade of drought, beginning in 1886. By 1896 Hammond declared, "We have just passed through the driest winter in the history of this county. . . . As a result, streams that were formerly large and springs that gave forth abundantly are now almost devoid of moisture as a tinder box."[18] Range grasses that were not eaten or trampled withered in the heat and drought. Frank Silvey, who lived through these times, tells of losing half his cattle to starvation. The days of large-scale, open-range cattle operations appeared to be coming to a close.

Ironically, as most large companies were suffering, John Albert Scorup, a Mormon cowboy who had settled in Bluff, started his own livestock operation. Hard work and good investments made his rags-to-riches story a lasting tribute to the dedication of the livestock industry in this difficult environment. He labored throughout the canyons and mesas, rounding up cattle for the Bluff Pool, saving his money, holding out during the economic slumps of the 1890s and post–World War I years, and always taking advantage of less-accessible rangeland. His company grazed from seven to ten thousand cattle each year on a two-million-acre range that extended from Blue Mountain to the confluence of the San Juan and Colorado. Scorup continued to supervise this operation until his death in 1959.[19]

Al Scorup, sitting astride his favorite horse, Ol' Booger, typified the tough cowboy required to herd cattle in the San Juan region. For many years, Scorup lived in Bluff and chased livestock throughout canyon country. (San Juan Historical Commission)

Thus, the ranges in the canyons and washes that stretched from mountain to river were of prime interest to everyone—settlers, cowboys, and Indians—in their search for grass, water, and shelter from cold winter storms. When all the horses, cattle, and sheep from the Anglo settlements combined with Native American livestock, the environment deteriorated rapidly.

Many variables must be considered when reconstructing the ecological effects of livestock from the historical record. Today scientific studies vary on the extent of these effects but agree on certain points.[20] For instance, cattle concentrate a lot of weight onto their four hooves. They have a tendency to cut and loosen the surface of the sandy soils of southeastern Utah, whereas other soil types compact. Sheep hooves, though smaller and bearing less weight, also cut deeply. If there is a slope or an embankment, hooves have a powerful mechanical ability to sheer off clods of earth.

Livestock also have a tendency to create trails by pounding and displacing soil. The net effect in a riparian corridor is that certain areas are overused, sudden rainstorms flush down well-trodden avenues, and erosion intensifies. Gullying from heavy animal traffic and overgrazing speeds the process, drying out the subsurface moisture. Thus, the entire landscape has less ability to support plant life.

Yet plant life is one of the main attractions of livestock to water. A recent study in semiarid rangelands showed that cattle favored riparian areas, which accounted for only 2 percent of the total grazing space but 81 percent of the damaged vegetation.[21] Food in these areas is more plentiful and often tenderer, water is close by, and there is added shade from the sun and protection from the wind. The drawbacks are that the riverbanks become badly trampled, grass and vegetation are removed, erosion increases, and the soil dries out. In the case of cottonwoods along the San Juan, cattle graze on the young, tender trees until the saplings are either dead or tall enough to keep the leaves out of reach.[22]

Arthur Spencer, known as Big Whiteman to the Navajos, displays rugs and pottery outside his trading post in 1917. Located near Mexican Hat Rock, this post, like many others, no longer exists and has left no trace. (San Juan Historical Commission)

Plant regimes also change. In a study conducted on the Boise River watershed in Idaho, for instance, bunchgrass, a desirable natural feed for livestock, was eaten first and replaced with downy-chess and needlegrass types of vegetation. Their root systems were shallower and less able to stabilize the soil, increasing runoff with all its erosive effects. Also much less water percolates into the soil with the type of plants in overgrazed soils.[23]

While early settlers and Navajos were largely unaware of these ecological factors, a few keen observers noticed the effects of increased livestock activity. Frank Hyde, raised at a trading post and on the ranges of the San Juan, came to the Montezuma Creek area in 1880. His description of the land in its relatively pristine condition is important, especially since he went into the cattle business shortly after his arrival:

> The river, when we moved into that country, was confined in a permanent channel, more so than it is now [1929]. There were willows and bullberry bushes on each side of it, [with] sloping grass banks. We could ride ponies across it most anywhere we came to without fords. As the country settled up, livestock tramped the grass down, made trails into the river; the timbers were cut off the headwaters, and the floods started to come. . . .[24]

Hyde then went on to tell that in 1884, his father's trading post, waterwheel, and farmlands were wiped out by the water-choked San Juan. His family remained for a short time, then moved down to Comb Ridge to open another post. This store also provided a ferry service, dependent on a cable system, at the mouth of Chinle Wash. And like others along the river, this post capitalized on the natural trail system used by travelers to cross the San Juan, water livestock, and trade outside the reservation.

Livestock were an important part of this operation. Hyde pointed out that cattle grazed as far away as Blue Mountain in the summertime but always returned to winter on the "sand flats" near the river and in adjoining washes. There was no doubt about the animals' importance to the trading-post economy: "My father traded a great many horses to the Indians for cattle and sheep, and we boys run them."[25]

To complicate the obvious competition among livestock owners over diminishing resources, two new economic possibilities arose: manufacturing wool cloth and running trading posts. One Anglo entrepreneur from Bluff wrote to the *Deseret News* in 1885 that the town was an ideal location for opening a woolen factory. He said, "Some of them [Navajos] own as many as twelve-to-fifteen-thousand head of sheep and goats. The wool can be purchased at the rate of 5 cents for white and 3 cents for black wool per pound."[26] While no factory materialized, trading posts purchased the wool, and this practice became an integral part of life along the river.

Thus, in addition to the San Juan serving as a magnet for the livestock industry, Anglo settlements drew Navajos to their trading posts. Beginning in the 1880s, posts became increasingly important for exchanging materials—wool, rugs, and silver—for products provided by Anglo-American society. Although they were usually friendly places, sometimes conflict flared. The stretch of river between Four Corners and Comb Ridge is a good example of the growth and problems encountered during this formative period between 1880 and 1895.

This was a colorful and important era, one when the reader can easily get distracted by details and lose sight of environmental trends. The first, and most important, is that posts attracted growing numbers of people into the region to trade. Also the stores were a nucleus for the development of communities later on: Montezuma Creek and Aneth are two of the clearest examples. And finally, sections of land along the riverine corridor were eventually added to the reservation because of conflicts over livestock. Unknown to the Anglos, some of this land was rich in oil, an economic boon to the Navajos starting in the 1950s. Thus, the posts that started as seemingly benign institutions actually had significant effects on the environmental history of the region.

Nine posts sprouted, bloomed, and died, some almost as quickly as they were created.[27] D. M. Riordan, a Navajo agent, complained in 1883 that "the Indians are persistently encouraged to leave the reservation by the small traders living around through the country surrounding the reserve. These men generally treat the Indians pleasantly and the Indians listen to them. It is `business' pure and simple with the trader."[28] From a purely environmental standpoint, the posts encouraged more and more human and livestock activity along the river, which was then the boundary between the Navajo and Anglo world.

Most of these posts followed a general pattern. The earliest structures were made of riverbed cottonwood logs and mud, eventually replaced by more substantial sandstone buildings, roofed with pine lumber from the mountains. Almost all the posts had skiffs, between twelve and sixteen feet long and five or six feet wide, flat bottomed and pointed on one end. Some of these were tethered by a cable system that prevented the boat from drifting downstream too far; to gain sufficient height above the river, the cable was suspended on a wood-and-rock crib, with sturdy poles protruding from the top.[29]

The largest of the cable ferries was at the Hyde Trading Post at Comb Wash. It was thirty feet long, twelve feet wide, flat bottomed, and held fifteen to twenty Indian ponies with their loads or two full-sized wagons. But even this sophisticated operation could not withstand the force of the river at full flood. Hyde reported that as the channel of the river filled with sand, the main current thrust against the banks, causing them to cave in; the river then took a new course until it was eventually forced back into the old streambed. The waves in the tumultuous river tipped over the cribbing that anchored the ferry's cable, causing all to be lost.[30]

Boats not attached to a cable drifted on the current and were loaded upstream from the desired destination, then angled to the far shore. Regardless of the boat's size and mooring, the river exacted its toll. During flood stage, the San Juan might claim three or four boats from one store in a season. Small wonder that some traders manufactured boats at their own posts.[31]

The Navajo reaction to riding in these boats was predictable. Many people mentioned their anxiety about crossing the river, with its supernatural power, at flood stage. Martha Nez, who

All that is left of the waterwheel system at the Hyde-Barton Trading Post, where the San Juan passes through Comb Ridge, is this anchor of logs and rocks. Many of the posts supplied some of their subsistence by cultivating crops. (James M. Aton photo)

lived on the south side of the river near Bluff, recalled, "Sometimes the river would have sand waves.[32] It was a scary experience crossing over it. The water sloshed against the boat and flopped over its edge. There was a man who lived close to the boat and rowed it. He was given five dollars for taking people's belongings to the other side."[33]

The four major trading posts—Noland's Four Corners Post, the Riverview-Aneth Post, the Montezuma Creek Post, and the Bluff Co-op—are good examples of the ebb and flow of commercial success. Each had its own history, continuous turnover of owners, and special personality, but all shared a common dependence upon Navajo trade and livestock for their survival. Each also drew its own clientele with their herds to foster economic development on the San Juan.

Claiming the title of the oldest-continuing mercantile business in San Juan County, the Aneth Trading Post began in 1885. Built of sculpted sandstone in an L-shaped configuration, the store sits on top of a bluff with a commanding view of the river. The Aneth post capitalized on a number of natural and man-made features. It was located on 160 acres of school-section property and so was exempt from becoming part of the reservation during the boundary changes of 1905. The wide floodplain at the base of the hill offered easy access, camping spots, and, at certain times of the year, fording sites along the riverbanks. McElmo Canyon provided a natural thoroughfare through the redrock country of southwestern Colorado and southeastern Utah, while McElmo Creek twisted its way to the San Juan, a water source for both agriculture and livestock. The road that followed the stream forded only shallow washes, effecting a year-round link with the Cortez-Mancos area beyond the slopes of Sleeping Ute Mountain.

Claiming the title of San Juan County's oldest still-existing business, the Aneth Trading Post has served Navajo customers for over a hundred years. This early photograph illustrates one of the essentials of a successful post—the ability to draw clientele. (San Juan Historical Commission)

Aneth, then, was at a communications choke point for travelers coming into the area.

An accounting of every trader at each post would read like a telephone directory. But a quick synopsis of the history of Aneth over a fifteen-year period illustrates the turnover rate. The first trader was an irascible troublemaker named Henry L. Mitchell. He hailed from Missouri, established a post on the floodplain below the present location, spent approximately six troublesome years antagonizing Navajo, Ute, and Mormon neighbors, and, eventually, with the help of other whites, killed or wounded some Navajo customers. That same year, 1884, the San Juan flooded its banks and took Mitchell's store, ranch, and crops, becoming one more inducement for him to leave. In 1885 he did just that.[34]

Shortly after Mitchell departed, Owen Edgar Noland, his son-in-law, relinquished control of the new post, built above the floodplain, to Peter and Herman Guillette, two brothers who also originally hailed from Missouri.[35] Both men freighted goods for a living and after they sold the Aneth post to Sterl Thomas, operated a flour mill in Mancos. Thomas in turn sold his rights to A. J. Ames and Jesse West.

Although Aneth by this time was suffering from a dwindling population and economic frustration due to the national depression of the 1890s, its importance lay in establishing a trade pattern. Navajos tell of ranging their herds on the far side of the river, then descending to a spot directly opposite the post, where they sheared their sheep. The traders provided ten-foot-long sacks to bag the wool, a boat to haul the goods across the river, and a burro to bring them up the hill to the store. There the owner weighed the wool, paid the customer around sixteen dollars a sack, and placed the goods in a nearby stone shed. Coffee, sugar, cloth, and flour were the staples of trade: to obtain these supplies, the Navajos took their pay in goods, at times never seeing the shine of a silver dollar or gold piece.[36]

Left Handed, a Navajo who visited the Aneth post in the late 1880s, described a typical transaction when the store was run by a man called Round (possibly one of the Guillette brothers). When Left Handed arrived, the trader came out, shook hands with his "friends," and helped carry the skins and hides into the store, where he weighed them. In exchange the proprietor gave them the usual commodities as well as a pair of overalls, shirt, red scarf, and box of .44 cartridges, after which the Navajos went on their way.[37]

In 1899 James M. Holley bought the Aneth post. Unlike many of the earlier traders, Holley took a great interest in developing the store as a center for the Navajo community and encouraging the Indians to adopt Anglo methods of farming and organization. One important contribution was his hiring Indians as workers. Not only did he pay them for improving the roads that led to his post, but he depended heavily upon their services as freighters. Holley hired Old Mexican, a Navajo whom he recognized as a hard worker, to freight goods between Aneth and Cortez, Mancos, or Durango in Colorado or Shiprock, New Mexico. For a three-day round trip to Cortez, he received ten dollars; for a six-day round trip to Mancos, twelve dollars.[38] He hauled sacks of wool, blankets, and hides and brought back two thousand pounds of flour and other supplies, including clothing and utensils for his personal use.

Another form of employment, derived directly from the river, was hauling Navajos and their goods from one side to the other. For instance, Jimmy Boatman, a Navajo, received his name for his long-standing service as a ferryman for wool, hides, and customers when the water peaked during spring runoff.[39] The traders, government farmers (men hired by the Indian Service to teach Navajos the latest agricultural techniques), and Navajos sponsored his entrepreneurial efforts by sharing the cost, which ranged between a quarter and two dollars. Jimmy started rowing on the south side of the river and tried to angle his wooden bark through the sand waves for the landing spot beneath the post. Occasionally he missed the mark and ended up downstream, much to the chagrin of the customers, who had to help drag the boat back. By mid-June, when the shearing season was over and the river level went down, he became unemployed, but when the waters rose, Jimmy was back in business.[40]

Owen Noland opened another post around the same time (1884) that Aneth started. Often referred to as the Four Corners Trading Post, it was located approximately four miles downstream from the current monument. After selecting a spot on the river with fordable access for both Ute and Navajo trade, Noland built a structure of large cottonwood logs. He soon replaced it with a sturdier edifice of rocks, quarried from the sandstone formations to the west on the San Juan. The western wall, the stem of an L-shaped configuration, was 117 feet long with walls more than two feet thick and eleven feet high, while the short stem was 65 feet in length. Eight large windows dotted the walls, with firing ports in places where there were no openings to see outside. Adobe covered the sandstone and coated the three fireplaces that heated the spacious rooms where blankets, silver, and trade goods lined the shelves.[41]

The San Juan Co-op was another post that started in the early 1880s but did not close until the 1920s. Founded on 29 April 1882, just two years after the Mormons settled the town, the San Juan Cooperative Company or Bluff Co-op reflected the leadership strengths of this tiny community. Platte D. Lyman was president; Jens Nielson, vice president; and C. E. Walton, Kumen Jones, and Ben Perkins, directors. These men founded the co-op for the purpose of "engaging in general merchandising" and divided their profits amongst its shareholders.[42]

Like other posts, it bought wool, pelts, and blankets from the Navajos and deer hides from the Utes and depended heavily on goods shipped from Mancos, Cortez, and Durango. But unlike the others, each stockholder took his turn at freighting and was paid accordingly. The first dividend came in five months, paying at a rate of 10 percent; within a year it had jumped to 25 percent.[43] The facility evolved from a roughly hewn log structure to a large, two-story rock building with a store and post office below and a meeting and social hall above.

Albert R. Lyman, a local historian raised in Bluff during its earliest years, believed that this store, with its freighting and stockholder revenues, allowed the town to survive economically until it made the transition from an agricultural community to the more profitable cattle industry.

The San Juan Co-op began in 1882 as a Mormon enterprise to trade with the Indians. Business success eventually gave rise to this building, which housed not only the store and post office but also a dance hall and stage in the upper story. (Charles Goodman photo, San Juan Historical Commission)

Lyman gave a colorful description of the post in its heyday:

> The Navajos came with their produce to trade in the little log store, which was generally surrounded with a motley tangle of cayuse saddle ponies, rawhide ropes, bundles of wool pelts, and snarling, mangy dogs. Trading was, to the Navajos, a rather festive occasion, deliberate and drawn-out. They camped nearby until it was finished to their satisfaction, crowding against the rude lumber counters in noisy talk and laughter, and always in a stifling cloud of tobacco smoke.[44]

The citizens of Bluff generally encouraged the trading business. In 1902 the co-op and citizens of the town held a fair, where Navajos exhibited their rugs, jewelry, silverware, and beads. It was such a success that the Aneth post followed suit.[45] This was seven years before the first Shiprock Fair, an institution that continues today.

The effectiveness of this type of business and the peaceful attitude of the Mormons are reflected in a report a few years later by a military group evaluating Navajo life in this area. Captain E. A. Sturgis visited Bluff and found that 950 adult Navajos had traded there in 1908, although only half of them lived within a sixty-mile radius.[46] As with other posts, the co-op ran a wooden boat, piloted, at one point, by a Navajo named Red Spotted Neck. The Bluff Co-op clearly served as a drawing card for the community.[47]

The Navajo Faith Mission, by 1901, showed the prosperity and attempted "civilization" that Howard Ray Antes desired for his charges. The prominent, lime-plastered building on the left served as living quarters for family and students; the schoolhouse is on the right. (San Juan Historical Commission)

Government policy also indirectly encouraged Native Americans to take up individual allotments on public domain. The Dawes Act of 1887 fostered the ideal of the Indian making his living as a farmer. Along the San Juan, Navajos had always used dry farming and small irrigation canals to water their crops on the floodplain and in tributary canyons. Now ever-increasing herds of sheep and a growing interest in farming intensified competition between Anglo settlers and Indians over resources. Commissioner of Indian Affairs T. J. Morgan watched the situation fester to a bursting point and claimed, "In the meantime I know of no other way to maintain peace between the non-reservation Navajo Indians who are on the public lands and the white residents except by the aid of the military."[48]

No one doubted the necessity for some type of controlling agency in this far-flung corner of the Navajo Reservation. Problems over land, water, trade, hunting, cultural values, and government control underscored the need for someone who could deal with issues in the Four Corners area before they became inflamed. Starting in 1903, the Shiprock Agency was founded by subagent William T. Shelton to address these problems.

Another person who settled along the San Juan to be an advocate for the Indian was Howard Ray Antes, a Methodist missionary. Antes and his wife, Evelyn, came to the Aneth area in 1895.[49] They built their first home of logs but soon started construction on a much larger and more elaborate sandstone structure. How much actual preaching Antes did to the few whites and numerous Indians is questionable, but the Navajos did name him Hasteen (Hastiin) Domingo or Mister Sunday. Facilities at the mission continued to grow. By 1904 the site boasted a large house, a smaller school building, and surrounding farmlands and orchards on the river's floodplain. Antes, however, never took up homestead rights on this property.[50]

At the time of his arrival, no real spokesman for the Navajos lived along the river, though the government owned a vast amount of territory in San Juan County available for settlement. The county commissioners oversaw activities and collected revenue for use, but land was open to any applicant. At the same time, reservation lands strained to feed the expanding livestock herds, unclaimed water holes were nonexistent or inadequate, and agents could not effectively patrol the boundaries. This situation, coupled with the attitude that Native Americans needed to become self-sufficient by taking out individual allotments, encouraged Navajos and agents alike to look for solutions across the San Juan.

By 1898 Antes took pen in hand on behalf of the Navajos. He accused Fred Adams, county tax assessor from Bluff, of locating Indian livestock north of the river and charging an inflated license fee of three or four sheep or goats per one hundred. Anglo livestock owners, on the other hand, paid only two-and-a-half cents per head. To Antes this was pure and simple extortion designed to force the Navajos with their large herds back on the reservation. He was told that the "interference of a missionary" was unnecessary, and so he wrote Secretary of the Interior C. R. Bliss, requesting that he intervene.[51]

Antes argued that the land was so barren and rocky it was suitable for nothing but grazing. He maintained that "fifty miles above us and twenty-five miles below us along the San Juan River, there are but two [white] men who have a few acres of cultivation" and a couple of trading posts; that Indian flocks would starve on the sandy, rocky wastes of the reservation; and so the Navajo should have untaxed access to the resources north of the river.[52]

Antes got what he wanted. Federal officials agreed that the Indians had the right to be there, they should not be taxed, and Adams had overstepped his legal bounds by using "false pretense."[53] Antes then assumed the responsibility of writing passes "on the authority of the Commissioner of Indian Affairs and the Secretary of the Interior of the United States" for Navajos wishing to graze livestock on the north side of the river.[54] The county commission was irate.

Yet for Antes and the Navajos, the time was right. Superintendent Shelton was well aware of this area's potential, too. While Antes was reporting the Navajos' wishes and championing their cause, Shelton wrote a letter to President Theodore Roosevelt on 10 April 1904, asking for an extension of the reservation. Chester A. Arthur had granted the first extension through executive order in 1884, which had moved the boundary to the San Juan River. These lands in the Aneth-Montezuma Creek area were the first that Navajos requested north of the river. Antes anticipated this additional land would lead to less friction between stockmen and Indians and more desirable economic conditions for the Navajos.[55]

Correspondence followed correspondence, but after much discussion and a few revisions due to survey problems, President Roosevelt signed Executive Order 324A on 15 May 1905, creating a new section of the reservation. Known today as the Aneth Addition, these lands encompass the region beginning at the mouth of Montezuma Creek, east to the Colorado state line, south along the boundary, then down the San Juan to Montezuma Creek. Lands previously claimed or settled were excluded from the reservation.[56] Antes had fulfilled his goal of annexation.

What were the implications for the environment? Navajos now controlled both sides of the river from Four Corners to below Montezuma Creek. Access to more land and a burgeoning population encouraged the Navajos to move north of the river and lay the foundation for two new towns—Aneth and Montezuma Creek. Conflict between Anglo and Navajo stockmen continued over the ranges north of the river, where every blade of grass and water seep grew in importance. Also the oil that lay beneath in what would later be called the Greater Aneth Oil Field was now a Navajo treasure waiting to be discovered. But much of this lay in the future.

By 1900 trading posts as an institution entered a golden era that would not decline until the livestock reduction of the 1930s. While stores along the river continued to flourish in Aneth, Montezuma Creek, Bluff, and Mexican Hat, the trend now was to expand from the borders into the heart of the reservation. Relaxing government controls and requirements that encouraged traders to live among their clients on the reservation initiated the new growth. A contributing factor was the expanding network of roads that started out as horse trails, upgraded to wagon roads, and eventually became maintained dirt highways by the mid-1950s.

Howard Ray Antes, "Mister Sunday," believed that the real salvation of the Navajo rested in the children. Perhaps that is why he named his location Aneth, a Hebrew word meaning "the answer." (San Juan Historical Commission)

The establishment of a transportation system had started much earlier on the river. The isolated posts along the San Juan required these economic lifelines that snaked their way across floodplains and through canyons to reach the market towns of Mancos, Cortez, and Durango.

As early as September 1882, the Bluff and Montezuma precincts spent $125 of their limited funds to pay settlers to improve what was called the Old Bluff Road.[57] This expenditure facilitated the mail service that started a month later, linking Bluff through Mancos to the outside world.[58] Even more important was the necessity to freight goods into and out of one of the roughest geographical parts of the Four Corners area. What with the steep hills and canyon walls, the mud and floods of the river, and the sand and rock in the washes, those responsible for pioneering and maintaining these fragile trails had their hands full.

Early descriptions of the roads are replete with the agony of those who traveled them. The old road to Colorado went from Bluff to Aneth and split after that. One fork wended its way up McElmo Canyon to Yellow Jacket Canyon, where Ismay is today, and then to Cortez and Mancos. Another road followed the river to the Four Corners Trading Post, then curled around the southern end of Sleeping Ute Mountain to Mancos. Another early road, with numerous modifications, curved its way up Recapture Wash from Bluff to Monticello, then continued to Thompson above Moab, where it eventually tied these towns into the Denver and Rio Grande Railroad in the 1880s.[59]

One group of visitors to the Aneth area lamented that for every step forward, one seemed to slide back two more in the sand; that every rock and bush had a rattlesnake behind it; that the water tasted poorly; and that the surrounding hills and cliffs were tedious to the traveler's soul. The spokesman concluded by saying, "I would rather walk five miles on an Ohio pike than one mile on any of the 'roads' in southern Utah or northern New Mexico. . . . [I] sigh for the green fields and shady woods of the East."[60]

This 1928 photo shows part of the abandoned Colorado-Utah road one mile below the mouth of McElmo Creek, looking upstream towards Aneth. Notice both the scarcity of vegetation along the river and the road's susceptibility to flooding. (Herbert E. Gregory photo, #558, U. S. Geological Survey)

Another traveler in that same year of 1892 felt differently, claiming that the thoroughfare "would do credit to a much richer settlement." He noticed that the road in the river bottom actually had to be carved out of the bluffs and riprap lined areas where the water could wash the bank away.[61]

Regardless of problems associated with weather, water, and terrain, interest in roads did not wane. The San Juan County Commissioners' minutes are filled with instances of leaders spending significant amounts of money, time, and effort to improve conditions, and often traders took the lead. For example, Henry Mitchell accepted the responsibility for maintaining the roads in the McElmo area. That same year, 1885, William Hyde and twelve other men secured four hundred dollars to build a route along the river to Four Corners, bypassing the road up McElmo and offering easier access to Noland's post.[62] Freighters continued to use the Old Bluff Road, marked with a stack of rocks every mile, starting in the town and proceeding all along the canyon.[63] Traders also paid crews of Navajos to improve roads near specific posts.

Later routes also linked the Four Corners region into a national network. By the 1920s many of the supplies for the posts came from either Grand Junction, Colorado, or Fruitland and Farmington, New Mexico, because of the railroad.[64] For example, the government built the first of a series of bridges across the San Juan, Colorado, and Little Colorado Rivers, eliminating the need for ferries and opening up the northern part of the reservation to more vehicles. The first bridge began construction in 1909 near the Shiprock Agency School, followed by one at Mexican Hat

Rush hour on the Mexican Hat Bridge, constructed in 1909–10. Oil, not sheep, was the economic boost that encouraged the building of this suspension cable bridge. (San Juan Historical Commission)

(1909–10), Tanner's Crossing (1910), and Lees Ferry (1925).[65] As the Anasazi ruins of Mesa Verde became increasingly popular, the government made arrangements to connect Gallup to southwestern Colorado. Starting in 1916, funds paved the way for a major construction project that opened the area to tourism, but the route was not completed until the summer of 1930.[66]

This expanding road network opened many resources to more people. What had been available to only a handful now became known to many. Since transportation is one of the keys to economic development, the growing sophistication in number, length, construction, and placement of roads spurred growth in the region. In turn, the entire process had a greater impact on the land and its resources.

In strictly economic terms, the burgeoning road network proved salubrious. As early as 1896, Colorado newspapers touted the effects of the San Juan trading posts on the economy, claiming that freighting outfits "loaded out from the Bauer Store [Mancos] often $1000 worth of goods a day."[67] By 1913 the *Mancos Times-Tribune* felt that trade "naturally gravitated" to this area, with sometimes as many as six or seven heavily laden wagons groaning their way to the river. This economic boom made Mancos the "recognized commercial and financial center" of Montezuma County, Colorado.[68]

Like many businesses, the trading posts hit a growth plateau and then declined. Some events were precipitous and others slow and inexorable, but all helped end the golden days of this institution. Most dramatic was livestock reduction of the 1930s, when the John Collier administration of the Bureau of Indian Affairs began a program that cut into the mainstay of the Navajo economy—sheep, goats, and wool.

Livestock was one of the foundations of traditional Navajo livelihood during the first quarter of the twentieth century. Horses provided transportation and food for the winter months, while goats and sheep were a continuing source of sustenance, blankets, and clothing and an entry to the barter economy of the trading post. Livestock also became synonymous with social status and emotional satisfaction, as Navajos watched their herds multiply and prosper. Suddenly, it all ended.

Indian agent B. P. Six had little understanding of the role of livestock in Navajo culture at this time. He did see, however, that the herds were expanding. During 1930 in the Montezuma Creek and Aneth area alone, 19,514 sheep and goats passed through dip vats filled with medicine to prevent scabies. The Oljeto and Shonto areas produced 43,623 more animals, while some Utah Navajos undoubtedly used vats at Kayenta, Shiprock, Dennehotso, and Teec Nos Pos. Still others probably skipped the process entirely, but if the totals from the Aneth and Oljeto areas are combined, at least 63,137 sheep and goats ranged over the reservation lands of southeastern Utah.[69]

What would soon end in the cold, hard statistics of lost livestock had its genesis in a scientific attempt to save the range from these herds. Depleted vegetation, soil erosion, silt accumulation at Hoover Dam, expanding herds, restrictions on off-reservation grazing, poor animal quality, and the faltering national economy were all part of the motivation to reduce livestock and modernize the Navajos' livelihood and management of resources.[70]

It should also be noted that range restrictions were not confined to Navajo herds. In 1934 the government placed controls on land use by Anglo stockmen through the Taylor Grazing Act. Science now dictated the carrying capacity of public land utilized by cattle, sheep, goats, and horses. The amount of feed was measured by animal unit month (AUM): the cost of feeding one cow, one horse, or five sheep for one month on a specific piece of land. Grazing districts subdivided the ranges and parceled them out in one-to-ten-year leases. The Grazing Service, later combined with the General Land Office to become the Bureau of Land Management (BLM), determined the capacity of the range, assigning permits and collecting fees from ranchers. In 1936 the Navajo Tribe adopted a similar system of livestock control and range management, bringing both sides of the San Juan into conformity.[71]

For older, traditional Navajos, however, who lacked a western cultural orientation, the whole process was difficult to understand and even harder to accept. Starting in 1933, Navajo goat herds were the first to be selected and gathered, then killed. A year later sheep came under the knife, followed by horses and cattle. The reduction that had started out voluntarily, as just one more incomprehensible government program, soon became a major threat to the Navajos' subsistence economy. Wealthier Navajos were more powerful and had better means of hiding their herds, so the poorer people, those who could least afford the losses and maintain self-sufficiency, were the ones who suffered. Impoverishment and dependency on the government became a part of reservation life.

In 1934 the Northern Navajo Agency reported that government officials had killed or sold seventy thousand animals and the Utah Navajos' herds were down to an estimated thirty-six thousand.[72] Because the nation's economy was wallowing in the depths of the Great Depression, the agent could price a sheep at only two dollars and a goat at one dollar. The annual report went on to say that "an excessive number of goats and sheep were slaughtered for food. There is every reason to believe that the next dipping record will show even a greater reduction than indicated by the number sold."[73] Horses and cattle suffered a similar fate.

What was the Navajos' reaction? Stunned disbelief and shock. Since cultural traditions taught that sheep bring rain and plants through prayer, it followed that there was plenty of vegetation before the sheep were killed. This was an exact reversal of the government's theory about overgrazing and soil depletion. One Navajo explained, "During the midsummer, vegetation, like sunflowers, colored the place. It grew in such abundance that the livestock walked in tunnel-like paths amidst it. . . . There is very little now for a sheep to take a bite of. All of this is due to the lack of precipitation. . . . Maybe they [Anglos] reduced that, too."[74] When Collier killed the sheep, he also affected rain and vegetation, one of the main things he was trying to protect. According to the Navajo, he used this excuse to "cheat" them at a time when there was abundant forage. Without the livestock's prayers for rain, the whole weather cycle collapsed.[75]

Since that time, everything has been different for the Navajos. The grass is gone. Russian thistle has become sharper and tougher, able to puncture a tire. It is so tough it can kill horses and sheep that eat it and make people ill if it scratches them. Weeds infest the soil, and droughts are common.[76] The land is desolate and reflects the older people's feelings about what happened to their way of life because of reduction.

Livestock loss not only forced Navajos into a wage economy but also pushed trading posts into a new system of cash and credit. The Civilian Conservation Corps, World War II factory work, seasonal migratory jobs, employment on railroad crews, and, in the 1950s, uranium mining offered the Navajos an alternative economy and lifestyle. The trading posts were a flexible-enough institution to struggle through these changes, and their final collapse didn't occur until the 1970s.

In a purely ecological sense, government control of the livestock industry on the reservation lessened the problem of overgrazing and gave the land a rest. Yet many of the elements associated with a wage economy, such as oil exploration and its accompanying industry, brought their own headaches.

Today important environmental issues concerning livestock and their effect on the riparian corridor of the San Juan remain. Increased tourism and environmental ethics have come into direct conflict with stockmen grazing cattle in more highly traveled areas, especially Comb Wash. Five side canyons—Arch, Mule, Fish Creek, Owl, and Road—were part of the range used by the White Mesa Cattle Company, owned and operated by the Ute Mountain Tribe. Beginning in 1991, a lawsuit charged the BLM with not properly enforcing grazing regulations to the detriment of water quality, wildlife habitat, soil stability, and scenery. The end result of the litigation is that grazing in the five canyons is forbidden, with a new watershed management plan now under development for the entire Comb Wash area. While this does not directly affect the San Juan, it once again highlights the continuing conflict over area resources in the river's drainages.[77]

There is, however, another side to the issue. Ever since the 1950s and 1960s, when the BLM began a series of field inventories and range surveys to determine the carrying capacity of the land, a mutual understanding that many of the problems of the past can be overcome with proper management has been growing.[78] Today there is seasonal (fall, winter, spring) grazing of cattle

Livestock reduction of the 1930s ended a way of life for the Navajos. No longer able to live through agriculture and animal husbandry, they were forced into a wage economy that took many off of the reservation. (National Archives, U.S. Signal Corps, #111-SC-89583)

(not sheep), managed by the BLM, along the north side of the San Juan River. Four allotments allow cattle access to the river so they can have water in a high-desert environment. Because the number of livestock and season of use are regulated, much of the harm done previously by uncontrolled access in a highly competitive environment is now an issue of the past.

Indeed BLM officials comment that the ranges are in generally good condition and welcome the presence of cattle. They point out that grazing forage plants stimulates growth, that cattle moving in the area help plant seeds by burying and covering them with soil, and that, with sufficient moisture, the ranges spring back better than ever. Riverbanks do not show any significant deterioration or sloughing, and the introduction of foreign weeds, such as camelthorn and Russian knapweed, is not a problem created by livestock; these invaders appear to be coming from upstream.

The number of livestock on the south side of the river is also theoretically controlled by permits, but in this case they are issued by the Navajo Nation. Unfortunately, Navajo lands on both the south and north (Aneth-Montezuma Creek) sides of the river are badly overgrazed.[79] The problem lies in enforcement of the allocation system introduced in the 1930s. Compliance with grazing laws is handled on the local level through the chapter system. Some range managers do an excellent job of ensuring that the required limit on animals is maintained. Other officials have a difficult time enforcing rules within their own community because of social pressures. Private transportation is the only means for these officials to get out into the canyons and mesas to check herd size, many of these district officials are not trained in soil conservation, and some Navajos rationalize that members of a growing family should have the same number of AUMs as their predecessors, thus increasing the actual number of animals on the same amount of range.

The end result is land that has been classified by experts as "low/fair to poor." Although the banks are stable because of tamarisk and other vegetation, there are extensive gullying and subsequent silt discharge into the river from eroded areas. The northern side of the river managed by the BLM is a one-way magnet that draws livestock to greener pastures. But what could be a troublesome conflict is being handled through peaceful cooperation. Both parties with livestock along the San Juan are sharing the responsibility of controlling the animals that range across the river when the water is low. This happens particularly around Recapture and Comb Washes. To prevent overgrazing from recurring, the BLM is providing individual Navajo families and Anglo livestock owners with materials to construct a three-mile fence in the Recapture area; the stockmen, in turn, supply the labor. The fences are built away from the river so they will not be washed out by high water. Thus, while livestock have been and will continue to be an issue in the San Juan region, there are strong indications that cooperation between the government and individual livestock owners is the key to defusing the problems.

A look at the past reveals that the early, most-intense period of grazing was associated with trading posts and road development along the river. This entire epoch revolved around the opening up of a relatively untouched reserve of grass. Navajo and Anglo herds of sheep, horses, and cattle descended upon the land from the north, east, and south to take advantage of open range. Trading posts and transportation followed, moving to market the products of this economy. Without vegetation for grazing and water to support life, there would have been little reason to establish the posts with their barter economy and the roads to feed the trade. There would also have been fewer reasons for Navajo and Anglo stockmen to expand as quickly as they did into peripheral areas of the San Juan. Each group came, saw, and acquired through various means what they could, precipitating a productive, yet destructive, period of history.

Not until government control divided and regulated land use based upon carrying capacity did the issue assume its present form. Now one of the big questions asked by local residents is whether there is too much control, too much intervention, for the livestock industry to exist. As the San Juan rolls into the twenty-first century, there will doubtless be other environmental issues, but never again will riverbanks feel the push, dig, and stomp of so many animal hooves in search of grass.

5 AGRICULTURE: *Ditches, Droughts, and Disasters*

The Southwest is known for its arid climate, dramatic beauty, and turbulent weather. To the inhabitants who wrest a living from this land, its unpredictability, especially supplying water, provides one of the greatest challenges. The Colorado Plateau and the Four Corners area are consummate examples. The San Juan River is the only major, continuously flowing source of water that courses through Colorado and New Mexico and then crosses into Utah at Four Corners. Melting snows in the spring and intense thunderstorms in the summer and autumn make the river rise and fall sharply. As the moisture pours off the San Juan and Sleeping Ute Mountains in Colorado, and the La Sal and Blue Mountains in Utah, dozens of tributaries swell the tide that scours the riverbanks and tears at the floodplains.

One of the most graphic examples of this phenomenon occurred in the fall of 1941. Between September 9 and October 14, the San Juan River changed from a placid, shallow stream 3 feet deep and 125 feet wide, flowing at 635 cubic feet per second, to a raging torrent 25 feet deep and 240 feet wide, gushing at 59,600 cubic feet per second.[1] The river ravaged hitherto protected floodplains, with only the highest banks able to contain the water. Few irrigation facilities and bridges survived the onslaught. The abrasive action of the stream's sediment load widened and deepened the channel, while the suspended matter swept down the stream, depositing its refuse as the waters receded. Eventually part of the streambed refilled as the river brought in new sand, silt, and rocks, but it took years to replace what had been removed so quickly.

The implications of depending upon a river like the San Juan are important. Until the government constructed dams to enact flood control, the river had its own say, exercised its own will. Although it could be destructive, at times it was also benevolent, bringing life-giving water and materials to those who came to its banks. The Anasazi used the river and its tributaries for two types of farming: pot and flood irrigation. The Navajos, much more so than the Utes, followed suit, locating their farms along the river bottom. Although the mean annual rainfall, only eight inches in the Aneth area, was lower than on other parts of the reservation, the river provided water continuously, while its lower elevation, 4,700 feet, offered a 161-day growing season.[2]

Getting the water onto the land was a whole other issue. Pot irrigation was inconvenient. Carrying water to fields was time consuming, yet more predictable than dry-farming techniques that depended on moisture in the soil and summer showers to keep crops alive. Irrigation was often more dependable but also entailed hazards. Navajos cleared and prepared their farms in April. Ditches from the river snaked across the floodplain, taking advantage of the natural slope in the land and direction of the river's flow. The Indians dammed arroyos and worked the waters over the fields in a process repeated once or twice during the summer.[3]

Alluvial fans extending from the mouths of intermittent or continual canyon streams, such as Recapture, McCracken, Montezuma, Allen, and McElmo Creeks on the north side, and Desert, Lone Mountain, and Tsitah Creeks on the south side of the river, encouraged settlements and farming there. Irrigation systems were also easier to install at these places because

Jim Joe and his family in their camp on the river near Butler Wash in 1921. Depending heavily on agriculture and livestock, he successfully utilized the resources in a number of ecosystems. (Hugh D. Miser photo, #564, U. S. Geological Survey)

the banks were lower, the soil was rich, and the water was less turbulent.

Other aboriginal farming sites below the Bluff area included the Comb Ridge/Chinle Wash vicinity, Butler Wash, Beaver Creek, and Paiute Farms. Obviously the farther downriver a traveler ventured, the fewer the farms because of steep canyon walls and difficult access. Although most of the farms which developed on these lower floodplains and alluvial fans were small, some were highly fruitful for a subsistence economy. For instance, Jim Joe, a Navajo friendly to the Bluff Mormons, reported in 1904 that he grew ten to twelve tons of corn at his residence at the mouth of Butler Wash.[4]

Paiute Farms, shared by Paiute and Navajo Indians at different times, was another favored agricultural spot. Tucked in the small valley bordering Nugget Creek, a tributary of the San Juan, Paiute Farms sat about a half mile from the river. It provided only a few hundred yards of planting space, with sufficient water to grow corn, pumpkins, and melons. White men passing through the area in 1894–95 noted Navajo homes scattered amid the stands of large cottonwoods. They also mentioned a conspicuous absence of willows. The flood of 1911 washed out these farms, leaving only the name to hint that agriculture had sustained life there.[5]

Obviously geography determined the extent and location of people's ability to sustain themselves in a sometimes-stingy land. Water in the desert is useless without access, and so the human drama that played along the San Juan pivoted around not only the presence of this precious resource but also the way to get it to the right place, at the right time, in suitable quantities to grow crops. Wide floodplains and a slower river flow offered the best chances for agriculture. Thus, the majority of Native American and Anglo-American agricultural ventures occurred on the upper end of the Lower San Juan near Bluff, Montezuma Creek, and Aneth. But even there, it was a nearly impossible struggle.

Actual planting by Native Americans began in early May and continued through the first part of July, when the "first fruits of the slim yucca

For centuries corn has been a major source of food for the Navajos. Its primary importance is reflected in religious teachings that tell that the holy beings created man and woman from this plant and compare the clan system to its growth. (Milton "Jack" Snow with Andy Tsinnijinni photo, # NA 4-14, Navajo Nation Museum)

burst open." The Navajos planted corn, then melons, then squash, and finally beans, based upon which had the longest maturation period. The gardener placed anywhere from five to fifteen seeds together in hills; those seeds that did not germinate were said to have been "eaten" by those that did. Men used digging sticks to create a hole approximately four to six inches deep, as women followed behind and placed the seeds.[6] Because livestock was an even more important part of their economy, the Navajos spent a lot of time ranging away from the plots on the river but returned occasionally to weed and water. Sometimes women, old people, and children stayed behind to tend the crops.

This general pattern changed according to specific conditions. Friction with Ute neighbors, demands of the livestock industry, shifting boundaries of the reservation, and a growing population base exerted pressures in different geographical directions. The overall effect was that the Navajos expanded outward from the heart of the reservation to the boundaries. At the same time, Anglo farmers and stockmen on the north side of the San Juan claimed that the public domain belonged to them, since the Indians had their own lands.

Yet Navajo agents did not agree. The government still wrestled with the idea of removing Indians from the reservation and nudging them into mainstream American society as farmers and mechanics. The Dawes Act of 1887 was designed to do just that, and reservations in other parts of the United States were broken into individual allotments. For the Navajos, however, the reservation not only stayed intact but expanded, while relatively few individuals took up allotments.

By 1892 government officials decided to build upon the already-established Navajo pattern of livestock and agriculture. Commissioner of Indian Affairs T. J. Morgan suggested a long-term approach to solving the problem of feeding

The L. H. Redd family visiting a neighbor, H. D. Harshberger, at his cornfield southwest of Bluff in 1898. The river in the background belies the fact that an artesian well was a necessary part of the operation. (San Juan Historical Commission)

and controlling this growing population. First, he believed the reservation should be carefully mapped with an emphasis on springs, water holes, and streams that could provide water for farms and livestock. Next, a system of dams, wells, windmills, and other water-procuring devices should be integrated into a program to make the Navajos self-sufficient. And finally, every effort should be taken to make Indian lands productive so that Navajos would not compete with Anglo neighbors.[7]

The commissioner charged the army with the task. Lieutenant Odon Gurovitz surveyed the south side of the San Juan and recommended that 260 acres near Bluff be turned into farmlands and that the Mormons supervise the project.[8] This decision was somewhat ironic because the last thing the Latter-day Saints wanted to do was attract more Navajos into an area where conflict with Utes and Navajos had already created bitter years of strife. Still, the river as a constant source of water could not be overlooked, especially since James Francis, a farmer in Fruitland, New Mexico, was already enjoying limited success.

Agent E. H. Plummer begged for money in 1893 to develop the possibilities. He argued that these improvements would act as the carrot to bring the Navajos home, since some people estimated that a third of the population was living off the reservation.[9] Plows, scrapers, wagons, and seed would be another inducement, and if three or four additional government farmers scattered to strategic locations where Navajos clustered, the south side of the river would become a magnet to draw back this transient population.[10]

Constant Williams replaced Plummer as Navajo agent the next year but continued to agitate for farming on the San Juan. On 11 December 1894, he went to Bluff, where he found the Indians "pitiable" because of crop failures over the previous two or three years.[11] It was ironic, however, for Williams to go to Bluff and suggest that large-scale farming was a viable means of livelihood, given the community's struggle to maintain itself through agriculture. Indeed, by this time, the Mormons had started to depend on the livestock industry, and many had settled away from the San Juan, where water was easier to control.

At this point, it is useful to pause and examine the Mormon pattern of experience that had started in 1880 and continuously faltered and failed up to this time. The settlers' struggle against the San Juan is a microcosm of what the Navajos encountered in a few short years. When the large Mormon contingent settled in Bluff, it started immediately to plow ditches and prepare for spring planting. Community cooperation and organization characterized this first year, but the ditches were unsatisfactory for a group of people who wanted to move beyond subsistence agriculture.

Some of the settlers cast about for a better solution to the ditch problem. William Hyde activated a large waterwheel, sixteen feet in diameter and twelve feet across, capable of sloshing twenty-three hundred gallons an hour onto the parched red soil of Montezuma Creek. This area was more fortunate than Bluff because it had rock shelves on which to anchor waterwheels, while the latter had to depend upon riprap dams and backbreaking shoveling to keep water on the fields and sand out of the ditches. Soon Harrison Harriman, James Davis, Frank Hyde, William Adams, Samuel Cox, and John Allen had each built a waterwheel on different sections of the river. Allen said of his wheel, "It's aya fine; I'd wish nothing better."[12] Adams declared that for less than three hundred dollars, a wheel could be built that would water two hundred acres of land and avoid the cost and labor of ditches. He believed, "These waterwheels are a success and cheaper to keep in repair and less liable to damage."[13] Indeed the only truly successful farming in 1881 that did not result from using a wheel occurred at the mouth of Recapture Creek.[14]

On the other hand, Frank Hyde later built a waterwheel on a twelve-foot ledge at Rincon, where its service was short lived. He awoke one morning to find the river one hundred yards away and only a dry sandbar where water had previously flowed.[15] The problem of high water/low water plagued both ditch and wheel operations, so it was only a matter of time before the river had its way. In 1884 the unpredictable San Juan claimed its share of wheels, sweeping all of them down the river in a torrential flood.

For a few years preceding this event, canals, ditches, and riprap dams seemed to hold the only possibility of success on stretches of the river where good rock foundations for wheels were not available. As early as 1879, the Mormon exploring party, looking for a place to settle, encountered Henry L. Mitchell in the Aneth area. While Mitchell had many, and would later create many, problems, one of his biggest at this point was his ditch, which he had "surveyed the wrong end up and the water would not follow."[16] To rectify the situation and raise the water, he tried to dam the entire river by building a barrier. The San Juan was determined not to be conquered and twice swept away the middle forty-foot section. Just when success seemed possible, the water level dropped, leaving the canal high, dry, and useless.

From an environmental standpoint, constructing these dams was ludicrous. The time—six weeks—and the effort—twenty-five men with teams—could not hope to harness permanently a river of that size, power, and unpredictability, given the materials at hand. What made the Mitchell attempt even more important to environmental history was where he got these materials. Eyewitness accounts estimated the dam to be two hundred to four hundred feet long and composed of "over 5,000 loads of young cottonwood trees and rock."[17] Cutting this many trees and hauling that many rocks from the riverine landscape did nothing but encourage the debilitating effects of erosion.

The Mormons took a similar approach two years later. By 1881 a new canal, costing from twelve to fifty dollars a rod, needed to be dug.[18] The headgate of this ditch was located four miles above the town at Walton's Slough, where the main canal passed over a long stretch of slickrock. The builders hauled logs, brush, rocks, and earth to construct the riprap channel that extended into the river to funnel the water. Three such walls controlled the water and allowed it to be turned into individual fields.

Uncontrolled flooding as well as scant river flow plagued farming efforts of both Mormons and Navajos. Large scale agriculture did not become a reality until dams and late twentieth century technology were introduced. (Milton "Jack" Snow photo, #NA 6-3, Navajo Nation Museum)

Men cut cottonwood trees from the riverbank and wove an estimated one thousand of them into the framework to hold tons of rocks and dirt. To encourage cooperation, the leaders sold stock for the new ditch, while church officials allowed some people to be rebaptized as part of the commitment to this new undertaking.

All winter long the men toiled. When April arrived, bringing thoughts of spring planting, the workers channeled the water down the ditch and watched it disappear through the porous walls of riprap. As the spaces filled with sediment, the water inched its way to the fields close to town. In May the river gnawed away the top of the ditch. The water then started to recede, so shovels deepened its course, and the crops succeeded.

The next year problems intensified. Banks broke, ditches filled with sand, crops withered, taxes increased to support the effort, and stockholders appointed new leaders in an effort to save the economy. Erosion also played a part, encouraging townspeople to take turns patrolling the ditch for "boulders the size of a wagon bed to a two story house that had recently rolled down."[19] The final straw, however, was the flood of 1884. The river carved up the canal, tore out the headgate, and covered what remained with sand.[20] A year later agriculture ceased to be the primary economic dream of the Bluff Mormons.

The discouraged settlers suggested that the community move away from the river and utilize a more-placid source of water. They considered Yellow Jacket Canyon until they learned that its owners wanted thirty thousand dollars for the land. F. A. Hammond, a newly arrived Mormon leader, decided that the anticipated twenty miles of floodplain farmlands would never materialize as he watched only three hundred acres being farmed successfully. He turned to the livestock industry and encouraged others to do likewise. Bluff blossomed as it shifted its attention away from the brown, roiling waters of the San Juan. Forty years after this farming project started, at an estimated total cost of $150,000 to $200,000, only 175 acres were still under cultivation and the ditch no longer existed.[21]

So when Navajo agent Constant Williams stood on the banks of the San Juan and insisted that its waters would be the economic salvation of the Navajos, one has to wonder who he had been talking to; it seemed history was about to repeat itself. Because the one farmer at Fruitland could not help all the Navajos, Williams requested one for the Bluff region. No one ever materialized to fill the position, so the government handed out seed and farm tools only to Navajos living along the Upper San Juan.

Most of these farming projects were small-scale, individualized efforts.[22] George M. Butler, superintendent of irrigation, had constructed several ditches on other parts of the reservation, the closest one to Aneth being the Carrizo Creek ditch. Sandoval, a Navajo from the Lower San Juan, rode one hundred miles to Fort Defiance during the winter of 1896 to solicit help in reclaiming some of the "fine tracts of land" near his home. Butler recommended a survey of possible locations.[23]

Little rainfall, cold springs, and early frosts discouraged the hardiest farmers, but government agents continued to call for surveys and ditches. In 1901 irrigation inspectors estimated that one-third of all the Navajos could prosper on the San Juan if they just had enough ditches.[24] In 1902 Samuel Shoemaker, supervisor of ditch construction near Fruitland, received orders from agent George Hayzlett to start a major ditch in the vicinity of Bluff. Shoemaker paid Navajo laborers a dollar a day as the ground thawed and work began. The agent supplied shovels, axes, mattocks, grubbing hoes, augurs, wrenches, hatchets, crowbars, and drills that he hoped would "soon make a mile of ditch in that part of the country."[25]

Hayzlett looked at lands above Bluff in the Aneth/Mancos Creek area, where farming would be "far cheaper per acre than any other part of the reservation." The Navajos there had repeatedly asked for help in creating ditches, and now that the Shiprock Agency brought government assistance closer in the form of William T. Shelton, their wishes could become a reality.[26] In 1904 newspapers reported that a prolonged drought had forced even more Navajos from the "interior of their reservation" to the San Juan "where they are farming all along both sides of our river."[27] The time was right for an even-grander scheme of government intervention.

Shelton analyzed the situation. He noted that Navajos often constructed ditches that washed out easily at the first high water because the trenches lacked headgates and protective barriers. The one exception in the Four Corners area sat at the junction of Mancos Creek and the San Juan. Eight men labored to build a two-hundred-yard-long, twelve-foot-deep ditch to bring water to fifty acres of a three-hundred-acre tract. Shovels and picks moved the soil, but the Indians had to carry rocks by hand for a quarter mile to create the riprap. The agent believed it was worth the five hundred dollars in labor to build, but "it will no doubt go out at the first high water, not being properly protected."[28] Government farmers could provide guidance to save ditches like this and teach ways to water four times the area.

Shelton looked at the Aneth region next. He realized that Navajos had successfully farmed with a number of small ditches around the mouth of McElmo Canyon. Old Mexican, a Navajo who worked the mouths of both Montezuma Creek and McElmo Canyon, provides a detailed account of what this experience was like. He tells about taking six days to dig a ditch a mile long to his field. Some passersby stopped to criticize his efforts, teasing that "all the people say water never runs up hill," but he persisted because the soil was good "to raise anything [he] wanted there."[29] It took six days to flood the level field, but by harvesttime, the corn had grown over his head.

Shelton understood the importance of this type of experience. Armed with three thousand dollars for irrigation projects in 1905, he appointed an additional farmer, James M. Holley, to supervise Navajo agriculture and livestock operations. Holley was no stranger to the area. He had come to Aneth in 1899 to open a trading post. During those six years, he had alerted government officials about the conflict over grazing land between white and Navajo stockmen and had even sought a position helping the Indians.[30]

Once appointed, Holley worked closely with Shelton, but of more importance was his impact on the Navajos as a government farmer.

Wagons, easily adapted to the Navajo lifestyle, often served as payment by the Indian agents. This wagon rests next to a bell-shaped hole, perhaps three to four feet deep and lined with cedar bark, used for storing produce. (Milton "Jack" Snow photo, #NAV 208, Navajo Nation Museum.)

Old Mexican again provides one of the most detailed accounts of what Holley tried to accomplish. One of his first tasks was to identify the best Navajo workers. The most deserving received tools, such as scythes, scrapers, pitchforks, hoes, and saws, as rewards for following the government program. Holley hoped to teach through example, while prosperous farms became the symbol. Thus, when Old Mexican harvested his foot-high hay field, obtaining a stack "eight steps wide and sixteen steps long, and higher than a hogan," the government farmer chose him for additional tasks and leadership opportunities.[31]

Holley marshaled community support for a number of projects. He built a riprap protective barrier to prevent the river from eating away the top of the Navajos' irrigation ditches. He started road construction to join Aneth to Four Corners and paid his labor in farm tools, some of which were very enticing. Forty-five-days labor got a worker a wagon; five days, a scraper; and one day a shovel, axe, or saw. If a person organized a group of men and helped feed them, the time to earn the reward decreased by half. Take, for example, the experience of one crew member:

> Slow had a wagon, but he wanted another. He took over a bunch of men; they were driving a horse and cow, and when they got to the place where they were going to work, they killed the horse and the cow to feed these men of his. In seven days they had earned a wagon for him.[32]

Four positive results came from this type of labor: These men built the road, improving transportation between Aneth and Shiprock; they worked as a team, creating greater community cohesion; the dispersion of tools ensured greater agricultural success; and the Navajos looked more and more to Holley and Shelton for advice, leadership, and equipment. In fact, Shiprock is still known to older Navajos by the name they gave to Shelton years ago—Nataani Nez, Tall Leader.

Much of Shelton's work on the river occurred in the fall when the water was low. In 1906 he requested a thousand dollars for men and materials to buttress his fight against the river. He estimated that it would take five hundred loads of brush (a dollar per load), juniper posts (twenty cents each), Indian teams (two dollars per day), Indian laborers (a dollar per day), and barbed wire (five dollars per one hundred pounds). One year later he received five thousand dollars for repairing ditches, headgates, and spillways along the San Juan. He reported there were twelve ditches between Shiprock and Aneth that, when kept in good repair, watered between six thousand and seven thousand acres of fertile land. While he also noted failures, Shelton was generally upbeat in fostering maintenance and development of these liquid lifelines of agriculture.[33]

Five years and one flood (1911) later, Shelton was not nearly as optimistic. He wrote to the commissioner of Indian Affairs that because of the "manner peculiar to local conditions" and "the treacherous nature of this stream [San Juan]," it was impossible to maintain a permanent "heading" on anything but the small ditches, and therefore it was impractical to encourage building permanent homes in their vicinity. He reported that the Indians had solved this problem by reverting to their old planting style: in the late spring, placing seeds in the areas they believed would receive high water and then waiting to see what happened. "If they succeeded in securing enough water for irrigation purposes, they usually raised good crops."[34] Yet Shelton would not give up his Anglo-American approach to agriculture. The government program persisted for an additional twenty years.

The construction of a government station by Holley on the terrace below the Aneth Trading Post was a symbol of this determined federal intervention. Joseph Heffernan bought the store from Holley after Shelton cautioned the farmer that being a trader and a government employee was incompatible. During the fall of 1906, Holley marshaled the aid of Navajo laborers and set to work. They made adobe bricks from San Juan soil, cement and lumber came from Shiprock, and Holley provided the floor plan.

Actual construction started in the winter. The bricklayer told Old Mexican to heat the water for the cement and keep him supplied with rocks and adobe. In two days, the foundation was completed, and in twenty days, the bricks laid. For his labor, Old Mexican received twenty dollars. That same year Holley built a barn for livestock and the bales of hay and alfalfa he had reaped from his own irrigated fields. The flood of 1933 swept both structures down the river.[35]

As a focal point for the community, the government farmer became increasingly prominent. In most areas of the reservation, the trading post became the center of community activity, but the government farmer, assuming that his personality and attitude were acceptable, became the bridge between Anglo and Navajo society, official policy and Navajo practices.

Holley's work as a farmer continued. Scabies, ticks, and lice infested the Navajo sheep, so he placed dipping vats at the mouths of McElmo, Montezuma Creek, and Recapture Canyons. At first the Navajos believed this medicine killed the sheep rather than helped them. Shelton tried his best to explain the benefits, but the Indians remained unconvinced. Old Mexican, showing his faith in Holley, suggested the group go talk to him. The farmer must have succeeded because larger and larger herds of livestock descended on the dipping stations.[36]

By the fall of 1908, Holley withdrew as government employee and returned to the life of a trader. J. H. Locke replaced him but lacked Holley's ability to speak Navajo. In spite of this drawback, the Navajos felt "the new farmer is kind to the Indians and gives them good advice."[37] He did not stay long, however; a little over a year later, W. O. Hodgson replaced him.

To support this agricultural program, Shelton called a number of Navajos from the Aneth area to Shiprock to ascertain what they needed. They requested help with their ditches, so the agent pointed to Hodgson and said, "This fellow will do the work." Then he told the Indians to collect a large group of people to get the project under way. The Navajos received a dollar a day for those who chopped trees and two dollars for those who hauled rocks and brush in their wagons. Hodgson had four hundred dollars set aside for the project. Fence posts and wire made the framework for the breakwater. The Navajos planted them in the

The Aneth Government Station, as it appeared in the 1920s, was a symbol of productivity and a gathering place for Navajos and Utes. In 1933 the river claimed it and much of the floodplain, ending large-scale agricultural attempts by the government. (San Juan Historical Commission)

shape of a triangle, in which they piled brush and rocks. Two weeks later eight of these structures were finished, but so was the money. The riverbank was only temporarily saved; by the next spring, the water had washed all the work downstream.[38]

Floods notwithstanding, Hodgson decided to build another ditch. He believed that if he dug deep enough, the water would flow better. Old Mexican cautioned against this proposal, but the farmer insisted that water would run twenty paces from the river. After the workers finished the ditch, the farmer ordered the headgates opened, but the water never went much beyond the entrance to the field. He commanded the workers to dig deeper, but the results were the same. Hodgson then turned to Old Mexican in exasperation and said, "Work it your way. You know more about it. Work it just as you like." And he did. The Navajo tied seven bundles of brush together, lined the bank with them, then spent three days piling rocks on top. He constructed a dam in front of the old ditch and forced the water over it into the field. His only comment when he was done was, "This Hodgson doesn't know what he is talking about."[39]

In the fall of 1911, disaster struck. People had flocked to Shiprock for the community fair. Heavy rains, however, bogged down the wagons and made setting up displays difficult. Participants watched the river rise, flood over its banks, and fill the fairgrounds. A reservoir upstream broke, adding to the torrent that battered at the adobe walls of the homes, school, and adjacent facilities. The new Shiprock bridge toppled as people moved onto hills nearby. Not until the water started to recede did the onlookers realize the extent of the damage to the school facilities, farms, and orchards.

When the Navajos returned to the Lower San Juan, they found their gardens and ditches obliterated and the ground covered with gravel. Many of the good sites no longer were worth

Herbert Redshaw standing next to his government car, a familiar sight to the Navajos of southeastern Utah. While he envisioned dams and irrigation systems to control the San Juan and its tributaries, there was never enough funding to make these dreams a reality. (San Juan Historical Commission)

farming.[40] The water had even undermined the foundation of the government station, making the house unsafe to live in. Hodgson withdrew to Shiprock until repairs were made, but by then he had developed heart trouble, left the Indian Service, and moved to Phoenix, Arizona.[41]

After the flood, life took up where it had left off. The next year Shelton reported that Navajos had built small irrigation ditches along the river from Farmington to Bluff, a distance of more than a hundred miles. He also pointed out that crops were rarely located near good grazing lands for sheep. Some Indians traveled up to thirty miles to get back to the river to weed and water crops.[42]

In 1914 Herbert Redshaw, the new government farmer, arrived in Aneth. Family members describe him as a "typical old English man." Dressed in bib overalls and a broad-brimmed hat, and sporting a mustache, he stood more than six feet tall. His steel gray eyes, large feet and hands, and rawboned build gave him a commanding presence, while a corncob pipe filled with George Washington tobacco jutted from his lower jaw. One man quipped that it took more matches than tobacco to keep the pipe operating, as the smoke curled around and colored his hat brim.[43]

To the Navajos, he was T'áá bíichį̨ndii. An exact translation of this name is difficult, but an approximation is His Own Devil. The Indians did not apply this epithet with rancor. Redshaw moved slowly and swayed slightly as he methodically swung his arms and walked; the name creates a feeling that he moved like a dead man returned to life.[44] The name is now applied to the Aneth Chapter, the place Redshaw struggled to develop.

Much of his life was filled with day-to-day, humdrum farming along the river. He lived in the government station, surrounded by forty acres of alfalfa fields and gardens, many of which Indians planted and maintained. He divided the produce among needy Navajos at harvesttime. His red barn and fences became a landmark to travelers, while his irrigation system proved ingenious. Redshaw not only used the waters from McElmo Creek, but when they weren't enough, he also drew upon the San Juan. His main ditch was four feet wide and two feet deep, with a headgate that returned much of the water directly to the river, and a smaller stream to flood his fields. This system alleviated the problem of silt buildup. Redshaw encouraged families to settle nearby as he made the government station a center of activity. He held community meetings under the cottonwood trees along the banks of the river and encouraged the Navajos to settle on the floodplains.[45]

Redshaw often spoke of his dream of damming the San Juan. He hoped to build a dam near Four Corners and eventually another at the mouth of McElmo Creek. The proposed structure would be as high as the surrounding hills with irrigation ditches paralleling both sides of the river. The dams would alleviate much of the danger of floodplain agriculture, which by then was becoming increasingly popular as Navajos farmed every available space along the river. Unfortunately, government shortages of funds and enthusiasm precluded the undertaking.[46]

Although Redshaw did not realize this dream, he methodically taught Navajos what he considered the proper method of agriculture. He spoke enough Navajo to get by, but for formal occasions, he used Eddie Neskaaii from Shiprock to translate. Harvey Oliver, a Navajo who worked for Redshaw for five years, explains his teaching style: "He would look at it. He did not just walk around, but he told us how to put watermelon seeds in the ground by counting them. Count the corn or the onions, this is what he said. There were distances between each onion that you should be aware of. He told me to learn all of this."[47] Oliver did learn, and by the end of his work with Redshaw, his salary had increased from one to five dollars a day.

By 1924 Redshaw had succeeded in persuading twenty-five families to settle around an irrigation canal that supplied water from the mouth of McElmo Creek.[48] He also kept track of the sheep dipping in the spring, though Evan W. Estep, superintendent of the Shiprock Agency, did not find him efficient enough. After telling how sheep dipping was progressing on other parts of the reservation, Estep commented that "Abba Chinda"—Slow Devil, as he translated it—was not ready and had left for Monticello just as a supervisor from Shiprock arrived. The agent threatened that if a quarantine occurred, Redshaw would be blamed and said that "no one ever knew Redshaw to do anything just when it ought to be done or when the other fellow wanted him to do it." Estep was anxious "to go down there and cuss him out right," but he also added that this farmer was a "good man . . . likely the best I could get in that out-of-the-way place, but he does get on my nerves at times, and no mistake."[49]

Redshaw played a vital role in many of the conflicts during the 1920s. One controversy important to the Montezuma Creek/Aneth area concerned range rights. Cattlemen and sheepherders vied for lands near the northern part of the reservation, and some of the ranchers slipped over the boundaries onto Indian lands. No fence separated property, so Redshaw told the Indians to herd the animals back onto the public domain, which did not sit well with the stockmen. Many of the Anglos thought talk of law and authority a bluff, especially the younger men who lacked "the fair attitude of the old timers." Redshaw pleaded for immediate government action.[50]

The agent agreed with the farmer's evaluation and added that some of these stockmen had been involved for years in stealing Indian cattle and making a handsome profit. White ranchers were also lobbying Congress to open Navajo and Ute lands to livestock grazing.[51] Tension increased. The end result of the conflict was the 1933 addition to the Navajo Reservation of the lands adjacent to Montezuma Canyon. What is important to realize is that Redshaw advocated for Navajos, and in some cases Utes, as they battled to maintain or obtain lands. He accompanied the Navajo agent, the Ute agent, and special investigators from Washington. In a few instances, he even retrieved livestock stolen from Navajos. He explained to one Navajo that he was an Englishman, he did not hate Indians, and he would not take their lands.[52] He was as good as his word.

In 1931 Redshaw retired. He stayed long enough to complete the census but avoided the trauma of Navajo livestock reduction in the 1930s. It was time for the government farmer to get out of the Indian Service. Redshaw moved to Ucolo, Utah, where he died in 1946. Almost as if the San Juan knew that Redshaw had left and livestock reduction had started, its waters gathered strength to undo what had been accomplished. In 1933 the river once again overflowed its banks, tore out the irrigation ditches, snatched away the headgates, wiped out Navajo farms, swallowed the government station, and forced abandonment of life on the floodplains. It also shifted from the south to the north side of the streambed and cut away every remnant of productive land. The Shiprock Agency withdrew its program of maintaining a resident farmer in Aneth and requested anybody desiring help to come to headquarters. The government's battle with the San Juan was over.

Was the government farming program a failure? Not really. It fit an era and a need that could not have been filled as successfully by existing programs. The Navajos adapted to it easily because agriculture was already an important part of their economy. The government contributed farm tools and equipment to a people who did not have the money to purchase them; it offered incentive to work as a community, yet rewarded individual efforts; Navajos were motivated to improve agricultural techniques and develop products comparable to those in the white economy; the program served as a vehicle to send children to school, produced a voice for law and order both on and off the reservation, and supplied men sympathetic to the Navajos' changing circumstances at the turn of the century. This last contribution was not measurable, like the vanished headgates and ditches along the San Juan, but was just as vital as any of the more tangible items. The river may have won the contest for agricultural lands, but settlement and development continued in spite of it.

In an environmental sense, this era and previous Anglo farming efforts concluded a period of "robbing Peter to pay Paul." While done with the best intentions, cutting down large numbers of cottonwood trees and tearing out implanted rocks on or near the river only sped the process of erosion from runoff. After removing the root structure provided by the trees, little remained to hold the soil together against the water and waves of the San Juan. While no doubt there had been flooding along the river before the introduction of Anglo-American agricultural methods, it had never manifested the degree of destruction that occurred later. Traditional Navajo farming practices were more capable of "breathing" with the mood or flow of the river; the Anglos wanted control so they could move beyond subsistence into a larger-scale market economy. Thus, the basic difference sprang from the philosophy and worldview of each group.

An important supposition underlying the Anglo-American attitude is that technology plays a primary role in forcing nature to comply with human plans. Brush, rocks, and wire were the basic tools that consistently failed to achieve people's goal. More sophisticated technology was needed to harness such a river. As early as 1899, Charles Spencer from Mancos announced his creation of a patented pump, run by a forty-horsepower engine, that could lift a continuous four-inch spray of water five hundred feet in the air. Claiming that it could be used for irrigating and placer or deep mining, Spencer believed he had solved the problem of fluctuating water levels by running his pump off an anchored boat. This was to be a "godsend" for farmers on the "San Juan, who on account of the sand and the ever-changing river bed have been obliged to see crops wither and parch for the lack of moisture."[53] Now the farmers could be rescued by the arms of technology. There is no record about how the pump was accepted, but it appears to have had little impact.

Not until around 1970 did technology provide an answer to the age-old problem of using river water to grow crops. That year the Bureau of Indian Affairs (BIA) completed a survey of possible farmlands bordering the San Juan River. Criteria included the plot being at least eighty acres in size and having a vertical water lift of less than five hundred feet, the soil being free from strong alkaline and saline content, and the land having a gentle slope of less than 10 percent. The BIA identified 52,984 acres that met these guidelines.[54]

Eventually ten different groups, including tribal, federal, state, and private agencies, participated in the project to turn the valley of the

The pumps, pipes, and motors used to push water on the land today are a far cry from this horse-driven irrigation pump at the Honaker camp in Montezuma Creek in the early twentieth century. (San Juan Historical Commission)

Lower San Juan into a lush agricultural zone managed by local Navajos as a private, cooperative enterprise. In 1971 only 65 acres were under cultivation; in 1974, 370 acres were planted; and when the project reached its height in 1976, about 1,000 acres were yielding crops from five sites on both sides of the river, ranging from the Utah Colorado border to Sand Island below Bluff.[55] Winter wheat, alfalfa, and oats were the main crops.

One site, the Tahotaile (A Wide Expanse of Land That Extends into the River) Farm Co-op, near Montezuma Creek, is an example of the way the program functioned. A 150-horsepower pump forced water from the river through pipes for 550 feet to a fourteen-million-gallon reservoir, where the silt settled to the bottom. Next a rolling sprinkler system of aluminum pipe traversed the graded farmland. Finally, families provided the necessary labor for weeding, harvesting, and marketing the crops.

Four major problems were encountered in the entire operation. The first was the silt that produced wear and tear on the equipment but was mostly removed once it reached the settling reservoir. Another was the meandering of the river, which had a wide streambed in which to roam. A third problem was the fluctuation in the water level, which could vary from day to day, and in some cases, hour to hour. But the fourth and final problem—individuals not being able to work in a cooperative effort—proved to be the final stroke that closed the project. Once the managerial system was relinquished to cooperative group control, individual differences hastened the abandonment of the farms.[56] Again the San Juan was left to run its natural course.

Today the best example of what it takes to utilize the San Juan on a large scale also illustrates the price that must be paid. On the Lower San Juan, some relatively small (one hundred to two hundred acre) Anglo operations employ hand-moved sprinkling systems to water fields of alfalfa and other crops. The vast majority of Navajo lands, however, lie agriculturally dormant. It is not until near Farmington, New Mexico, that reservation land on a large scale is under the plow.

The Navajo Indian Irrigation Project (NIIP) began in 1970 as part of the tribal-sponsored Navajo Agricultural Products Industry (NAPI). Part of the NIIP infrastructure consists of the Navajo Dam reservoir and a seventy-one-mile canal and pipeline water-delivery system that puts 508,000 acre-feet of water on 110,630 acres of farmland. The project has swallowed $370 million with an estimated $260 million more needed to complete future development. Fiscal reports suggest that it is "profitable" and annually pumps $35 million into the tribal economy.[57] How profitable it will be in the future with rising costs and less-available water remains to be seen. What is important is to understand that it took a huge investment in time, money, technology, and materials to harness the San Juan—something that government agents in the past could hardly have comprehended in their wildest dreams. For the Lower San Juan, this dream has remained out of reach.

One issue that lies beyond the scope of this chapter but bears mentioning because of its effect on the Lower San Juan concerns further use of river water by the NIIP. Although the Winters Doctrine *(Winters v. United States*—1908) provides the guiding principles concerning Native American legal rights to water, major questions still beg to be answered. The doctrine states that the establishment date of a reservation guarantees preemptive appropriative use, nonuse does not justify loss of the water rights, and sufficient water will be available to irrigate agricultural land. While this has not been a major issue along the Utah portion of the San Juan, the NIIP is now being challenged by the Endangered Species Act that prevents further utilization of the water to protect Colorado pikeminnows (squawfish) and razorback suckers, even though only half of the intended irrigation lands are under cultivation.[58] More litigation is in sight, directly affecting how much water can actually be taken from the river.

As the San Juan rolls into the twenty-first century, there will probably be just as many efforts to utilize its water for agricultural purposes as there were in the twentieth. There will also be more voices demanding their rights: for a fair share of water, greater recreational use, environmental concerns, or development of other economic schemes. Whatever the river's eventual fate, it will rest upon a historical legacy of trial and error, boom and bust, reflecting the way that Native people and Anglo-Americans tried to wrest a living from the water and lands along the San Juan.

6 City Building: *Farming the Triad*

Today a traveler, coming in sight of Bluff from the desert and canyon country to the west, is struck by the contrasting redrock cliffs and gnarled, green cottonwood trees. Indeed the trees are implausible until one sees the sinuous bend of the San Juan River, snaking its way against the bank that abuts the southern bluff. The cottonwoods suck their life from the brown waters and high water table, then give it back through an exploding tangle of leaves and limbs. Certainly nothing is more pleasant than a shady retreat, leaving behind the sun, heat, and dwarfed desert growth.

More than one hundred years ago, Mormon settlers, completing their six-month trip over the Hole-in-the-Rock Trail, felt the same emotions of relief. They had traversed some of the most inhospitable terrain, starting from Parowan and Cedar City in southwestern Utah, then moving across the desert of southern Utah to that narrow cleft called Hole-in-the-Rock that overlooks the Colorado River. The epic travail of building a road through a cliff and down one side of the escarpment, floating the wagons across the river, and continuing the road out of the canyon is a tale that has been told elsewhere.[1] The Mormon faith in the leadership of this church-directed colonization and tenacity in facing the elements have become legendary.

When they finally arrived in Bluff, they did not have the strength to journey any farther. Bluff, with its numerous stands of cottonwoods, level floodplain, and sheltering canyon walls, was just too inviting. Trees for shade and building materials, the wide floodplain for crops, water from a continuous source—what location could be more perfect for an agricultural community? Why continue another twenty miles upstream to the Montezuma Creek-Aneth area, where eighteen non-Mormon families from Colorado had already made their home? Better to stay put and use the resources lining the banks of the river than get involved with outsiders.

And so on 6 April 1880, the main body of Mormon pilgrims from southwestern Utah settled in the southeast, attracted to a desert land of promise, made visible through trees, land, and water. Members of this religiously based community were accustomed to the idea that covenants, with visible signs, expressed intangible relationships. The life that appeared so possible could be shaped at this spot where resources spelled more than survival. Some of the people felt this was the appointed place where their deity wanted them to settle.

The ensuing one hundred years proved disappointing. The water that had enticed them to settle became one of the main drawbacks. The trees that offered shade from the blistering sun proved unsatisfactory for much of anything else, though large numbers were harvested from the riverine corridor. And much of the floodplain, so amenable to agriculture, washed down the river during a series of unpredictable floods. If a covenant of cooperation between people and land ever existed, it broke fairly regularly. One geologist summed up the situation this way: "The San Juan River has eroded its banks at Bluff during catastrophic floods to a greater degree during the past 100 years of historic agriculture than it had in more than 1400 years prior to this."[2]

This chapter examines what part this triad—river, cottonwoods, and settlement—contributed to this region's history. Each one had an impact upon the other two as, year in and year out, they carved out and maintained

Looking southwest from the cliffs north of Bluff, this picture captures the desolate feeling of the landscape beyond the town. The San Juan River (left), Cottonwood Wash (behind town), and white-tipped sandstone bluffs (right), frame the world the settlers encountered upon their arrival in 1880. (Hugh D. Miser photo, #560, U. S. Geological Survey)

their ecological niches. Each seemed to have a will of its own. Whether during the dramatic floods of the San Juan or when the river faded to a trickle; during times when cottonwood communities proliferated or in years of decline; during the infusion of settlers or the years of steady exodus, all three elements struggled to follow individual paths.

When the settlers arrived along the banks of the San Juan, they encountered a variety of plant communities. On the five-to-six-mile-long river bottom near Bluff existed everything from microscopic spores of cryptogam to cottonwood trees averaging sixty feet high. Erastus Snow, a Mormon apostle visiting the newly founded community in 1880, remarked about the land and its variety. He estimated that the bottoms along the river varied in width from one-half to one mile, with some places upstream as wide as a mile and a half. He went on,

> extensive cottonwood groves in places, and generally [the ground is] covered with sunflowers, greasewood, rabbit brush, sagebrush and other luxuriant growth. Deep rich alluvial soil. The bench lands and adjacent hills covered with grass not a very heavy growth and in places extensive forests of cedar [juniper] and pinion pine. . . .[3]

Other plants found along the river included cattail, reed, cane, willow, arrowweed, serviceberry, Mormon tea, spiny aster, milkweed, Indian paintbrush, broadleaf and narrowleaf yucca, scrub oak, and a variety of cacti.

Into this comparatively untouched region came 230 people, eighty-three wagons, and more than a thousand head of livestock. The group encountered three families who claimed small tracts of land in Cottonwood Wash. Now, however, true civilization (i.e., city building) had arrived in one day.[4] Division of the land and the start of an irrigation ditch were top priorities. After some abandoned schemes and heated debate, fifty-nine men each drew an acre lot in town and a field of from eight to twenty acres, depending upon the quality of the land.[5] Since the planting season was already upon them,

The early home of the Wayne H. Redd family in Bluff illustrates building with what Albert R. Lyman called "that rams-horn breed of trees . . . whose walls bowed in and out with wonderful irregularity." (San Juan Historical Commission)

work started immediately on the ditch, whose entrance was placed four miles above the town.

Church leaders counseled the pioneers shortly after their arrival to remain nearby. A half-dozen families continued to their original destination and established a community at Montezuma Creek; another thirty families had already had enough and moved on to Colorado or returned to southwestern Utah and a brighter hope. For those who stayed in Bluff and Montezuma, the order of the day was erecting a fort for protection. They built the Bluff fort with houses around a four-hundred-foot square and all doors facing inward. Stockade fences stood between the houses, a well within the courtyard provided water, and a meeting house was the first completed public edifice.[6]

While materials for later construction relied on the straight ponderosa pines of the Abajo Mountains, forty miles away, lumber for early building came from local cottonwood trees. Results were less than gratifying. Albert R. Lyman, who lived in Bluff beginning in 1881, described the results of using this wood that was so "determined to warp and twist like a thing in convulsions, [it] would not lie still after being nailed down." Fences were made with "crooked stakes and riders of crooked cottonwood limbs into a hocus-pocus barrier," which he blamed for the "breach cows that have pestered Bluff ever since." And here is his classic description of life in a high-desert environment under a cottonwood roof:

> More still, it [invincible attitude] undertook from that same rams-horn breed of trees, to select logs and build houses, whose walls bowed in and out with wonderful irregularity and chinks ranging from nothing to a foot wide. It roofed them with thick coats of sand, which feathered out into a crop of runty sunflowers and stink-weeds, if the weed seed had time to sprout before the wind carried the sand away. But whether it raised weeds or blew away, it never turned the rain, which dripped dismally from it long after the sky was clear. These houses had doorways without glass and floors which

required sprinkling at intervals to lay the native dust and tempt the soil to harden.[7]

Lyman recorded other uses of native cottonwoods. The hundred cribs built to line the banks and irrigation ditches and prevent the sandy soil from washing away came from these trees. He estimated that thousands of these logs were harvested from the banks, jammed together, then backed with brush and stones. "Even then these tortuous members lay ready, with a little help from their kinsman the river, to come writhing from their prison and go twisting and rolling in glad somersault down the streams."[8]

No one kept track of exactly how many trees were cut and what comprised the brush that fortified the banks. But the project went on year-round for years. The winter was a particularly good time for construction because of the low water level before the spring onslaught. Settlers removed stone from above the riverbank and cut cottonwoods from both banks and sandbars. Lyman recorded in March 1897 that a recently exposed twenty-acre island was quickly cleared, stripping all the brush between Bluff and Recapture Wash, a distance of about six miles. The product: a mile of riprap.[9]

Although the "cottonwoods sprang up again like so much big hay," making the settlers believe they had an endless supply, the river continued to whittle the riprap away about as fast as it was erected. Entire "forests of new cottonwoods were gnawed away," and when the river "surrendered" access to a previously unavailable source of trees, it was "promptly skinned."[10] Little wonder that when Lyman writes about the community's struggle with the river, the rhetoric is steeped in war metaphors. The desperation seemed comparable.

What was taking place in Bluff occurred on varying scales in other settlements upstream. In addition to the Montezuma Creek community and the non-Mormons living in the Aneth-McElmo area, small, family agricultural operations and trading posts were scattered along the river. Each demanded something of the local resources. Take for instance, John Holyoak, who in 1882 established a village at what became known as Peak City because of a prominent feature nearby, possibly Peter's Nipple. Described by Platte Lyman as twenty-five miles upstream from Bluff, it was probably located near Rockwell Point on today's maps. The "city" included a home and store that doubled as a post office, which was soon augmented by the cabins of John Robb and James Dunton. By November of that year, however, these two men had pulled out, and Holyoak eventually followed suit. Like so many who had seen the promise of financial gain in the land, he found looks deceiving. Later a passerby described the remnants: "Its lonesome cabins and rude chimneys became the doleful abode of rats and chipmunks, until the pestulent [*sic*] river whittled the sand from under them and scattered their logs along winding banks."[11]

The land paid a price for supporting the efforts of this pioneering generation. Over the first twenty years, the cost included materials to sustain settlements, trading posts, large herds of livestock, a gold rush, agricultural efforts, and the start of an oil boom. In a relatively fragile, high-desert environment, where recuperation is slow and scars long lasting, the sudden onslaught exacted a heavy toll.

A good indicator that the land was ecologically challenged was the explosion of weeds. These plants are themselves pioneers; they take advantage of unsettled or "new" conditions where the land and plant communities have been disturbed. Windy, sunbaked areas subject to erosion were prime candidates for the new invaders. Weeds followed a marked succession. The first rooted quickly but also didn't last. Their seeds arrived in massive numbers, laying the foundation for sturdier, more slowly developing plants. Some of these pioneers provided shade, retarded wind flow over the ground, produced organic material, and stabilized soil. One author has described them as the "ecological Red Cross" that helps heal and prevent further damage.[12]

While most people view weeds as bothersome and unworthy of notice, a few settlers recorded their presence in Bluff. In 1885 Jens Nielson, the local patriarch and bishop, wrote that the community's greatest drawback in planting crops was a "heavy growth of weeds that have sprung up on our cleared land. Sunflowers grow large enough for fence poles and as close together as it is possible for them to stand."[13] In the same breath, he mentioned that beekeeping was a success, no doubt because of the profusion of flowers that accompanied the weeds.

Built in the fall of 1880, this structure for fourteen years served the pioneers as a church, school, dance hall, and public meeting place. Old veterans of the Hole-in-the-Rock ordeal and settlement of Bluff are pictured (left to right): Kumen Jones, Platte D. Lyman, Jens Nielson, James B. Decker, and Francis Hammond. (Charles Goodman photo, San Juan Historical Commission)

Albert Lyman made a similar observation, saying that the sandy roads of the metropolis of Bluff in 1888 were nothing more than a "narrow pass between two forests of stinkweed [purple bee balm]. They grew ten feet high, loaded with rich purple blooms, and always full of the buzz of bees wild and tame."[14] He went on to tell that a neighbor had spent two days scouring the countryside for a missing cow, only to find it in the weeds between Bluff streets. In other areas, where large herds of livestock grazed, there was no such luxurious growth.

One pioneer plant that grew quickly but hurt the livestock was pigweed *(Amaranthus retroflexus)*. In the early spring, redroot pigweed contains high concentrations of nitrates which can poison horses. Lyman recalled it growing in profusion on the benches above Bluff on a heavily grazed winter range. Later the settlers deduced that this plant caused their horses to go blind.[15]

If the horses had trouble with their eyesight, so, too, did the settlers, but for a different reason. Sandstorms that "make your eyes look and feel like kidney sores on a cayuse" arose in the spring.[16] Lyman reported one sandstorm was so bad that he could not see five feet in front of him.[17] Ernest Hyde, another longtime resident of Bluff, related that the gulches west of town filled in level with the sand blown by these storms. Heavy rains later loosened the sand and silt, then dumped them in the river.[18] Fish sometimes found it difficult to breathe as they choked in the sediment-laden water. The settlers, on the other hand, capitalized on the situation, scooping the fish out by hand from the shallow eddies, throwing them onto the bank, then later collecting the catch.[19] While all of the problems of wind, sand, and sedimentation cannot be laid at the feet of the settlers and their livestock, certainly the removal of grass and trees was a significant factor.

Undoubtedly the most powerful of all the antagonists in this struggle for dominance was the San Juan. It is no small irony that the river that drew the settlers to its waters also proved to be the biggest challenge. Poor crops, sand-filled ditches, and destruction of dams and channels discouraged the heartiest souls. By December of 1882, church authorities in Salt Lake City gave an official release for those who wanted to quit the San Juan "mission," but encouraged all to remain. Three years later Francis A. Hammond, the stake president (ecclesiastical leader) of the Four Corners region, made a progress report. Of the 150 men called to serve initially in the San Juan mission, only 25 had remained.[20] The primary reason for many was the river.

Before examining the historic record, it is useful to look at the general interplay of forces that affect the San Juan River, both locally and long distance. For instance, one side canyon may consistently deposit far more silt than another, and local storms may be more prevalent in one area than another, but there are always exceptions.

The San Juan has two different sources of runoff. The first occurs at lower elevations (from five to six thousand feet) and consists of the snowmelt and rain that flow down the canyons and washes and dump their load during the late winter and early spring. Local summer thundershowers also contribute. The second, higher-elevation runoff appears in April through June and originates in the mountains of Colorado, New Mexico, and Utah. The flow at a specific time depends upon the available snowpack and how quickly it melts. High temperatures lead to sudden release. The entire drainage area of the San Juan is twenty-three thousand square miles.[21]

Another aspect to consider is how much precipitation above or below normal an area receives over an extended period of time. While data were not scientifically collected, some interesting weather patterns started in the Southwest about the same time that the Mormons arrived in Bluff.[22] What had been a fairly placid and predictable seven-hundred-year period of precipitation and scouring of streambeds suddenly changed into a series of violent storms that delivered large amounts of water over short periods. A parallel to what occurred in Bluff has been well documented by Richard Hereford, G. C. Jacoby, and V. A. S. McCord in their study of the Virgin River, Utah.[23]

Gullying moved at a rapid pace into the twentieth century as flash floods tore away at the landscape. A few examples corroborate that what was occurring in southeastern Utah was happening elsewhere in the Four Corners region. Oraibi Wash at the southern end of Black Mesa, Arizona, was only 20 feet across and 12 feet deep in 1897; by the 1930s, it averaged 150 to 300 feet across and 30 to 35 feet deep. In 1880 Keams Canyon Wash did not exist; by 1930 it was 25 feet deep.[24]

The culprit was not necessarily increased rainfall but the type of storms and suddenness of water delivery. Certainly the removal of trees and brush, the grazing of livestock, and general settlement activities (both Navajo and Anglo) had an impact. It is, however, interesting to note that a similar phenomenon of erosion and gullying occurred during the twelfth and thirteenth centuries (the Anasazi Pueblo period), long before modern activity could have caused deterioration.[25] Beginning in the 1880s, violent, heavy storms tore at the landscape with a frequent ferocity that ravaged the river corridor.

Besides the water, what and how much flowed down the river? Between 1970 and 1979, more than 18 million tons of sediment (averaging a daily load of 5,000 tons) made their way through the channel.[26] Compare this to the 1930–39 period, when 395 million tons flowed down the river past Bluff.[27] Just as impressive was the contribution made by Cottonwood Wash, which drains only 205 square miles or about 1 percent of the total upstream area. During one six-month period in 1968, it contributed 10 percent of the annual sediment load of the entire river.[28] In addition to the soil sluiced down from canyons and washes, sandstone, siltstone, and shale underlie the channel, contributing to the suspended particle load. At flood stage as much as 75 percent of the river's volume can be silt and sand.[29]

The actual flow of water varies with the season, but an annual average is 2,542 cubic feet per second. This capacity can increase to 62,300 cubic feet per second during the highest flood stage, an event that can statistically happen every fifty to one hundred years.[30] Although floods are often viewed by those living along the river as highly destructive events that should be controlled, scientists who study riverine habitat now believe that flushing and scouring are actually healthy for the river. By flooding the banks

This 1909 photo shows the broad floodplain, wide river, and shallow banks that easily allowed flooding to occur. Consequently, crops and irrigation ditches were frequently lost. (Stuart Malcolm Young Collection, Cline Library, Northern Arizona University, #NAU.PH.643.25)

and carrying materials downstream, the water adds nutrients to the bottomlands and washes old, spent soil and debris away. Since the impoundment of water by Navajo Dam in 1962, extreme floods have been reduced. Now that the river is more confined, the banks are stable and heavily infested with tamarisk, Russian olives, coyote willows, and other undergrowth.[31] Flooding and replenishing soil, however, have also slowed down. In the lower canyons of the San Juan, because there is no floodplain and the river drops more steeply with high rock walls to maintain the channel, flooding has had minimal impact. Understanding this general behavior of the river makes what happened to Bluff starting in the 1880s clearer.

Kumen Jones, one of the original settlers, noted that at the time of his arrival, the San Juan River coursed through the middle of the bottomlands and was confined to what appeared to be a permanent channel "with cane and willows and cottonwood trees up and down."[32] He also noticed piles of driftwood some distance from the river, indicating big floods in the past.

Past phenomena, however, soon became part of the present. Starting on 22 December 1883, Bluff received the first of many prolonged dousings of rain and snow, lasting for forty-eight hours.[33] One storm followed another. Heavy showers in February 1884 continued into March, raising the river seven feet above normal. Cottonwood Wash added to the melee, spewing forth a torrent of water "loaded with drift and stinking loud with filthy sediment."[34] Worthless white sand spilled over and covered some of the best agricultural fields; ten inches of water pooled on the floors of homes in the southwest corner of the community and the fort; the river badly mauled irrigation ditches.

By May the San Juan, swollen by continuing heavy showers, gained more momentum from the melting snows of Colorado. The precious headgate at Walton's Slough, key to the entire irrigation system and symbol of sacrifice, was now threatened. Samuel Rowley recalls that the community mustered everyone into service, but the people's efforts were like "pitching straw against the wind."[35] The men camped away from the river that night, listening to trees that had stood for "centuries" crash into the water. In the morning, the headgate bobbed in the current, entangled in a cottonwood. As it broke free, some of the men tried to lasso it, but to no avail. It disappeared, taking their hope with it.

The flood peaked on 18 June, sweeping everything except the settlement of Bluff before it. All of the buildings in the Montezuma Creek-Aneth area, as well as the individual homes along the river, were first flooded, then washed away. The experience of Jane Allen is typical. Jane and her small children tried ditching to turn the river away as it flooded one side of their bottomland farm. Montezuma Creek contributed its share to the problem, so soon water stood a foot deep in their home. Bob Allen, one of Jane's sons, came from Fort Montezuma, where things looked generally bleak, and tried to rescue the woman and her children with a buckboard. The wagon quickly mired in the mud and sand. He then lassoed a molasses boiler, placed as many family members in it as he could manage, and pulled them to safety. After a number of trips, the family sat on high ground, watching home and belongings disappear. All three log cabins on the property, the fields and orchards, Fort Montezuma, and the entire community of Aneth flushed down the river, except for the Harriman home, built too high on a rock for the river to snatch. "The site of Montezuma was a yawning gap of sand. . . ."[36] The flood was too much. Many people left the San Juan to return to southwestern Utah or greener fields in Colorado. A few went to the battered city of Bluff, where friends and relatives helped them. No one stayed in Montezuma Creek.

In cool retrospect, what can be said about the flood of 1884? Most obvious is that it followed a classic pattern. First, local precipitation produced flooding, followed by a rising river from more-distant spring snowmelt. This was also a year with an inordinate amount of precipitation, accompanied by cooler global temperatures, associated with the explosion of the volcanic island of Krakatau in the Indian Ocean and probably an El Nino off the west coast of South America. The ash and debris sent into the atmosphere affected major weather patterns, increasing rain and snowfall around the world.[37] The combination proved fatal to most of the settlements in southeastern Utah.

On the other hand, Bluff, one of the lone survivors, was not without fault. As early as 5 September 1880, church authorities counseled the settlers to avoid building communities close to a bend in the river or "near the mouth of any wash" that might be subject to cloudbursts or "mountain floods."[38] Bluff sat right next to Cottonwood Wash, one of the main contributors to its problem. It is also interesting that the white sand that covered the fields washed down from the bench above town, a site of intense livestock grazing. No doubt most of the grass and other plants had either been trampled or eaten so they no longer stabilized the soil. If these lessons were difficult to grasp, the future would present many opportunities to relearn them.

Floods were one type of problem; droughts were another. Although the 1880s were generally characterized by above-normal precipitation, the 1890s proved to be the opposite. By 1893 a prolonged drought was taking its toll on the farms and ranges of southeastern Utah. Streams that had run full now dried to a trickle; springs that had consistently gushed water were now as "devoid of moisture as a tinder box."[39] Three years later there was little relief. Presaging what would occur in the Oklahoma dust bowl of the 1930s, the elements exacted their dues.

Albert Lyman, with typical detailed observation, recorded that the "hideous specter of drouth came stalking over the whole country." All of nature worked in concert to undo the settlers. "Dry winds drove clouds of dust fiercely along from the southwest, drinking up moisture like a sponge, leaving weeds and grass dry and withered. Crops failed. Loose soil on newly plowed land was swept from hilltops, leaving naked markers of the plow running across the hard earth."[40] The San Juan River ceased to flow, so now people could cross without getting their shoes wet. Intermittent pools in the streambed contained barely enough water to support fish. Settlers and Navajos, however, descended on the river with spears in hand, filling their sacks before heading to town to sell their catch.[41]

In September 1896 the cycle temporarily broke with three continuous days and nights of precipitation. Hammond later reported more rain and snow pelting to earth that winter than anyone in Bluff could remember. Again Cottonwood Wash played havoc with the town. The "boiling mass" crested the eastern bank after attaining a depth of twenty and a width of one hundred feet near Hammond's house. Closer to the San Juan, it was half-a-mile wide. Two to three feet of sand and mud washed over the orchards, suffocating some of the trees; sand and silt again buried the fields; water and sediment tore at, then filled the ditches; and green cottonwoods, sixty to eighty feet in length and one to two feet in diameter, floated, then settled in "great piles" upon the land.[42] It was time once again for the town to dry out.

By 1898 and 1899 the drought had resumed. Now the people of Bluff had a new idea: Pump the water out of its diminishing, wandering bed and send it down the ditch. The machine—a large steam engine—looked promising. Once in operation, it pumped a "fair quantity of muddy water through its pipe."[43] There was one problem: Where could the operators find enough wood to keep the old engine going? Wood haulers searched "up and down the river for many miles . . . nor did it take very long to complete the skinning." Soon the enterprise was abandoned; damming and riprapping continued.

In 1902 citizens from Bluff wrote to the Navajo agent, saying that his wards were starving because of nine years of drought.[44] Upon investigation, the condition of the Indians was far better than reported; there was, however, no denying the stressful climatic conditions. Louisa Wetherill, trader to the Navajos, recalls 1902 as the year the San Juan dried to its lowest stage of six inches deep and three feet wide outside of Farmington, New Mexico.[45]

The fall and winter of 1904–05 again reversed the sequence. Just like a serial on television, the newspapers carried the latest word about the ongoing struggle of Bluff with the San Juan. October: The river was on a "big spree," fed by the heavy rains in the "upper country"; March: News had just arrived in Mancos, Colorado, that "Bluff was washing away"; May: The San Juan had sliced around the dam, threatening to "cut a new channel right through the town."[46]

Bluff was not alone. William T. Shelton, Navajo agent at Shiprock, had been fighting a similar battle ever since he founded the agency in 1903. Shelton recognized that every year the river cut away hundreds of dollars of valuable land and was very "shifting in its nature . . . from first one side of the valley to the other. Hundreds of fine trees were swept away by the high water this spring that should have been used in protecting the banks and to prevent encroachments of the river on the farming lands."[47] The San Juan continued to antagonize anyone wishing to settle its banks.

Shelton had other plans. In the fall of 1906, he considered buying either Bluff or the Navajo Faith Mission (near Aneth), owned by Howard R. Antes. Either site could accommodate a boarding school for the children of the estimated two thousand Navajos living on the Lower San Juan. Agriculture would be the school's main curriculum. By June of 1907, Bluff appeared to be the strongest candidate, since the bottomlands of Aneth were in the process of washing downstream.[48] After the government considered its options in obtaining Bluff, interest cooled. The rough land south of the river, the danger of crossing during high water, and the large amount of quicksand along the banks made it impractical to establish a school there. Cottonwoods for fuel were "none too plentiful," raising the issue of heating in the absence of coal.[49] As far as Shelton was concerned, there was no place between Shiprock and Bluff where topography, water, and Navajo needs could successfully merge. The town remained a private enterprise.

It is pertinent to ask at this point just how much land was actually washing away. A rough estimate is provided by two newspaper items from 1907. The first, published in July, tells that the river line was approaching the historic landmark known as the Old Swing Tree. Under this cottonwood, Bluff settlers had held their first church services. Initially it had been situated approximately halfway between the town and the river on the northern floodplain. Now the water was gnawing at the bank fifteen feet away. By September the cottonwood was gone, but not

Citizens gather for a last picture and farewell to the Old Swing Tree before it is swept downriver during the summer of 1907. This symbol of the Bluff settlement marked the place where the settlers held their first town and church meetings. Its loss also represented the antagonistic relationship between the settlers and the San Juan. (Charles Goodman photo, San Juan Historical Commission)

before a crowd of residents paid local photographer Charles Goodman to take their picture with the "doomed sacred tree."[50]

There is little wonder that two years later, one of the townspeople took pen in hand to let the world know how desperate the situation was. He wrote that Bluff's hay fields were being transferred "down into the Gulf of California" and that if something was not done to stop the river, "we will have to take to the cliffs and become cliff dwellers."[51]

As the river "licked up lucerne patches, barbed wire fences and ponderous old trees with a fluency which would sicken a saint," the people determined to launch a war to make Bluff safe for habitation. The river's main avenue of approach was up Walton's Slough, east of town. Community members donated time and money, while the LDS Church opened its coffers to support the fight. Workers hauled pine logs from Blue Mountain, set them in the ground with pile drivers, and backed them with rock and brush, making "all other riprapping campaigns dwindle to insignificance."[52] After two years of extensive labor, the dam met the river's onslaught successfully, turning away the flood of 1911.

Other communities were not as fortunate. Starting in July and August, rain deluged the Four Corners area. In October the precipitation intensified as one two-hour storm dumped 4.8 inches of rain. The weather bureau later reported that between September 1911 and March 1912, 27 inches of rain fell in San Juan County, twice the normal amount for even the wettest areas. Bluff averages almost 8 inches a year.[53]

Water from both local and distant sources coursed down the river, sweeping everything before it. Shelton reported that, starting in Shiprock, the entire valley flooded, in many places "from hill to hill." He estimated the depth as twenty times greater than he had ever seen it; he knew that parts of the school lay submerged under six feet of water, nine adobe structures had "melted" away and all the larger buildings held water, and he had "sent ten to twelve thousand fine melons down to the people living along the Gulf of California."[54] The river also replanted the recently placed steel bridge a quarter of a mile downstream.

In Utah, Navajo homes along the river near Aneth washed away, as did the two-year-old steel bridge at Mexican Hat. The bridge had cost the state four thousand dollars, a sum willingly paid for the anticipated wealth from the oil fields.[55] In the lower canyon, Otto Zahn, a miner, returned to his camp to find only the top of his home protruding out of a mass of mud. After estimating the low water level of the river, the height of his home, and what was left, he believed the mud flow was seventeen feet deep.[56] Once again, the San Juan emerged the victor.

But nothing was ever final with the river. There would be periodic floods for the next fifty years until its turbid waters were finally brought under control. There was the flood of 1927, which, according to one eyewitness, raised the water to thirty-three feet above its normal September level. Debris from Gypsum Creek, opposite the town of Mexican Hat, was so plentiful that it almost dammed the river.[57]

After the flood of 1941, the Soil Conservation Service assumed the responsibility for forcing the river back into its original channel. "Lovely cottonwood trees," a report tells, were cut down for riprapping to protect the land and seventeen families residing in Bluff. By now it was all a familiar scene: "The water rolled and boiled, cottonwood trees fell, the banks melted like sugar until 96 acres of irrigable pasture had disappeared in two weeks. Two days later an additional 18 acres of alfalfa sluiced away."[58] Five-foot waves swung the river from one side of the streambed to the other. At the end of June, the waters finally abated.

In 1948 the Army Corps of Engineers linked arms with the Soil Conservation Service, county officials, and Bluff residents to raise money for a joint venture in erosion control. The project, costing an estimated fifty-five thousand dollars, attempted to prevent more acreage from washing away. The plan included a large rock crib southeast of Bluff, with a stretched cable securing pole jacks or large cedar trees to a protruding bank. When the river washed against this breakwater, slowed velocity made the sediment drop, while the jacks caught the floating debris. The structure helped but never affected the extreme fluctuations during flood times.[59] It was not until 1962, with the completion of Navajo Dam, that the cycle of torrent and trickle took on any semblance of managed uniformity.

After eighty years of combat with the river, what conclusions can be drawn about wood, water, and people? There are no simple answers. The life of the river is complex, and its environment depends on many factors, some far distant and others very localized. Trees, both near the river and farther away, have definitely played an important part in the history of the Lower San Juan. The trees cut along the river and its headwaters had an impact further downstream. Frank Hyde, who traveled through areas such as Dolores, Mancos, and Arboles in Colorado, remembered how destructive clear-cutting timber was to the ground cover. Once the yellow pines and blackjacks covering the base of the mountain had been harvested, the trees on the slopes were the next to fall beneath the axe and saw. The heaviest cutting of low timber occurred before 1896 and was associated with the settlement of towns and construction of the Denver and Rio Grande Railroad. Big sawmills continued to operate for another thirty years. Hyde recalled, "I saw a great many of those forests before they were cut and I have seen the places where they have been. . . . There were great forests in there for miles. The sawmills took out all, cleaned it up. . . ."[60]

A. L. Kroeger, a civil engineer and resident of the area in Colorado for more than forty years, was familiar with the lumber companies and corroborated Hyde's astute observations. Kroeger stated that 782,000 acres of Colorado and New Mexico forests were harvested for timber.[61] The result was no underbrush, pines, thistles, or leaves remained on the mountains to slow the wash of water and subsequent erosion into the San Juan. Hyde verified the impact: "Since the timber was cut down, my observation has been that the water in the river flows off quicker."[62]

This riprap barrier in Bluff suggests not only fear of flooding but the tremendous toll exacted from the environment to combat it. Multiply the quantity of trees necessary for this structure by the miles of riprap dam; then multiply that by the number of years riprapping was constructed. The amount becomes staggering. (Charles Goodman photo, San Juan Historical Commission)

Removing cottonwood trees on the lower San Juan also had a debilitating effect. Large trees provided shade and slowed surface evaporation, decreased wind, added organic materials to the soil, reinforced the riverbank, and rooted the ground. Since local precipitation could arrive in sudden, violent, downpours, streaming off the sandstone cliffs and naked slopes in a deluge, cottonwoods helped stem the flow.

Along the banks and bars of the river, mature trees did not stand a chance fending off floodwaters that first undermined, then toppled the stately monarchs. They were just too inflexible to withstand a direct onslaught. In some instances, large trees, caught in the tide, created a horseshoe vortex upstream that channeled the water around them and destroyed the bank even more.[63] A large cottonwood might also lodge and dam a part of the river, sending the water and sediment in another direction and creating a new streambed. The shallower banks on the broader bottomlands soon flooded, changing the course of the river.

On the other hand, younger trees, especially those on sandbars, were limber enough to bend with the floodwaters, slow and catch the sediment, and scour the stream bottom, which eventually changed the course of the channel. The more plant cover existed on a floodplain, the greater chance of withstanding the ravages of a flood.[64] So as the settlers cleaned the land of any vegetation that could be used for riprapping, they ironically destroyed the most beneficial element in counteracting erosion: natural cover.

But it was also the river that fostered the life and regeneration of cottonwood stands. Besides providing necessary moisture for growth, the San Juan played a crucial role in planting the trees. Cottonwoods have adapted to the ebb and flow of water in the Southwest. Their seeds are viable approximately three to four weeks, peaking in mid-May. This period corresponds exactly with the high water on the river, so that when the level drops, the seeds have moist soil for germination. Indeed, one method of seed dispersal is floodwater that washes over banks or sandbars and plants a new tree in a safe spot. Sufficient shade also prevents the soil and germinating cell from drying out too rapidly.[65] Understanding this cycle of growth explains why there are fewer cottonwood communities along the river. As the settlers cut the trees to build their riprap walls and homes, they not only removed the source of new seeds but also altered stream flow that encouraged regeneration.

Early settlers' descriptions of the riverbanks make it clear that the San Juan was confined to a well-established course lined with mature growth. Kumen Jones's observations on the banks and riverbed, cited earlier, bear closer examination. He wrote,

> The channel was fixed and definite when I arrived and these were lined on each side [with] old trees and old willow patches and the river had definite banks and the channel was confined in the original position of the river as I saw when I first went there; and that condition continued until the first flood [1884] changed it some by running over the old channel in some places, and after the flood subsided, the channel almost entirely resumed its position in the old channel. The position of the river changed during the second flood. After the flood was over it did not come back in most places. The [first] time that the channel changed in any substantial degree was during the flood that occurred in 1896.[66]

That was sixteen years after the settlers arrived, sixteen years of intense tree and brush cutting that left the banks and sandbars bare. Add constant livestock use and the effects of drought and wind, mixed with periods of extensive precipitation, and there is little wonder why the San Juan ignored its earlier boundaries and started eating away at the bottomlands.

Many variables must be considered when examining the geologic characteristics that affect the river. Stream depth and gradient, sediment loads, volume of water, texture (roughness) of bank and bottom, soil consistency, vegetation, and tributary washes are all factors. A few general points, however, can be made about the Bluff experience. The town sits on a wide floodplain, but across the river stands a four-hundred-foot sandstone cliff. During floods there was no direction for the river to go but toward the settlement. Water follows the path of least resistance. The high rock cliffs, barren slopes, and Cottonwood Wash just compounded the problem when intense showers or rapid snowmelt overloaded the waters of the San Juan.

The loosely packed, sandy soil of southeastern Utah did not retain this gift of extensive moisture very long. The water selectively eroded gullies, then fashioned mud flows and sandbars from the materials. Frank Hyde recalled a time when Cottonwood Wash deposited so much sediment that the north side of the San Juan River was "choked off," and it took another five or six days to flush the debris downstream.[67] The river also created mud balls, some as large as a wagon wheel, from cobblestones, clay, and sand. The spheres rolled down the river channel, sometimes collecting and damming the flow. Quicksand, either blown by the wind or carried by the water, accumulated in shallow bends, ensnaring livestock that ventured into the mire. This was particularly true of the thirty-five-mile stretch of river between McElmo Canyon and Comb Wash.[68] Once submerged in quicksand with only their backs showing, horses and cattle had to be dislodged by ranchers.

Even more dramatic than pockets of quicksand and moving mud balls was the braiding of the stream. This phenomenon was caused by the decreased velocity and capacity of the water to transport its bed load and sediment. The deeper, narrower, and faster the river, the more capable it is of transporting large objects. When the moving silt, sand, and rocks hit objects or entered still water or a broad floodplain with unstable banks, they dropped and came to rest. As velocity decreased, the finer sediment settled out.[69] On the San Juan, this meant that gravel bars were more common above the narrow canyons of the lower river and the open area by Paiute Farms became a multichanneled series of sandbars. Other factors like sediment load and amount of water also caused the channel to shift dramatically, sometimes in a relatively short period of time.

Historic testimony supports these occurrences. Eyewitness accounts tell that the main current "would shift from one side to another," the water would be "four feet deep on one side, and coming back three or four days later, the deep channel would be on the other side and it [Clay Hills] would be impossible to cross on account of quicksand," and, once the channel was filled with sand, "it had a tendency to throw the heavy body of the current against the banks, undermining them and caving them in. It cut in a half circle until it cut the bottom entirely in two," then returned to its old channel.[70] The more sand was added to the channel, the greater the possibility that the water would rise and spill over the banks. Thus, the dramatic swings in the river's channel from high and dry to submerged and deep were all a result of the equation of water, velocity, bed, sediment load, and makeshift dams.

There were also seasonal variations. Sudden showers in the fall often washed windblown sand out of gullies. Fall floods characteristically deposited more sand, which remained in the river all winter. The water in the spring rose and fell more gradually, eating away at the sandbars and rearranging the load downstream.[71] Paiute Farms provides a good example of the result on a grand scale. Bert Loper, a miner and river runner with a long history (beginning in 1893) on the San Juan and Colorado, estimated the riverbed at Paiute Farms was between three hundred and four hundred feet wide. Years of flood and deposit changed it dramatically. By 1921 the actual Paiute farms were gone, and the river measured thirty-three-hundred-feet wide. Nothing but three or four shallow streams of water were braiding through a landscape of sandbars.[72]

Layers are still being added to the environmental history of the San Juan River. The stabilization of the river, the introduction of salt cedar, and the government's plans to control the San Juan as a resource will be discussed later. But even since Navajo Dam was built, there have been problems with flooding. Cottonwood Wash continues to be a nemesis. In 1968 a summer flood carried large cottonwood logs down the wash in a flow metered at twenty-three thousand cubic feet per second. The steel-girder bridge spanning the wash was badly damaged, and a number of homes were flooded.[73] In 1973 Cottonwood Wash repeated its performance, washing out a steel bridge, splashing ten-foot waves along its banks, and flooding Bluff so that some people could paddle around in boats.[74] Nothing is new under the sun.

But most of the riparian landscape has changed. Only faint traces of the extensive irrigation ditches once important to Bluff's survival still exist. The shores stripped for riprapping are now covered with tamarisk, Russian olive trees, and other vegetation. Cottonwoods persist and tower above the lower growth, harkening back to the time when they dominated the banks and floodplains. And the San Juan, partly restrained by Navajo Dam, winds its way to Lake Powell. Much like the country it passes through, the river can only suggest the freedom it once enjoyed.

7 MINING: *Black and Yellow Gold in Redrock Country*

Once a beautiful, well-dressed woman visited the home of a powerful stranger. The master of the house invited her inside, asking who she was. She replied that she was the goddess of wealth, which pleased the master, who in turn entertained her with kindness. Soon another woman appeared, but this one was ugly and dressed in rags. The master of the house inquired her name, and she answered that she was the goddess of poverty. The man became frightened and tried to drive her away, but she hesitated to leave. She explained, "The goddess of wealth is my sister. There is an agreement between us that we are never to live separately; if you chase me out, she has to go with me." Disregarding this advice, the master evicted the ugly woman, only to have the woman of wealth also disappear.[1]

Wealth and poverty have always been close relatives, as this Buddhist fable points out. There is no better historic example of this truth than the exploitative attempts in the nineteenth and twentieth centuries to wrest resources from the Lower San Juan River. When obtaining riches seemed possible, the desert and tortuous rocky canyons along the river became a welcome Eldorado for the miner and oil man. When mineral wealth literally did not pan out, the ugly and desolate wretch was abandoned to her own devices. The outcast river wandered along its course uninterrupted, waiting to be rediscovered.

Until the early 1890s, few Anglo-Americans had ventured into the canyons below Bluff. Two prominent early prospectors, Ernest Mitchell and James Merrick, searched for a rumored Navajo silver mine.[2] Instead, they found death at the hands of a Paiute-Ute band in Monument Valley; they ensured, however, that others would seek the same fabled riches.

Most notable among them was Cass Hite, who during the 1880s and 1890s wandered along the San Juan and Colorado in search of gold and silver. He eventually found deposits of copper at the head of Copper Canyon and small amounts of gold near the mouth of White Canyon at a place which now bears his name. In 1883 a mild rush followed, when several hundred miners sought placer gold along the banks of the Colorado River in Glen Canyon.[3] Prospectors staked their claims on the gravel bars at the river's edge, and by 1889 twenty-one sites were distributed from the mouth of White Canyon to Lees Ferry.[4] Miners drifted in and out to try their hand, though none achieved dazzling success. The Colorado was still the ugly woman of poverty, proving more tenacious than her flirtatious, wealthier sister.

Obtaining gold, that symbol of easy riches, rested upon the three-legged throne of environment, attitudes, and machinery. These elements were central in luring large numbers of miners into a difficult, trying business venture in the hopes of becoming instantaneously affluent. Only after the price in men and machines was tallied against what the river and its environs had to offer could a decision about pursuing mineral wealth be reached.

The first large gold rush on the San Juan started in December 1892. Bluff was the jumping-off point for claims scattered from the Four Corners to beyond the confluence of the San Juan and Colorado. Most activity was concentrated in the region around and below Mexican Hat. The rush started when a trader named Jonathan P. Williams showed some entrepreneurs and

Dwarfed by the landscape, this pair of placer miners suggests the enormous efforts required to wrest wealth from the San Juan. Rock, sand, and water comprised the environment at the foot of the Honaker Trail. (Charles Goodman photo, Manuscripts Division, Marriott Library, University of Utah)

railroad men samples of coal and other minerals found where the two rivers meet. Word leaked out that gold, not coal, had been found.[5] Since silver, in the early 1890s, was no longer the basis for United States currency and a nationwide economic panic was then under way, many silver miners from Colorado saw the discovery of placer gold as an enticing antidote for unemployment.

How much and what type of gold actually existed along the shores and in the waters of these desert rivers? In geologic terms, gold and silver are closely associated and occur in igneous rock formations, not the sedimentary sandstone that characterizes the vast majority of topographic features in the river corridor. Therefore, no veins of gold existed along the San Juan. Nuggets, the next most profitable size, were deposited by the river as they tumbled their way downstream. Obviously the more water and the stronger the current, the greater the possibility of finding nuggets farther away from the mother lode. Ancient gravel terraces, sometimes two hundred feet above the high-water level, held some larger pieces of gold.[6]

Finally, there was placer or "flour" gold, small particles the size of coarse grains of sand. The vast majority of the gold found in the San Juan was this type, indicating that its source was the San Juan Mountains in southwestern Colorado, the Carrizo Mountains in northeastern Arizona, and other mountains whose tributaries flowed into the San Juan. Geologists have not determined a specific origin for this flour gold; both dependable rivers such as the La Plata and Animas, as well as intermittent streams, contributed.

The miners in the 1890s knew much about the origin of gold, but facts did not dampen their hopes. A quick perusal of reports from the goldfields shows not only the newspaper "boosterism" of the times but also the way distant perceptions differed from reality. One of the earliest notices of the strike in Salt Lake City's *Deseret News* came from a Flagstaff, Arizona, dispatch

From the simple gold pan to the rocker and improvised wheelbarrow, technology became increasingly important as men worked the land to yield its riches. This staged photo at "Dempsey's claim" in July 1894 illustrates the equipment used in the earlier years of gold mining on the river. (Charles Goodman photo, Manuscripts Division, Marriott Library, University of Utah)

stating that two hundred "locations" were spread along fifty miles of river, with "gold fields reported as being the richest ever found." That was on 13 December 1892. A few weeks later the *Salt Lake Herald* started announcing amounts and types of gold. Small bottles of the precious mineral began to appear, each valued at around fifty to sixty dollars, along with reports of "small nuggets the size of peas." While "five ounce nuggets are not plentiful, some have been found," but most of what was being scraped off the gravel bars was around an ounce.[7]

Cass Hite, speaking with a voice of experience, did not believe these accounts. In a letter to the Denver *Republican,* he tried to set the readership straight by pointing out the impossibility of these claims. Hite believed that any gold in sedimentary formations was characteristically fine, flour gold; coarse gold did not travel far from its point of origin, and any gold that had moved a considerable distance might have started out soft and heavy but before long would have been ground to a very fine consistency.[8]

Others joined in trying to stop the "senseless stampede," as one paper called it, while another attacked the "San Juan fake." Eyewitnesses told of spending days on the river with little to show for their efforts because the gold was "so fine and light that so far it has been impossible to gather it."[9] One person testified that many of the articles he had read about the goldfields ended with the statement: "The San Juan is no country for a poor man." He went on to refute this: "It is the greatest place on earth for a poor man, and the longer he stays there the poorer will he become."[10]

How many people actually mined gold along the San Juan will never be known. Miners poured in from every direction with jumping-off points in major cities of the Four Corners region. Flagstaff, Arizona; Durango, Colorado; Salt Lake City, Utah; and Farmington, New

Mexico, advertised transportation lines that led to the mining district. Newspapers estimated enough gold and land for ten thousand miners to remain employed. Near the height of the rush in January 1893, an estimated two to three thousand men, "with more arriving every day," worked their sites. Other figures vary from a low of seven hundred to a high of five thousand, with one person claiming that one thousand miners passed through Bluff on New Year's Day alone.[11] By March most of the boomers had left the diggings to the more determined and affluent miners.

Somewhere between boosterism and bleak reality lay partial success. Although many left the fields disappointed, stories circulated about some who enjoyed limited prosperity. William Hyde gauged the amount of gold by pointing out that typically a "pan of dirt [would] wash out in which he counted seventy colors with his naked eye."[12] A few months later Walter Mendenhall averaged a dollar for each yard of gravel that he ran through his gold-saving machinery. But he also estimated that within a year, he had taken out four or five thousand dollars worth. And Bennett Bishop believed that during this same period, fifty men along the river were pulling wages of ten to twenty-five dollars a day in gold.[13]

There appeared to be just enough gold to sustain a level of enthusiasm for hopefuls and diehards, with reports of wealth continuing for years to come. The shifting sands and gravel beds of the river always held possibilities. In 1898 an article in the *Mancos Times* reported that "397 pennyweight of San Juan Gold" was shipped in from Bluff. In 1904 headlines announced that a "half pint of gold" was brought in by James Hyde, a Bluff merchant. A year later mining engineers publicly announced that in "over two hundred tests made in bars covering a distance of nearly 20 miles along the San Juan River, not a single barren pan of dirt was found."[14]

But most incredible of all was a newspaper article entitled, "How Gold Nuggets Grow," which stated that when gold was left in its natural environment, it attracted other particles and grew in size. Examples of this phenomenon had been observed in mines in California. The article closed by speculating that people might abandon regular farms to establish gold ones, where they would grow nuggets for a crop.[15] In the goldfields, hope sprang eternal.

The question of how to wrest flour gold from the San Juan was the real issue. Traditional panning tantalized but never produced sufficient wealth to interest the freelance miner. One ingenious individual went back to an expensive, yet ancient, technique of placing cattle hides fur side up in the shallow part of the stream. Water, laden with silt, washed over the hide, depositing its heavier load. After a few days, the miner removed the skin, dried it, then burned it and recovered the gold from the ashes. Reportedly a full pound of gold resulted from this process, but it cost thousands of dollars. Both expense and effort proved too much to sustain this ingenious operation.[16]

As in countless other stories about extracting wealth from the landscape, the situation called for better, more sophisticated technology and more investment capital. In a land that provided nothing, it took green dollars to milk wealth from the brown waters of the San Juan. Beyond the traditional pan, rocker, and sluice box, a second phase of mineral extraction began in early 1894. By this time the get-rich-quick boomers were gone. It was now technology's turn to pit itself against the resources of a stingy land.

Two men, D. H. Lemmon and Major J. W. Hanna, exemplify those who put their money where their faith was. The two used a Kennedy machine that required an engine both to pump water from the river fifty feet away and separate gravel from the high bars above the bank. Sand and silt in the water clogged the filters and quickly wore out the packings on the pump. Yet the miners wanted this sand because it held the gold. At one point, Hanna believed "the finest sand is no more nor less than a gold quartz." To retrieve it, he considered a cyanide mill, but there is no further mention of this deadly chemical being utilized for mineral extraction.[17]

Still, the experimenting went on. By August 1894 Lemmon and Hanna had completed extensive testing of the large gravel beds along the river and chosen a spot near Bluff. There they placed two Kennedy machines and "an electroplate of large capacity" that were said to save all the gold that funneled through them.[18] The men believed they could process between 200

Sluice box and waterwheel were a part of more-permanent mining operations. A. L. Raplee owned this camp at the foot of Mexican Hat Rock. A large community of more than a thousand people settled briefly in this area because of the gold and oil industry. (Charles Goodman photo, San Juan Historical Commission)

and 250 cubic yards of gravel a day at a value of more than fifty cents a cubic yard.

In September Major Hanna pronounced himself a success. In a newspaper interview, he declared he had spent seven thousand dollars experimenting over a year's time. He estimated that there were now two hundred men working both the high and low banks for seventy miles down the river. "A man with a rocker made out of candle box makes $3 to $7 a day. With improved machinery he can do better."[19] Hanna went on to describe his operation on the high sandbars. Wheel scrapers pushed boulders and sand to the machine, where the large rocks then fell through to the river while the sand passed through three sets of screens. A copper plate with quicksilver amalgamated the gold, saving it for further processing. According to Hanna, at this point in the venture, he had recouped his expenses.

Hanna and Lemmon would mine for a few more years but eventually gave up. Others, such as Charles Spencer in the early 1900s, then took up the banner, swearing that crushing rock, dredging the river, and chemical amalgamation held the solution to the problem.[20] However, more-expensive schemes and better technology still could not wrench enough precious metal out of the river to make it pay. Machines which processed fifty cubic yards a day were to be replaced with groups of them that could do a hundred cubic yards. At one point, Spencer estimated that eventually these river sites would be churning out five thousand dollars worth of gold a day. Spencer, like Hanna, was good for a few years before he also climbed out of the business.

One aspect of the rush that illustrates the determination to overcome physical odds was the transport of men and supplies to the fields. Some methods were ingenious, others dangerous, still others labor intensive, but all pitted man against the land. Take, for instance, the network of roads that crisscrossed the rough canyon country skirting the river and slicing across the high desert of southeastern Utah. To get to the goldfields, roads for horses, pack mules, and, in a few instances, wagons started to appear in some of the most impossible places. Paths snaked up Comb Wash, crossed over Lime Ridge, then branched off to Mexican Hat or continued to Clay Hills or the mouth of Slickhorn Canyon.

One engineering feat known as the Honaker Trail serviced miners at the turn of the century. Its most dramatic section began about eight miles below Mexican Hat, where it traversed down the face of an escarpment for two-and-a-half miles. Although people hoped to bring pack animals over the trail, that proved impossible.[21] Other miners resorted to lowering supplies by rope down the cliffs to the banks below. Bert Loper, who came to San Juan in 1893, describes the difficulty in using this system.

> We freighted our stuff to the rim of the canyon and there two or three of us young fellows at that time carried the stuff down from ledge to ledge until we got to the last big ledge [which] was about 130 or 140 feet down from the rim. We had to let our stuff over with ropes, and then when we got our stuff let over the cliff, we would go out to the point where the trail now goes over and climb down a rope ladder and then come back to the ledge and down to the river.[22]

In the contest of man against the land, the land often threw down an interesting series of challenges.

Miners also brought their equipment and supplies down the river in boats. Local entrepreneurs built and sold many of them to transients anxious to get to the fields, but that increased the already-heavy demand for lumber. Rockers, sluice boxes, flumes, waterwheels, and general construction all depended on wood in a generally treeless area. Boats came into service to haul it down to the work sites. A. L. Raplee, a miner and oil explorer, recalls these boats were about eighteen feet long, with a four-and-one-half-foot beam, flat bottomed, and drew about six inches of water. Trips were made all times of the year, but in the summer, it could take fifteen hours to travel twenty-five miles because of the lower channel, slower current, and numerous sandbars.[23]

Slowness was the last of the problems that challenged Otto J. Zahn in the late spring and early summer of 1905. He built a fourteen-foot raft of bits and pieces of drift timber, secured with baling wire. He then loaded a 250-pound hopper and launched off, alone, into the floodwater for his placer camp, fifteen miles downstream. His only steering device was a thirty-foot rope that he pulled against when the current ran the raft ashore. After the boat was dislodged, he hopped aboard until it ran aground again.[24]

A few of the miners even managed to go upriver using poles, oars, and ropes. Frank H. Karnell remembers building a boat in Bluff, filling it with one thousand pounds of supplies, and sailing it down to his camp above Mexican Hat. He used the same boat to make his way upriver, two to three miles at a time, to work various placer sites on the gravel bars. At one point,

These two sections of the Honaker Trail illustrate some of the problems in servicing mining camps. Lowering supplies by rope over cliffs was time consuming and dangerous. Eventually, if a mining site proved profitable, intensive labor might forge a trail. Traversing the cliff meant numerous, narrow switchbacks. (left: Charles Goodman photo, Manuscripts Division, Marriott Library, University of Utah; right: E. G. Woodruff photo, #171, U. S. Geological Survey)

he and his partners had three boats plying the San Juan; they were not alone. Frank recalls twenty groups sailing past at different times during this period.[25] Few, if any, of these travelers ever brought their boats back to Bluff, their point of origin.

In addition to the gold seekers on the San Juan, crews also sifted the sand in Glen Canyon. Some continued to use the standard pan, rocker, and sluice, but as on the San Juan, others invested in more complex, expensive machinery. Entrepreneurs introduced more than one hundred different types of patented apparatus designed to extract the gold, but none proved successful. Robert B. Stanton supervised the Hoskaninni Company that, from 1898 to 1901, operated a gold dredge on the Colorado. The system was expensive to run, went aground on the shifting sandbars beneath the water's surface, required continuous repair, and turned no profit. The total venture may have lost as much as a hundred thousand dollars before the owners abandoned it.[26]

A few conclusions can be drawn about gold-mining operations on the Lower San Juan in the two decades straddling the turn of the century. The first is the role that machinery played in heightening hopes. People, ever

Flat-bottomed boats, manufactured locally, brought supplies to the camps downriver. Few came upstream for a return trip. (Charles Goodman photo, Manuscripts Division, Marriott Library, University of Utah)

desirous of controlling their surroundings, allowed technology to dupe them into believing more was possible. Once the easily obtained gold was gone, technology offered the solution. The shifting ratio between the cost of machinery and the amount of gold extracted became the measure of success.

Another point is that in spite of all of the digging, scraping, crushing, sifting, and sluicing, the actual impact on the course and flow of the river was negligible. A few rip-roaring seasonal floods erased most evidence of any activity except on the highest banks. It was as if no one had ever been there.

Oil—an increasingly get-rich product—soon replaced the gold frenzy on the San Juan. The earliest discovery of oil harkens back to 1882, when a band of prospectors in search of the lost Merrick and Mitchell mine crossed the river and noticed the strong smell of petroleum. Cass Hite, Ernest B. Hyde, and other members of the party camped on the north side of the San Juan, where they dipped pieces of bark in the oil floating on top of the water, then burned them. Hyde returned and staked the spot but allowed his claim to lapse. The 1892–93 gold rush lured others, such as Melvin Dempsey, A. L. Raplee, Charles Goodman, and Robert Mitchell, into the area where they prospected. These men became intimately familiar with the land and its wealth and participated in the subsequent oil boom.[27]

Serious drilling for oil started around 1904 and continued near Mexican Hat into the 1920s. By 1909 various oil companies had eight drill rigs in operation, had punched twenty-five holes—80 percent of which were producing—and had established a field that eventually encompassed the area between Bluff and Slickhorn Canyon. The home of E. L. Goodridge near present-day Mexican Hat became the freighting terminus and post office for the majority of businesses sprouting along the river.[28] Promoters in 1910 proclaimed Goodridge (Mexican Hat) a "thriving village" that boasted a platted town site; a telephone line that would soon connect through Bluff, Blanding, and Monticello to Thompson with its Western Union terminal; a water system with eight hundred feet of pressure; a hotel and restaurant; and a "goodly number of citizens [estimated by one author at fifteen hundred people at its height] who propose to make their residence permanent."[29] Once the boom ended, much of this infrastructure fell into disrepair.

Approximately twenty-five miles above the fields, Bluff organized a board of trade to care for the new arrivals. As during the gold rush, this struggling community saw an opportunity to boost its economy through promotional advertising, road development, and sale of goods. As one newspaper explained, "Bluff has a number of the finest homes in the state and these will be thrown wide open to visitors."[30] In the same breath, the writer spoke of the developing wagon road to

The Atwood mining camp and boats represent one of the largest and most-sophisticated investments in technology to procure gold from the San Juan. When the enterprise proved unprofitable, the boat went to Lees Ferry, where it transported people and equipment for years. (San Juan Historical Commission)

Goodridge, a new "gusher" capable of pumping an estimated six to seven hundred barrels a day, and outside experts from Pennsylvania, Ohio, New York, Washington, Illinois, and Colorado, who saw nothing but promise in the growing fields. While a gusher was highly unlikely (perhaps staged for promotional purposes), given the substrata rock formations, and hundreds of barrels of oil from this field even more unlikely, it was an era of rampant salesmanship.

Gushers were reported to have pushed oil from forty to seventy feet in the air, symbolizing both the hope and promise of the economy. A well drilled at Goodridge, about two hundred yards from the river, was said to have gushed an unbelievable 287 feet, and when a second well nearby began producing the same amount of oil, the boom started.[31] The well flowed for four months, then went dry. Another well not far from the first produced ten thousand barrels before it, too, was spent. Frank Hyde, an Indian trader temporarily turned oilman, drilled nineteen wells in the Mexican Hat-Goodridge area. The most successful were the seven lying within a mile-and-a-half radius of this general vicinity.[32]

Engines to drive the drills were initially fired with driftwood from the river. A lot of the petroleum from the field was consumed locally, burned in the engines that drillers hoped would produce more oil. Yet even with this need met, other problems existed. Finding the pockets of oil was often accidental; moving heavy, deep-drilling machinery into difficult-to-reach locations proved impossible; the closest railroad lay more than one hundred miles away; and lack of wood, good water, and a convenient road network made work in the oil fields expensive and labor intensive.[33]

The excitement generated by wealth and the hope of obtaining it provided impetus for

Is seeing believing? This oil gusher, photographed at the Goodridge well on 8 March 1908 is a difficult phenomenon to explain geologically. The hope for oil in this region was not truly fulfilled until the Aneth field opened in the 1950s. (Courtesy of Doris Valle)

road and bridge building. In 1909 the Midland Bridge Company received a five- thousand-dollar contract from the state to build a cable bridge capable of supporting light traffic at Mexican Hat. There were also budding plans for a railroad to move oil to the bustling markets in the East. Unlike the bridge, which was completed within a short time, the railroad never moved beyond planning maps in the offices of the Denver and Rio Grande and Southern Pacific Railroads. As the prospects of substantial oil deposits diminished, the enthusiasm for laying track waned.

By 1912 the boom had largely ended, though a few believed the canyons would still yield a rich treasure in "black gold." (Today the twenty-five oil wells around Mexican Hat usually each pump one to two barrels a day.)[34] Some oil prospectors looked to the reservation, hoping that Navajo boundaries would change. The history of the Paiute Strip, an area south of the

This image captures the get-rich-quick hopes for the San Juan. Neither the gold miners in the background nor the oil workers in front realized the anticipated profits. (Charles Goodman photo, Manuscripts Division, Marriott Library, University of Utah)

river and west of the 110th meridian, indicates the success and failure of that notion.

In 1892, eight years after President Arthur set this region aside for Navajo use, the gold rush pushed the strip back into the public domain. Because it was sparsely settled by both Navajos and Paiutes and rumors of gold were rampant, the government easily removed the strip from the reservation. By 1908 things had quieted down so it returned to Indian ownership, this time for the Paiutes living under the supervision of the Western Navajo Agency in Tuba City.[35]

In 1921 economic forces called for a new determination of the land's status. Paradise Oil and Refining Company, Monumental Oil Company, and traders such as John Wetherill and Clyde Colville from Kayenta sought the right to locate and pump petroleum from an area they believed was unsettled. Once again, the Paiute Strip became public domain; however, the oil companies realized little, if any, profit from their ventures.

Between 1930 and 1932, two large areas of land—the Paiute Strip and the region around Aneth on the reservation—became points of contention between Navajo and Anglo stockmen. On 19 January 1933, the incessant finger pointing stopped, and the sought-after solution became law. Those representing Anglo interests agreed that the Paiute Strip and additional lands, now called the Aneth Extension, would become part of the Navajo Reservation.

Of equal import for the future was the agreement that as the tribe allowed oil exploration and leasing of the land in the Paiute Strip and Aneth Extension, 37½ percent of the revenues would be used for "Navajos and such other Indians" living on this section, with the remainder going to the tribe.[36] The money would fund education, road construction, and the general well-being of the Navajos. The law expanded in 1968 to include all Navajos living on the Utah portion of the reservation.

This background information makes it clear why oil from this region became critical to both local and reservationwide Navajos. From an environmental perspective, the land was about to produce black gold, wealth that the early gold miners could never imagine. The land, used previously for grazing, was now going to be punched and prodded to yield as never before. Roads, oil derricks, pipelines, and storage tanks became symbols of the unparalleled exploration and exploitation of the land.

Both sides of the river in the Aneth-Montezuma Creek area were dramatically affected by what was about to begin. Starting in 1953, Humble Oil and Shell Oil initiated agreements with the Navajo Tribe and the State of Utah to unlock the rich petroleum reserves beneath the land. By February 1956 the Texas Company (predecessor of Texaco) was hard at work in Aneth. In its first full year of production, the field yielded nearly 1.3 million barrels of oil, which increased to more than 30 million barrels by 1959. Soon the area became known as the "giant" Aneth field.[37]

Beginning in November 1956, the Navajo Nation officially opened its doors to general bidding on the 230,000 acres of oil-rich reservation land in southeastern Utah. An estimated five hundred to six hundred oilmen attended the first session in Window Rock, Arizona. Two days later they left behind more than twenty-seven million dollars in lease money and an agreement that the Navajos would receive rentals plus 12½ percent of the gross value of any oil produced.[38] Thus concluded what the Bureau of Indian Affairs termed the largest sale in its history.

In 1956 alone, long before its peak, the Aneth Oil Field produced $34.5 million in royalties to the tribe.[39] With a population of more than eighty thousand, the Navajo Nation decided against making a per capita distribution, which would only amount to an estimated $425. Instead, the leaders invested the royalties in services such as education and economic development. Much of this money, however, remained on the central part of the reservation and not in the periphery, where the wells producing the wealth were located. As time progressed, this situation became increasingly inflammatory to the Utah Navajos, especially those living in Aneth.

The exact amount of income from oil-field royalties is difficult to calculate, since wide discrepancies appear in the available literature. It can be determined, however, that the tribe received more than $10 million annually in royalty income during the early peak years of production, after which the amount dropped to $5 million a year. The Utah Navajo administration averaged $1,352,821 from royalties between 1960 and 1991.[40] According to tribal sources, the royalty money from oil and gas rapidly became the "backbone" of tribal income, contributing anywhere from 50 to 80 percent annually.[41]

Production peaked in 1960. The next year the oil companies began injecting water into the wells to enhance productivity, but the decline continued. By 1972 Aneth's oil output had dropped by 74 percent, down from a high of 32.4 million barrels in 1960 to 8.3 million barrels, and the tribe's royalty income was cut in half, from $10 million in 1961 to $5 million in 1972.[42]

Despite this decrease, Aneth remained the largest oil field in Utah, with business people still describing it as "huge." In an effort to conserve the oil and prolong the life of the field, the Conservation Commission established eighty-acre spacing for the wells. This also prevented the area from becoming crowded with equipment as in other oil fields.[43]

Yet, Aneth was one location where the tribe could hope for further development. Because the oil field already had long-term leases in place and was located in a sparsely settled part of the reservation, the tribal offices saw an opportunity to involve the oil companies in expanding drilling operations. More development ensued, and oil-well spacing decreased from one per eighty acres to one per forty.[44]

This change initially increased production until it stabilized. The drilling did not discover any new oil but only drained the resource faster than the original development scheme. In fact, the tribal offices admitted that the future for Aneth oil looked "dismal." But they had no recourse, since, as they consistently mentioned in their economic reports, tribal income depended on this oil field.[45]

The people of Aneth were not unaware of these events. They witnessed firsthand the developments in their backyards. As keen observers of their environment, based upon religious beliefs, the livestock economy, and agricultural pursuits, many older Navajos became increasingly concerned. They watched oil-company workers, accompanied by tribal representatives, locating new wells. There were also social issues for those who lived in the midst of the boom and bustle of the oil field. The road improvements and the addition in 1958 of a $300,000 bridge across the San Juan River still did not endear many of the people to those they considered white interlopers.[46]

Some Navajos complained that they did not know they had "oil under [their] feet" and the land had been "given out at [their] leader's office [in Window Rock]."[47] One man explained the oil exploration in this manner:

> They came and it just happened. For ninety-nine years they would drill for oil and pump it out. . . . The Anglo put up ribbons to outline what they were going to do. They were driving all over the place in automobiles and drilling . . . but no one bothered them. Then we found out it was not a good thing. . . . It was after this that the water was not good any more. They drilled and let whatever came out drain into the wash. Then the horses and sheep drank this water. It was from this time that things started to go bad. It was because of this that there were gripes against the drilling of oil.[48]

Another person remembered how beautiful the land had been, with vegetation in abundance, before all of the destruction started:

> The prairie dogs stood on their hind legs and chattered as the tall grass made waves in the breeze. It was a beautiful sight. Then came the oil wells. Bulldozers tore up the land. . . . We could not get a drink of cool, unpolluted water anywhere without getting sick. It [pure spring water] did not cause heart problems, bone disease, headaches, or cramps like it does today. All these health problems began when the oil wells were put up. It has all been polluted and ruined.[49]

This observation is very much in keeping with what the oil companies have been investigating. Although there has always been fear of an oil spill from a broken pipeline into the river, the major environmental concern in the Aneth area is the creation of more saline water. Based upon recent reports from the U.S. Geological Survey, salinity is increasing in the wells; how much depends upon the location.

In simplest terms, underground water flows from neighboring mountains and higher elevations toward the river. The Navajo Aquifer, which provides drinking water in shallow wells, has been contaminated with salts from the lower Paleozoic Aquifer. Natural seepage is responsible for some of the increase, but active drilling for oil through the layers of rock accounts for most. The water table has also been lowered throughout the area over the years. Thus, the Navajo elders, although unfamiliar with the complex chemical and physical imbalances being created far below the earth's surface, can see in very practical terms that the water and land have been hurt by activities in the oil field.[50]

In addition to harm to the springs, pools, and vegetation, some sacred sites on the mesas could no longer be used to pray for rain, plants, and livestock. Fumes permeated the air, "galloping" pumps dotted the land where horses once trotted, and machines sucked oil from the earth's bowels. At night the grinding noise of the pumps kept people awake, there were fears that the carbon dioxide injected into wells added to the general contamination, and livestock suffered from continuous incursions on the ranges.[51] All of this was taking place at local people's expense, while the government at Window Rock and the oil companies appeared to be getting rich.

Out of these conflicts and others on the reservation arose a self-styled champion known as the Coalition for Navajo Liberation. The organization defined itself as an advocate for the "rights of the grassroots people." It proclaimed a desire to foster "the protection of our natural resources against white corporations, the protection of our Mother Earth, and the protection of individual rights."[52] Thus, the coalition said white corporations were exploitative at others' expense. While not everyone in the Aneth-Montezuma Creek area subscribed to all the activist sentiments of the coalition, many wanted answers about who was responsible for the problems in the oil field and what could be done to solve them.

Ella Sakizzie, an older Navajo resident, remembers the situation well. Her problems with the oil companies go back to the 1950s. Like most local people, she had not been informed

Navajo protestors at the Aneth Oil Field in April 1978 sought greater benefits from Texaco. Environmental degradation was one of their complaints. (Utah Navajo Development Council)

about the tribe's or oil companies' intentions. Suddenly she saw trucks cruising over the rangelands, drilling rigs punching holes in the earth, and "smoke stacks popping up here and there." The white workers became very "careless" and "ignorant," running over with their vehicles dogs, goats, and sheep. At one point, a driver plowed through part of Ella's herd, and the goats "came rolling out from under the truck like balls." Another time she went toe-to-toe with a bulldozer operator who was clearing rangeland for another oil pad near her trailer. She explained, "This greasewood pasture is where I take my sheep every morning, but now look at what you have done! You have completely stripped my land. Turn that bulldozer off right this instant! I'm not kidding you."[53]

On 30 March 1978, a group of forty to fifty Navajos seized the Texaco pumping station in Aneth and stopped the oil company's operation for two-and-a-half weeks while they expressed their environmental, economic, and social concerns. The people stated that generally the underground wealth had not meant a better or easier life. In fact, it had killed their cattle and sheep, destroyed their environment, and disrupted their lives. The environmental concerns included the emission of noxious gas fumes and pollution of waters by spilled petroleum.[54] By the end of the takeover, the oil companies promised some improvements but not enough to truly stop the degradation of air, land, and water.

Since then, a new company, Chuska Energy Corporation, has entered the Aneth area. For the elders, who have watched events unfolding and couch their understanding in religious values, the problems have still not been solved. One resident said, "Because of this mist of gas that hangs over us, the good rain clouds do not come over us any more. . . . I feel like our place [represents] the total destruction of 'Mother Earth.'"[55]

Dozens of interviews with older people in Aneth paint the same picture, and feelings are intense. One person observed, "'Life' [oil] is being pumped out of the earth. It helps the earth function in its natural way, but it is being removed. All these natural things recycle themselves, returning back to rain and this was what our people prayed to." Another noted, "It [oil industry] ruined our environment, polluting our water supply everywhere. We had natural water springs, but they have all been destroyed by the oil." Still another said, "When you compare yesteryear with today, it's more dangerous to eat a sheep now than back then. It is because of these 'injection' oil wells. All the chemicals and explosives used to drill for oil have contaminated our vegetation so that it will not grow anymore."[56]

As a postscript, history continues to repeat itself. On 24 February 1997, the companies again reacted to local complaints concerning the environment. They allocated more money for postsecondary scholarships and community-to-company liaison positions, as well as making promises to monitor air, land, and water quality more closely.[57] A year later the federal government filed lawsuits against Texaco and Mobil for polluting the San Juan River. Oil spills of various sizes had occurred since 1991 and been recorded as the bulk of the violations. Improper procedures and faulty or missing equipment comprised the rest. Because of errors, Texaco has leaked eighty-five barrels or 3,570 gallons of oil during this period; Mobil has had seventy-three spills, losing 2,000 barrels of contaminated "production water" (used to pump oil out of the ground) and 450 barrels of oil.[58] The government is seeking fines of twenty-five thousand dollars per day for each violation. The oil companies feel there are no grounds for these charges. How much improvement is necessary to change these conditions, if they exist, remains to be seen.

Thus, for the older Navajos, polluted air and water, lost vegetation, sickly animals and people, and impoverished human relationships are some of the primary products derived from the oil field. From their perspective, "The oil wells have killed our land."[59] The oil companies, on the other hand, tend to deny and gloss over the complaints because they lack a Navajo ethical basis of understanding. Money speaks loudly. Just as economic survival of the fittest characterized the philosophy of the miners working the banks of the San Juan at the end of the nineteenth century, it also underlies many businesses today.

In summarizing the past one hundred years of mining along the San Juan River, the following points are clear. The first and most obvious is that underlying the changes in human values, perception, and technology during this period, the emphasis remained on wealth. Mining for gold and digging for oil are the only extractive industries that have affected this riparian corridor. But whether one looks at the struggles of the turn-of-the-century gold miner or the besieged Texaco executive facing environmentally concerned Navajo elders and 1970s hippies, the conclusion is that money or lack of it was a determining factor in maintaining the operation.

The attention paid to ecological damage by either group depended as much upon which part of the San Juan was involved—the more heavily populated upper portion in the Aneth-Montezuma Creek area, or the wilderness in the lower canyon section below Bluff—as it did on the environmental ethics of the times. Where people were directly affected, political forces were activated to curb harmful effects. However, mineral wealth decided the fate of the sparsely settled Paiute Strip, as it bounced back and forth between Indian agents and entrepreneurs.

As more and more wealth disappeared from the land, business invested larger amounts of money to gain greater profit. A general premise running throughout American culture is that technology is the savior for any failing enterprise. Therefore, it is only logical that machinery seemed to hold the answer for diminishing amounts of gold and decreased flow of oil. Eventually the expense outweighs the profit, leading to abandonment of the endeavor.

On the positive side of the ledger, a lot of new territory was explored and developed for further use. The canyon portion of the San Juan was one of the least known areas of the entire river. With the discovery of gold, practically overnight people traveling along its banks mushroomed from a handful per year to hundreds a week. As the river became more familiar, people started visiting it not because of the mineral wealth but for its scenic beauty. Bert Loper, a famous river runner, is perhaps the best example

A reminder of the value of good roads and what it took at the turn of the century to move freight. A. C. Honaker clears a path on the way to his San Juan trading post. (San Juan Historical Commission)

of someone initiated into river navigation through his experience in the 1893 gold rush.

A significant portion of the road network that services remote canyons today, as well as more populated areas, had its start during the gold rush. The first cable bridge at Mexican Hat was one result. The bridge in Montezuma Creek as well as the miles and miles of paved and dirt roads near Aneth attest to the power of extractive industry in the region. And the general boost to a local economy as men and money flowed into the area, was an "exciter" for cash flow, employment, and sales. Wealth begot wealth.

However, returning to the analogy at the beginning of the chapter, wealth is accompanied by poverty. No miners ever became independently wealthy from San Juan gold. Indeed most of the hopefuls from the late 1890s and early 1900s ended in poor financial circumstances. And although tremendous wealth has poured off the Aneth Oil Field, many of the local Navajos feel greatly impoverished in both a monetary and environmental sense.

Thus, the sisters—poverty and wealth—are still companions. Today they wear political cloth and are wrapped in red tape, but they remain as inseparable as they were a hundred years ago. Future miners of wealth, take note: Both women wait just around the bend on the San Juan.

8 The Federal Government: *Dams, Tamarisk, and Pikeminnows*

The federal presence on the San Juan appears in the khaki-and-green uniforms of the Bureau of Land Management (BLM), the National Park Service (NPS), the Fish and Wildlife Service, and other public-land agencies that have jurisdiction over parts of the river. Of all public-land issues, water development has loomed like the four-thousand-pound gorilla and had the greatest impact on the San Juan landscape in at least two fundamental ways. No single human activity along the river has wrought so much change in ecological processes as Navajo and Glen Canyon Dams. Moreover, concern about water was the first manifestation of the turn-of-the-century conservation movement and eventually led to the post–World War II environmental movement, which ironically gained its voice, strength, and momentum by defeating a major Colorado Basin dam at Echo Park in Dinosaur National Monument. A flood of environmental laws in the 1960s and '70s followed in the wake of the Echo Park victory. That, in turn, encouraged scientific study and a deeper understanding of the river's ecology. Water's story particularly is the subject of scores of books, monographs, and articles. To understand federal water development on the San Juan, however, a bit of background on the Colorado Basin is necessary.

Many western historians regard the Reclamation or Newlands Act of 1902 as the most far-reaching piece of legislation enacted by Congress regarding the West. The law was the culmination of decades of debate over ways to facilitate irrigation and encourage settlement. Some see the Reclamation Act as the climax of nineteenth-century Manifest Destiny and the creation of a hydraulic empire, controlled by the "iron triangle"of the modern federal state: science (the Bureau of Reclamation, Corps of Engineers, and U. S. Geological Survey Water Resources division), the state, and capital.[1] Other scholars think the act represents the West's failure to formulate a unified water policy, causing individual states to gain considerable influence at the expense of federal control.[2] Whether one believes that the federal government or local politicians are wagging the western water tail, clearly national laws and agencies have built, funded, and maintained these projects during the twentieth century and will continue for the foreseeable future. The San Juan River, as part of the Colorado Basin, exemplifies everything that happened in the big dam-building era; it has a large dam, Navajo, near its headwaters, and the Glen Canyon Dam sits below its confluence with the Colorado. Before examining the effects of these two dams on the San Juan's riparian and social life, let's look at the way they came about and how they are managed today.

The Reclamation Service (changed to the Bureau of Reclamation in 1907) set out to construct dams to aid small farmers in living the Jeffersonian agrarian dream. But the bureau had trouble from the beginning making reclamation pay for itself, which the law required. Few farmers could afford water at the bureau's costs. Moreover, the agency failed to develop a social program to teach farmers how to work arid land. Engineers know how to build big dams but have no clue about creating reclamation societies. Always in debt and often under attack, Reclamation began to see its way out of the red in the 1920s, when hydroelectric power came along as a cash cow that would help finance projects.

Considerable debate arose during the first two decades of the twentieth century about hydropower development in the West. For a number of reasons, discussion focused on the Colorado River Basin. First, California was the fastest-growing and most-powerful western state, and the Colorado River was its closest source of water and power. Second, Bureau of Reclamation Director Arthur P. Davis had an absolute fascination, perhaps obsession, with building a giant dam in one of the Colorado's canyons. Exactly how those dams would be built and who would regulate them were eventually worked out in the Federal Water Power Act of 1920, a compromise between total federal control and private monopoly. The law created a Federal Power Commission with authority to determine who could build hydroelectric dams on public lands like the San Juan. It also allowed, as discussed in chapter 3, private utilities such as Southern California Edison to help fund government surveys like the Trimble Expedition in 1921 to look for dam sites in the San Juan canyons.[3]

The real disagreement over water and power, however, did not concern public or private control. The crucial issue revolved around *which* states would get *what* out of the Colorado and its tributaries. Seven states—Wyoming, Colorado, Utah, New Mexico, Nevada, Arizona, and California—were contending for water and power. In 1922 all but Arizona signed the Colorado River Compact, which divided the waters into an upper basin—Wyoming, Utah, Colorado, and New Mexico—and a lower one—California, Nevada, and Arizona. The dividing point was Lees Ferry, Arizona.[4] California received the first benefits because it had the political clout, means, and perceived need for all the river could provide. The result was a series of small dams on the lower Colorado, followed by Hoover Dam in 1936, just outside Las Vegas.

In the late teens and throughout the twenties, USGS water geologists had combed the canyons of the Colorado looking for dam sites for a massive storage unit. E. C. LaRue, a cigar-chomping, outspoken hydrologist, had made a number of trips through Glen Canyon and on the Lower San Juan; he lobbied strongly for a big dam just four miles above Lees Ferry.[5] There were two problems with the Glen Canyon site: upper-basin states were not ready to buy its power, and California would derive little benefit from it. Some believe that LaRue argued so stridently for the Glen Canyon site that he lost his job.[6] The Boulder Canyon Act was signed into law in 1928, and construction of what was then one of the greatest building projects in world history began.[7] Hoover Dam ushered in an era of unprecedented dam building in the West that utterly changed the waterways of the Colorado and tributaries like the San Juan.

During the construction of Hoover Dam (New Deal Democrats like Franklin Roosevelt preferred to call it Boulder Dam, even though it sat in Black rather than Boulder Canyon), a potentially significant lawsuit arose. *United States v. Utah* (1931) threatened federal water development in Utah and perhaps elsewhere in the basin. Known as the Colorado River Bed Case, the court proceedings had to decide whether the Green, Colorado, and San Juan Rivers were navigable at the time Utah became a state in 1896. Navigability meant that Utah owned the rivers; otherwise, all the rights of ownership rested with the federal government. At issue were oil-drilling permits and fees as well as power sites and mineral leases.[8] After thousands of pages of testimony before Special Master Charles Warren, the Supreme Court decided that most of the rivers were navigable and hence Utah's. The San Juan below Chinle Wash and hence through the canyons, however, was deemed nonnavigable. In 1960 the United States brought suit again in *United States v. Utah* (1960) to obtain title to the San Juan between Four Corners and Chinle Wash. Utah lost the case as well as control of mineral leases.[9]

These cases actually had a negligible bearing on the San Juan's environmental history. They are monumentally important, however, for the wealth of historical information they provide about the use of these rivers. For the San Juan in particular, everyone from old gold miners like W. E. Mendenhall to Mormon pioneers like Kumen Jones and old river hands like Bert Loper testified. While attorneys for both sides tried to focus the testimonies on navigability, all sorts of other information crept out. For example, it became clear that the Mormons changed the course of the San Juan at Bluff, cut down many of the cottonwoods and, unknowingly, hastened erosion along the river. They were aided by their counterparts upstream in New Mexico and Colorado. Reams of information about land use along the San Juan lie in the testimony of the case, offering a wealth of understanding for river historians.

With the court case settled, water development in the upper basin had to wait for World War II to end, but in 1946 the Bureau of Reclamation was ready with a thick, lavishly illustrated, almost-hyperbolic report, *The Colorado River: "A Natural Menace Becomes a National Resource."* This report, which ultimately became the Colorado River Storage Project (CRSP), proposed nothing less than utilizing Colorado Basin water "to the very last drop" and was significant in a number of respects.[10] First, in planning to use every ounce of upper-basin water, it outlined scores of dams, big and small, on all the rivers. Four sites were considered for the San Juan—the Great Bend, Slickhorn, the Goosenecks, and Chinle Wash (sometimes called the Bluff Dam). Second, the report's language was full of unbridled optimism and hubris. It depicted the Colorado River in language close to the propaganda used to describe the "Nazi hordes" the Allied powers had recently defeated in Europe: "a natural menace," "tore through deserts," "ravaged villages," "man was on the defensive," "He sat helplessly . . . in vain to halt its destruction," and so on. Further, the report concluded that controlling the river's "terrifying energy" would build "beautiful homes for servicemen" and "bulwarks for peace." In short, the bureau's grand proposal emphasized empire, wealth, and absolute control of the whole basin. It projected images of clean, orderly houses; plentiful fields of crops, prosperous cities, healthy livestock, new dams, and happy recreationists and painted a glowing picture of America's future, to be delivered by the "total use" of the Colorado River.[11]

On environmental, even aesthetic, concerns, the report was strangely silent. It described the spectacular beauty of the many canyons the bureau hoped to dam, but in the next breath discussed the way that beauty would be somehow enhanced and made more accessible to the public by dams and reservoirs. It mentioned fish and wildlife, but only to the extent that reservoirs would increase sport fishing and hunting.[12] In all *The Colorado River* was an extraordinary blueprint that outlined another kind of conquest for a nation flushed with victory after World War II but also wary of threats from communist Russia and China. What the bureau proposed, and largely accomplished, was what water historian Marc Reisner calls "the most fateful transformation that has ever been visited on any landscape, anywhere."[13]

That transformation did not happen immediately. First, the upper-basin states had to agree on water allocations. The Upper Colorado River Basin Compact was signed in 1948 and ratified by Congress the next year. The Bureau of Reclamation began issuing planning reports the same year, but the Korean War put CRSP on hold again. By 1953 the first CRSP bill finally arrived in Congress, and hearings began the next year.[14] Two House and two Senate subcommittee hearings on irrigation and reclamation convened in 1954 and 1955. After thousands of pages of testimony and a major renegotiation in Denver by upper-basin congressmen, CRSP passed both houses in 1956, and President Eisenhower signed it into law. Two dams that affected the San Juan—Glen Canyon downstream and Navajo on the upper river—formed part of the original CRSP law. The way these dams made it into the final legislation has been the story of many recent publications, but here is a brief summary.[15]

Glen Canyon Dam had always been a high priority for the bureau, going back to 1916, when E. C. LaRue first proposed the site. As the CRSP legislation proceeded, however, the act ran into opposition from conservation groups over its other top recommendation, a dam at Echo Park in Dinosaur National Monument. Conservationists like David Brower of the Sierra Club, Howard Zahniser of the Wilderness Society, and others effectively blocked the construction of a dam there, which moved Glen Canyon to the sole number-one spot. This does not say, as many environmentalists have over the years, that the conservation groups traded Echo Park for Glen Canyon.[16] It would have been built anyway. Conservationists opposed Echo Park because it violated the integrity of a national park. The Glen Canyon Dam site, however, sat in no such park.

Moreover, Brower and the Sierra Club did not want to appear opposed to all "progress." Although Brower later regretted not opposing the whole CRSP project, defeating a dam in a national park was probably the best conservationists could do at the time. A few faint-voiced souls, however, did argue strongly against Glen Canyon. Calling themselves "the Utah and National Committee for a Glen Canyon National Park in Opposition to the Proposed Glen Canyon Dam," they managed to have their petition read into the record of the 1954 Senate and 1955 House hearings on CRSP.[17] But their statements had no more impact than a small gust of wind on an ocean. No senator or congressman even acknowledged them.

What is interesting about the opposition to both Glen Canyon and Echo Park Dams is the nature of the arguments. Both the major conservation groups and this small group of river runners cited aesthetic and spiritual values to oppose the dams. Again and again, running through the testimonies of Howard Zahniser, Joe Pinfold of the Izaak Walton League, and Sigurd Olson and Fred Packard of the National Parks Association are statements like this one from writer Olson: "We also believe that these areas are for the education and spiritual rejuvenation of all people. . . . unspoiled nature has a greater significance than any other and . . . any change in these areas which depreciates the spiritual values is wrong."[18] Surprisingly enough, little discussion centered on environmental values—wildlife, riparian ecology, and so on. The closest any conservationist came was in statements by Richard Pough of the American Museum of Natural History and George Fell of the Nature Conservancy. Both argued for preserving Echo Park for scientific study, saying that biologists needed "untouched areas" where they could set up outdoor laboratories to study "undisturbed communities of wild plants and animals." Pough and Fell justified these nature labs because they might yield new antibiotics, medicines, and other "new uses for obscure organisms."[19]

To contemporary ears, already full of talk about endangered species, habitat conservation, and ecosystems, these arguments may sound quaint and inadequate. To the western senators and congressmen, however, they were ridiculous and certainly not mainline arguments against dams. Although conservation groups formulated other arguments against Echo Park, such as pointing out math errors by the bureau and advocating "alternative" energy from coal and atomic power, they primarily stuck to the same arguments that Muir and the Sierra Club had used unsuccessfully earlier in the century to fight Hetch-Hetchy. Times had changed more than the terminology.

Glen Canyon, then, sailed through the CRSP hearings unscathed, but the law included a provision to protect Rainbow Bridge National Monument from the encroaching waters of Lake Powell. The bridge sits below the confluence of the San Juan and Colorado. After passage of CRSP and their victory at Echo Park, conservationists pressed the bureau and Congress to abide by the provision to protect the spectacular 278-foot sandstone span. It soon became apparent, however, that saving it required construction of at least one dam downstream from the arch to prevent Lake Powell from lapping at its base. Another dam upstream would divert water around the bridge. Congress decided in 1960 that this was too expensive and construction would mar the area more than doing nothing. They continued to refuse funding through 1971. Conservationists, who successfully posed the precedent argument at Echo Park (one incursion into a national park would open the floodgates for more), suddenly found themselves on the defensive. They pressed their case through the courts in the 1960s and '70s, contending that Congress must uphold the law. The environmental groups won in federal court, but the U.S. Court of Appeals overturned the decision. In 1974 the Supreme Court declined to review the appellate decision, effectively ending the bid to protect Rainbow Bridge National Monument.[20]

Now, given easy access via Lake Powell, as many as one thousand tourists arrive daily during the summer to view the sandstone span. Recently, however, the Park Service, responding to Navajo, Paiute, Ute, and Hopi claims that the bridge is sacred, have posted signs asking tourists to refrain from walking on or under the bridge.[21] The idea of safeguarding the bridge has evolved, and many federal agencies and jurisdictions have been part of the story. The

Navajo Dam is New Mexico's main source of upper Colorado Basin water. It does not, however, generate hydrodollars for the Bureau of Reclamation. It is now being operated to mimic historic flows to help the endangered Colorado pikeminnow. (Bureau of Reclamation, Upper Colorado Region)

controversy over the meaning and application of protection measures has obviously not ended.

Navajo Dam did not generate any of the controversy that surrounded Rainbow Bridge or Echo Park. And unlike Glen Canyon, it was authorized through the efforts of New Mexico Senator Clinton P. Anderson and a chorus of constituents, including members of the state delegation, Navajo tribal leaders, the commissioner of Indian Affairs, and local politicians and water-board officials. Navajo Dam, in fact, had not even been part of the original bureau proposal in 1946. By the time CRSP appeared before Congress in 1953, however, a number of smaller dams in northern New Mexico had coalesced into this one big project.[22]

New Mexico's only real chance at Colorado River water allocation rested in the San Juan. Fortunately or unfortunately, depending on one's opinion of the dam, the state had a very able and influential advocate in Anderson. At one point in the 1954 hearings, he responded to a question from Utah Senator Arthur Watkins by saying, "If the Navaho [*sic*] Dam is not included, I will say to my distinguished friend from Utah that the bill [CRSP] will pass over my dead body."[23] Actually Watkins had asked Anderson a different question. Anderson's blurted, misdirected response clearly indicated the fervor he felt for Navajo Dam. He ultimately engineered the agreement to drop Echo Park, which left room for one or more major projects to be included. Anderson saw to it that "his dam" was one of them.

Construction began on the earthen structure shortly after CRSP's passage. The government dedicated the dam, the first major

The Bluff dam site, a mile down river from the mouth of Chinle Wash, was first proposed in 1916. A dam here would have flooded the town of Bluff and destroyed countless prehistoric and historic sites. (E. C. LaRue photo, #827, U.S. Geological Survey)

accomplishment of CRSP, in September 1962.[24] Although the bureau originally estimated the dam would cost thirty-six million dollars, the final bill rose to forty-two million. Unlike Glen Canyon and other "cash registers" spilling out hydroelectric dollars, the cost of Navajo Dam is completely nonreimbursable. It is, as one political scientist has written, "a charge completely on the federal treasury."[25] In the 1970s the bureau proposed adding a power station at Navajo, but the Navajo Tribe filed suit, claiming all money from the sale of power should go to them. That effectively shut down the bureau's project. Later in the 1980s, Farmington, New Mexico, settled with the tribe for millions of dollars and constructed a power facility at the dam, which came on-line in 1987. Farmington maintains the plant and derives all benefits from it.[26]

Power generation or not, the inclusion of Navajo Dam in CRSP indicates that dams need friends, preferably ones in high places with lots of connections. Dams proposed for the Lower San Juan, especially the so-called Bluff Dam, had no such friends, but its story sheds light on the way water politics work.[27] The first plan for a dam just below Chinle Wash came in 1914. Bureau of Reclamation surveyors had suggested the site, and E. C. LaRue pushed it in his 1916 report, saying, "Unquestionably the Bluff reservoir site . . . will prove of value in connection with the control of the Colorado River." The dam, he noted, would submerge the village of Bluff, but it was needed for irrigation, flood control, and silt retention.[28] LaRue continued to push this site through the 1920s, while also advocating Glen Canyon Dam. It is not known exactly what Bluff residents thought of losing their town, their houses, and their cemetery, but the *Montezuma Journal*, reporting from a safe distance in Cortez, Colorado, observed that it would cost the government a lot of money to compensate the citizens of Bluff for their loss. The paper quickly added, "The benefits derived will more than offset the vast initial expenditure."[29]

The Bluff Dam, as well as the Slickhorn, Goosenecks, and Great Bend ones, continued to be discussed when the bureau published *The Colorado River* in 1946. The report indicated that the Bluff Dam was "urgently needed to prevent floods and retain silt" but later admitted that the other dams would adversely affect "scenic values on this impressive section of the San Juan and . . . would flood a number of sites of archeological importance." The bureau estimated the costs for the San Juan dams as Bluff ($30.5 million), Goosenecks ($8.3 million), Slickhorn ($10.1 million), and Great Bend ($16 million) for a total of $65 million.[30]

The Bluff project stayed in the plans through the signing of the Upper Colorado River Basin Compact in 1948. It appeared in the bureau's planning report in 1949 but had mysteriously disappeared by 1950.[31] By the time CRSP legislation appeared in Congress in 1953, the Bluff Dam proposal was nowhere to be seen. Ironically, it was resurrected by a conservationist: General U. S. Grant, III.[32] A former Corps of Engineer staff member knowledgeable about dams, Grant then served as president of the American Planning and Civic Association, a conservation organization opposing Echo Park. The grandson of the former president had carefully studied the bureau reports, looking at cost-benefit ratios as well as technical aspects of proposed dams. At the opening House hearings in January 1954, he urged the subcommittee to reinsert the Bluff Dam in the bill, contending that it could retain silt and store water. Grant also suggested other sites upstream on the Colorado and Green Rivers that, as a package, could substitute for Echo Park. Grant made the same argument later that summer before the Senate. Senator Arthur Watkins tried to trip Grant up by questioning his figures. Anderson and New Mexico also did not like Grant's proposal because it threatened Navajo Dam. In the end Grant's idea went nowhere, and the Bluff Dam proposal finally died in Congress, where it met stony silence from both the House and Senate.[33]

In the 1960s the bureau proposed a dam one mile downstream from Mexican Hat, which would have flooded up to Cottonwood Wash at Bluff and cost fifty-one million dollars, but it remained a dream.[34] Only Navajo Dam and Glen Canyon Dam became realities for the San Juan. The former changed stream flows and riparian life along Utah's section of the San Juan. The latter drowned the lower part of the river, effectively killing riparian life.

Ultimately what saved Utah's San Juan from all dams was its inaccessibility. Relatively few people who could advocate for a dam lived around the river. Utah's politicians were clearly preoccupied with Echo Park and the Central Utah Project, both of which promised water and power for the populous Wasatch Front. Tiny Bluff didn't amount to much in the political long run, so a good stretch of San Juan canyons was saved from the dam-builders' concrete. Only one leg of the iron triangle of western water development, Reclamation, stood for the Bluff Dam. The other two, Congress and capital, walked away.

The story of Glen Canyon Dam's construction is well told in Russell Martin's *A Story That Stands Like a Dam.* No such story exists, however, for the noncontroversial Navajo Dam. Nevertheless, both dams came on-line within a year of each other, Navajo in 1962 and Glen Canyon in 1963. Flows on the San Juan were immediately affected since Navajo Dam cut major floods by half. The waters behind Glen Canyon Dam slowly backed up into the San Juan arm of Lake Powell, all the way to the foot of Slickhorn Canyon by 1980. In the process, Lake Powell drowned the Lower San Juan between the confluence and Paiute Farms and the last twenty miles of the Middle San Juan between Mexican Hat and Paiute Farms. Both these dams had a major impact on plants and animals in the river corridor.

Another dam which threatens to have a major impact on San Juan River ecology is the Animas–La Plata Project (A-LP), proposed on a major tributary, the Animas River, below Durango, Colorado. First authorized by Congress in 1968, A-LP has seen many modifications as proponents have refashioned it to adjust to the shifting realities of San Juan basin politics. A major change in 1988 involved agreeing to include the Ute Mountain and Southern Utes in the complicated water-delivery scheme. This would satisfy the tribes' water claims as granted by the 1908 Supreme Court decision known as the Winters Doctrine. Opponents, however, now estimate the dam's cost at close to one billion dollars. They cite fiscal irresponsibility and major environmental impact to native fish and the San Juan River corridor as reasons to kill a project which has survived decades beyond the bureau's big dam-building era. Unlike the Bluff Dam, this project has many powerful backers: the southwestern Colorado ranching and farming communities, nearly all the Colorado politicians, and the local Indian tribes. But opponents are well organized and armed with various environmental laws. Although A-LP seems to keep rising from the dead, it appears less and less likely that Congress will ultimately fund it.[35] Either way, congressionally mandated environmental studies of the river's ecology have had a major influence on the project.

The first of these riparian studies was conducted by Angus Woodbury. The increased federal presence in the Colorado Basin and the San Juan area has intensified scientific study and netted a wealth of understanding about river systems.

In his trail-breaking survey of Glen Canyon-San Juan flora and fauna, Woodbury identified, described, and catalogued three plant zones: streamside, terrace, and hillside. He noted that once the canyons were flooded, all streamside habitats would be completely wiped out. Depending on their location in the canyon, terrace and even hillside vegetation might also be covered by water. What concerned Woodbury the most, however, was the loss of streamside vegetation. Because that lush, narrow strip of plants growing along the river was the direct source of food for many mammals and indirectly involved with nearly all others, Woodbury predicted that its loss would drastically affect populations. "Beavers," he believed, "are doomed." And he went on to predict "that the mammalian fauna will become drastically reduced both in kinds and number." He concluded that the banks of the filled reservoir would "in no way provide the number and types of ecological niches which presently exist in the canyon. These banks will possibly become even more sterile because of fluctuations of the water level."[36] Clearly the few banks left on the San Juan arm of Lake Powell are sterile compared to the lush variety of the old riparian habitat. Studies of the environment around the lake since inundation have largely borne out Woodbury's predictions, although the news is not entirely bad for the fauna, as the Lake Powell Research Project (LPRP) has shown.

Begun in the mid 1960s, the LPRP in many ways picked up where the Glen Canyon surveys had ended. Multidiscipline in nature, the project ranged far and wide, looking at everything from the concentration of heavy metals in game fish to the social consequences of boomtowns (like Page, Arizona); from the shoreline ecology of the lake to the prehistoric and historic trails in the Lake Powell area. The LPRP found, for example, that habitats for birds generally decreased as a result of flooding. Their range is now restricted mainly to the flowing part of the San Juan above Paiute Farms. But the lake has increased the number of waterfowl, shorebirds, peregrine falcons, and bald eagles. The peregrine population at Lake Powell has grown to be one of the largest in the United States. It appears that the falcons have shifted their diets from the swifts and swallows that used to nest along canyon walls to the waterfowl and shorebirds that have taken up residence at the lake.

As for the beaver whose doom Woodbury predicted, the LPRP said, "Little is known about the fate of this animal." Although it has been seen burrowing into the banks of the lake, it appears to be largely gone from the San Juan arm. Nor has the lake been especially kind to native fish, of which more will be said later.[37] In summary, some native and nonnative fauna have flourished with flooding, but most have not. The riverbank flora basically were wiped out in the flooded section of the San Juan, with the exception of native willows and nonnative tamarisks, which have flourished in the postdam environment.

A number of interesting changes have occurred as the result of the shoreline ecology that has developed around the lake. Dominant species like Indian ricegrass, broomgrass, Mormon tea, goldenweed, and sand sagebrush have markedly increased on sandy slopes. Fluctuating shorelines have caused a significant surge in the exotic species, Russian thistle. Accidentally brought into South Dakota in 1886, this aggressive pioneer, commonly called tumbleweed, colonizes sandy, disturbed soils. Its rapid seed development, quickly growing taproot, and resistance to periodic flooding make it a highly successful invader; only tamarisk has adapted better. Near Paiute Farms, Russian thistle has grown as large as Volkswagens.

Conversely, decomposition of snakeweed and other plant debris accounts for increased mercury levels in the lake. Carp, introduced into western streams and lakes in 1875 as a source of protein for local people, register the highest levels because they feed on the bottom where plant debris accumulates. Other game fish like walleye, bass, and trout also have higher concentrations of mercury per kilogram as they grow larger. Besides an increase in mercury, other heavy metals have concentrated in the lake because of discharge from the coal-fired Navajo Generating Station near Page.[38]

Perhaps the major effect of both dams on the shoreline and the riverbank ecology of the

San Juan is a proliferation of tamarisks and, to a lesser extent, Russian olives. Gallons of ink have been spilled in trying to tell the story of tamarisk. From the tamarisk's point of view, the narrative would surpass even the old stories of frontier conquest in its triumphs and completeness. But from the opposite standpoint, the narrative would look like a B-grade, space-invaders movie. This exotic species excites high emotions.

Tamarisk, or salt cedar, as it is commonly called, is an Old World plant from the Middle East. Its history goes back to biblical times. Although some attribute its introduction into the Americas to the Spanish, this appears to be due to mistranslating the Spanish *taray* as "tamarisk." Probably a more accurate translation is "willow," which is a native riparian species. The first mention of tamarisk was in an 1823 Old American Nursery catalogue from New York City. The catalogue offered it for sale, and then several East Coast nurseries followed suit in the 1820s and 1830s, advertising it as an ornamental and describing it as "hardy," "beautiful," and "much admired." The first record of tamarisk in the West was in 1856, when A. P. Smith listed it in the Pomological Garden and Nursery catalog in Sacramento. Botanists assume that tamarisk escaped from cultivation; its first naturalized occurrence was on Galveston Island, Texas, in 1877. It was noted in the 1880s in St. George, Utah.

Between 1890 and 1920, tamarisk spread slowly at a rate of about twelve miles per year from the lower to the upper Colorado Basin, yet it largely escaped the notice of most sportsmen, ranchers, and farmers until the 1920s. The Roaring Twenties was a period of relative dryness for the Southwest, which hastened its multiplication. Using historic photos, botanists have pinpointed 1922 as the probable time tamarisks arrived at the mouth of the San Juan River. A 1921 photo by Hugh D. Miser of the Trimble Expedition shows no tamarisk at the confluence of the San Juan and Colorado. Herbert E. Gregory's 1923 photo at the same place shows tamarisk. Presumably it migrated upstream thereafter.[39]

Contrary to contemporary appearances, however, tamarisk did not immediately dominate willows, cottonwoods, and other riparian species. As late as 1958, Angus Woodbury, Stephen Durrant, and Saville Flowers calculated that willow *(Salix exigua)* covered roughly three acres of streamside beach for every one acre of tamarisk *(Tamarix pentandra)*.[40] In most places along the San Juan now, the opposite ratio exists. What, then, facilitated tamarisk's eventual dominance of Southwest waterways? The answer, as already suggested, is dams.

Prior to dams, rivers like the San Juan and Colorado had very high-volume spring and early summer floods, which scoured banks and deposited large beaches of sediment in their wake. Floodplains remained wide, and willows and, later, tamarisks were unable to colonize the unstable sands. One reason salt cedar flourished in the 1920s is that relatively low spring runoffs allowed it to invade and colonize previously unstable beaches. It was in the right place at the right time, a factor that became even more pronounced on the San Juan when Navajo Dam began controlling spring floods coming out of the San Juan Mountains in southwestern Colorado. Prior to Navajo Dam, floods of thirty thousand cubic feet per second were common on the San Juan, and ones of near or more than a hundred thousand cubic feet per second, like in 1884 and 1911, came swooping down periodically.[41] Since Navajo Dam's completion in 1962, the largest flow at Bluff registered fifty-two thousand cubic feet per second.[42] Clearly the river channel has narrowed considerably, and tamarisk moved in to colonize the beaches that had previously been flooded yearly.

To appreciate its success, one must understand a bit about the way salt cedar reproduces. A large, mature tamarisk tree sheds 250 million seeds per year. An individual plant may live up to a century. It is highly tolerant to varying amounts of moisture. Once established, it can withstand the fiercest drought the arid Southwest throws at it as well as prolonged flooding. Some tamarisks along the Colorado River below Glen Canyon Dam survived five hundred days of inundation by cold reservoir water. Part of tamarisk's ability to endure such extremes rests in its taproot, which may reach down more than 100 feet, with lateral roots spreading out 150 feet. The roots grow quickly, well ahead of drying surface sands. Tamarisk is also tolerant of high levels of salinity, and southwestern rivers have become saltier because of irrigation.[43]

In 1924 the San Juan near Aneth was wide and braided. Note the man standing in river. A 1995 rephotograph shows constricted river bottoms and nonnative Russian olive trees in the foreground. (C. H. Dane photo, #160, U.S. Geological Survey; 1995 photo by Lin Alder, Ecosystems Research)

Tamarisks have spread along the shorelines of the San Juan arm of Lake Powell for the same reasons: sand banks are covered with water when the reservoir fills in late spring and early summer and left dry when the lake lowers in mid-to-late summer. Tamarisk's long germination period allows it to flourish under such conditions where willows and cottonwoods cannot.[44] It has proven to be a biological superplant. Like many species that evolved in the Mideast, it found a perfect opportunity to colonize and dominate when Americans dammed rivers in the Colorado Basin.

One thing that ecologists have studied since the advent of tamarisk along western waterways is its effect on wildlife, especially bird populations. In some places like the Grand Canyon, where vegetation was traditionally sparse, tamarisk growth has led to an overall increase in lizards, small mammals, some new species of birds, like Bell's vireos, summer tanagers, hooded orioles, great-tailed grackles, and native birds like yellow warblers and ash-throated flycatchers. At the same time, cliff swallows have been largely extirpated.[45] At other places like the lower Virgin River, where studies were conducted and vegetation was relatively dense, tamarisks provided poor wildlife habitat, and bird densities were substantially lower than with native riparian plants.[46] Although no studies have been made of the free-flowing sections of the San Juan, based on these two studies, a Grand Canyon-like scenario appears most likely. Historic photos of the river from Four Corners to the Colorado show wide, sandy bottoms with sparse vegetation except for cottonwoods. Tamarisk has now invaded the entire corridor, and increased wildlife habitat seems logical. The thick, mature tamarisk stands at places like the mouth of John's Canyon, for example, must inevitably have increased habitat for lizards, small mammals, and some bird species. It is interesting to note that beavers gnaw on small branches, deer and bighorn sheep occasionally browse on tamarisk, and honey- producing bees use it as a source of nectar.[47]

Wildlife biologists, then, face the perplexing situation of managing a *naturalized* ecosystem—"a blend of the old and the new, a mixture of native and introduced organisms and natural and artificial processes."[48] Federal land agencies, however, have been slow to shift their thinking about dam-changed, tamarisk-dominated river corridors.

Many have damned tamarisks (and Russian olives) as the "scourges of the West." River runners hate tamarisks (or "tammies") because they are not native, and they choke off camping beaches, make hiking a skin-tearing, eye-poking bushwhack, and often harbor billions of blood-sucking mosquitos. Farmers, ranchers, land managers, and dam operators also loathe tamarisks because they consume extraordinary amounts of water. LPRP ecologists estimate conservatively that the West has one million acres of tamarisks, which consume twice as much water as California's major cities.[49] For this reason and others, scientists and land managers have tried to control or eradicate the plant. Some have used flood and fire with no success. Tamarisk comes back even stronger. Others have combined plowing it up with applying herbicides like garlon.

One researcher, Jack DeLoach of the Department of Agriculture, has found what he believes are effective biological controls—a Chinese leaf beetle and a Dead Sea mealybug, which only attack tamarisk. DeLoach was set to introduce his insects in 1995, when the U.S. Fish and Wildlife Service halted his project. They feared reducing tamarisks would harm a recently listed endangered species, the Southwest willow flycatcher. This has set off yet-another round of debates about tamarisk. Some ecologists, however, have questioned whether these programs are economically feasible or ecologically desirable. They say that a sick, static river system and irrigated saline soils are the problems.[50] Thus far, no tamarisk eradication programs have been attempted on the San Juan. Part of what the tamarisk debate has done, however, is begin public discussion about what western rivers should look like and ways to manage them.

Recent events in San Juan country point to an additional, completely unexpected consequence of dams and tamarisk thickets. On 29 May 1998, three men from the Cortez-Dolores, Colorado, area—Alan Pilon, Jason McVean, and Robert Mason—killed a Cortez police officer. They apparently fled first to the Hovenweep area in June, then moved down to the tamarisk-Russian olive bottoms near Montezuma Creek. A

In 1914 the San Juan at the confluence with Chinle Wash (right) was a wide, cross-bedded stream. In the post–Navajo Dam era, the river is confined to a relatively narrow, tamarisk-choked course (facing page). (Herbert E. Gregory photo, #243, U.S. Geological Survey; later photo by Lin Alder, Ecosystems Research)

large police force from the San Juan County Sheriff's Office, the Navajo Tribe, and the FBI converged on the area in July 1998, searched the thickets, then tried to burn the fugitives out. They failed in their efforts, even though they found numerous clues that two of the men had hid out in the bottoms; Mason apparently committed suicide shortly thereafter. On 31 October 1999, Navajo deer hunters found McVean's body in Squaw Canyon near Hovenweep National Monument, another apparent suicide. Pilon is still at large, though police expect to find his body in the area in the same condition. The cost for the river operation totaled nearly two million dollars. Before Navajo Dam and tamarisks, criminals trying to hide along the San Juan near Montezuma Creek would have been as exposed as slickrock.[51] These kinds of social and economic costs were never factored into the Bureau of Reclamation's projections.

Besides questioning the ecological and economic value of tamarisk control, some scientists and land managers are beginning to see the plant in a new light. They are even espousing a new tamarisk aesthetic and ethic. Instead of viewing it as an alien species that must be driven out because it is "not natural," a few ecologists have suggested that tamarisks and Russian olives are nature's way of coping with the different water regime since the advent of dams. Weeds like tamarisks, after all, are colonizers, the "Red Cross" that rushes in to stabilize soil when something has disturbed the natural order. Rather than being an aggressive exploiter, the plant can be seen as a first line of defense against soil erosion. Because tamarisk has only been around the upper Colorado Basin for seventy years, one needs to take a longer view, they argue, and see how these new, naturalized areas evolve. Some anecdotal evidence even suggests that once

tamarisks invade and successfully stabilize an area over a few decades, native plants like hackberry, cottonwoods, and redbuds will begin to reestablish themselves. Ultimately, if the amount of flow remains stable for a century or so (a mere wink of time), southwestern rivers may even see at least a limited return of the native streamside vegetation that ecologists like Angus Woodbury described along the Lower San Juan in 1958.[52]

Some "tamarisk philosophers" like Eben Rose have even pointed out that Euro-Americans and tamarisks share a special kinship. Both are "introduced species vying for a place in what will become the new balance of biota" in changed Southwest riparian ecosystems. Since Anglos have played the major role in that change—building dams and introducing tamarisk—no amount of "tamicide," he argues, "will halt these forces of change and evolution's relentless effort to reach a balance." The wisest position right now, Rose believes, is just to sit back, watch closely, and learn.[53]

The same period that ushered in phenomenal expenditures of public money on dam building in the West also saw the emergence of a newly invigorated environmental movement and subsequent passage of a slew of regulatory laws. The post–World War II era clearly marked the transition between the old and new environmental values. Those values reflected social changes in America in the postwar period. The explosion of nature writing as a literary genre, which will be discussed in the next chapter, exemplified that shift. Where the conservationists of the first half of the century had stressed efficient resource development, the environmental movement stressed the quality of life and human experience. While conservation ideas and actions rolled down to society from government and scientific leaders, the popular environmental movement spread from the middle of society outward, continually forcing reluctant leaders to respond.[54] The recent movement sprang from many places in society and has produced thousands of laws, policies, and consequences for the American landscape.

The Lower San Juan River especially, since it is administered by at least three government agencies—the BLM, the Bureau of Indian Affairs (BIA, the Navajo Tribe), and the NPS (Glen Canyon National Recreation Area)—serves as an excellent case study for working out the debates, laws, and policies of the environmental era.[55] The Navajo Tribe governs both sides of the river from Four Corners to just east of Bluff. From there, the BLM administers the north side to the Honaker Trail; the Glen Canyon National Recreation Area (GCNRA) picks up from there to the old confluence. The Navajo Tribe continues its jurisdiction along the south side of the river, more or less to the old confluence. All of these federal entities, however, operate under the same set of environmental laws passed by Congress during the 1960s and 1970s.[56]

The first major environmental law of the era, and the one that Congress spent the most time on, was the Wilderness Act of 1964. First conceived and proposed by Wilderness Society President Howard Zahniser, the act went through nine separate hearings, six thousand pages of testimony, and sixty-six rewrites between 1957 and 1964, when President Lyndon B. Johnson signed it into law. Originally a big, bold proposal by Zahniser, the act which passed made wilderness designation a long, cumbersome process of federal-agency reviews and separate acts of Congress for each state's system. More than thirty years after the fact, Utah has two separate wilderness bills before Congress, and the process has divided state and local communities as few issues before.[57] Environmental groups want almost five times as much acreage as the BLM has recommended or the Utah congressional delegation has proposed. In San Juan County, commissioners have stated they want no wilderness at all.

Along the San Juan, most of the Grand Gulch plateau and Cedar Mesa area have been proposed as wilderness, as has the north side of the river in GCNRA. The Utah Wilderness Coalition, however, wants almost twice as much land in the San Juan area as the BLM and Utah delegation (400,000 versus 200,000 acres).[58] Moreover, it wants both Comb Ridge and Lime Ridge, north of the San Juan, designated as wilderness, while neither the BLM nor the Utah delegation has included them. No Navajo land, to the south or in the upper stretch, figures into any wilderness proposals. At this point, it appears that a good part of the Grand Gulch complex of drainages into the San Juan, as well as the GCNRA sections, will become wilderness if a bill ever passes. Sections of Cedar Mesa remain questionable.

Whatever happens about wilderness along the San Juan River, the debate has been extremely contentious, even though most wilderness study areas (WSAs) along the river have not generated as much controversy as other areas of San Juan County. Most of the San Juan has been a de facto wilderness for Anglos since the Spanish and fur trappers nipped around its edges. Generally the region has fewer mineral or agricultural resources than other parts of the Colorado Plateau and so has proven less of an issue for wilderness advocates and opponents.

Consequent with the Wilderness Act was the Wild and Scenic [River] Act, signed into law in 1968, again by President Johnson. Like the Wilderness Act, wild and scenic designation involves a slow, cumbersome review and legislative process which is even more difficult. Aiming specifically to protect river ecosystems from inundation by dams and safeguard other ecological, historic, and recreational values along riparian systems, the wild and scenic system has languished because it is so difficult to get competing users of waterways to agree. Moreover, major environmental organizations look upon rivers as secondary concerns.[59]

The San Juan has been considered by the BLM in its Resource Management Plan (RMP) as a wild and scenic river. From Bluff to Clay Hills, all but two miles around Mexican Hat have been judged "wild," with those other two categorized as "recreational." A National River Inventory judged the San Juan as "having outstandingly remarkable science, recreational and geologic values."[60] The RMP also declared that a formal study of the San Juan for wild and scenic designation should be conducted within five years. That was in 1991. As of 2000, however, all studies have been suspended indefinitely. Until

Norman Nevills (front, center) was the first commercial boatman on the San Juan. His trips attracted passengers such as the future Arizona senator, Barry Goldwater (back, far right). (Manuscripts Division, Marriott Library, University of Utah)

a study can be conducted, however, the San Juan is being managed as a wild and scenic river.[61]

This temporary designation means that no dams or other construction can be built along the San Juan. The river will remain as it is, but that does not indicate an absence of management. Boating on the San Juan has been popular since the early days of commercial river running in the late 1930s with Norman Nevills.[62] As will be discussed in the next chapter, Nevills and subsequent commercial outfitters popularized the San Juan as a kind of "river for Everyman." The big boom began, however, in the mid 1960s, when the national campaigns against the Echo Park and Glen Canyon Dams publicized the wonders of canyon river running. By the end of the 1970s, both the BLM and NPS, the two agencies administering the north side of the river, realized that a permit system was needed to limit the huge numbers of people running the San Juan. Along with the limit, there should also be safety inspections and waste and garbage-disposal requirements. The San Juan, in short, was getting trashed. In 1974 approximately two thousand people floated the San Juan. By 1995 that number had risen to more than thirteen thousand, where it remains today, fixed by BLM rules.[63]

In 1979 the NPS and BLM signed an agreement that allowed the bureau to issue permits, collect fees, enforce rules, and patrol the river. The agreement also called for developing a river-management plan, which BLM recreation planner Jerry Ballard drew up that spring. That plan,

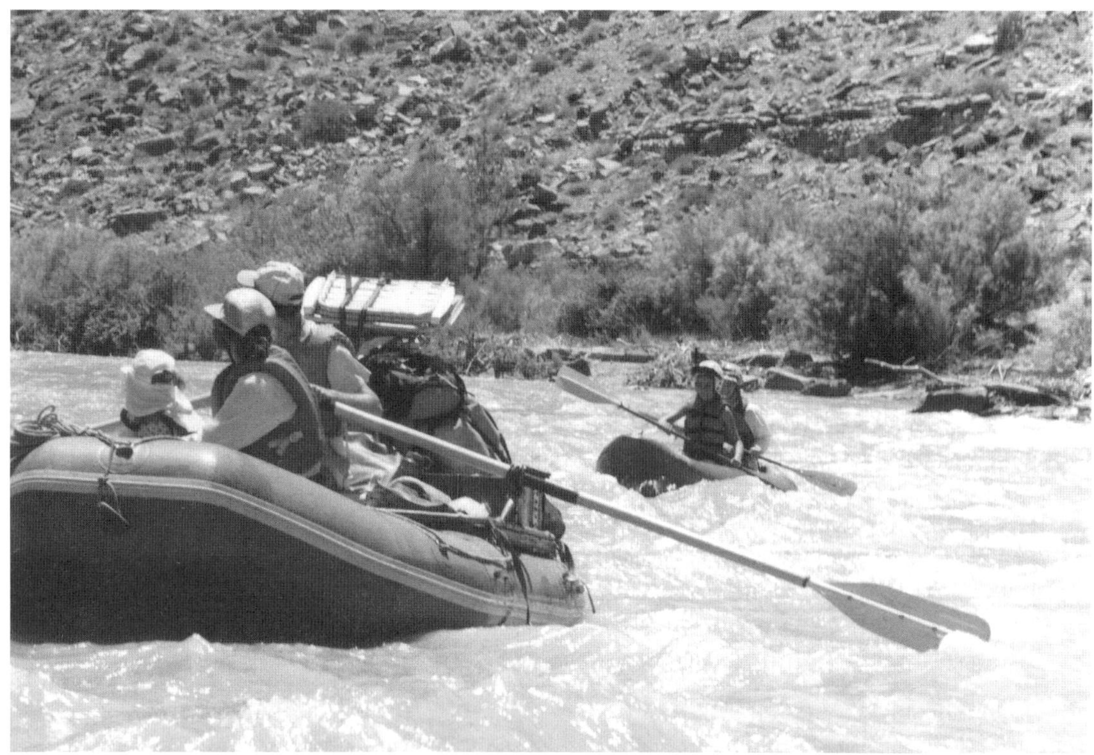

More than thirteen thousand boaters now run the San Juan annually. The Bureau of Land Management regulates the river through a lottery-and-permit system to protect the riparian corridor from being loved to death. (James M. Aton photo)

after some tinkering, has formed the foundation for managing the thousands of people who use the river every year. It controls garbage and waste, educates the public about cultural and natural resources, and protects camps along the river. The Navajo Tribe, which owns the land south of the San Juan and the south half of the river itself, has recently begun requiring an additional permit to camp on their land. Although the presence of thirteen thousand people has undoubtedly had an effect on such popular places as River House Ruin, the Kachina Panel, the Honaker Trail, and Slickhorn, river runners, following BLM guidelines, have been responsible recreators.[64] In fact, an independent study of the beaches and campsites along the river revealed "little biophysical damage."[65] Still, the BLM would like to do much more patrolling the river, educating river runners, and minimizing human impact on the ecosystem. The problem has been, and continues to be, funding.[66]

A law which has had a profound effect on the San Juan and federal land administration is the National Environmental Policy Act of 1969 (commonly known as NEPA). Pushed through Congress by Senator Henry "Scoop" Jackson of Washington with remarkably little debate, NEPA fundamentally changed the way the nation did business when it came to land, water, and air. It required environmental impact statements for projects that might potentially harm natural resources. NEPA recognized environmental quality as part of American life. Together with its cousin act, the Federal Land Policy and Management Act of 1976 (FLPMA), NEPA has completely changed the way federal agencies like the BLM, BIA, and NPS administer the San Juan River and its drainages. For example, the BLM's recent RMP was mandated by FLPMA. Moreover, the agency had to employ NEPA's principles and consider environmental issues. The same ideas were inherent in the BLM's original wilderness inventory, or draft environmental impact statement, in 1980. NEPA and FLPMA obligate the BLM and other federal agencies administering lands along the San Juan to study environmental impact before any project moves forward. This could mean a permit to graze cattle, drill oil wells, or build a new road. No longer may federal agencies make decisions about land, water, and air based solely on economic principles. Ecological principles have entered the discussion.

Another law with an equally important effect on the San Juan—perhaps even greater than NEPA and FLPMA—is the Endangered Species Act (ESA) of 1966, 1969, and 1973. The ESA of 1973 not only sought to identify species that were endangered or threatened by extinction but also proposed plans to reestablish "critical habitat" for species designated by the U.S. Fish and Wildlife Service. Along the San Juan, the Southwest willow flycatcher and peregrine falcon have been on that list. The falcon, whose eggs were decimated by the petrochemical DDT, has recovered so well on the San Juan and throughout the West that it soon may be delisted.

The Fish and Wildlife Service has mostly concentrated its efforts, however, on two native fish: the Colorado pikeminnow *(Ptychocheilus lucius)* and the razorback sucker *(Xyrauchen texanus)*. The resulting seven-year study (1991 to 1997), known as the San Juan River Basin Recovery Implementation Program (SJRIP), was the most thorough analysis of the river ever undertaken.[67] The SJRIP looked not only at endangered native fish but also examined the entire fish community, which included an intensive look at geomorphology, hydrology, and habitat.

The pikeminnow, whose ancestors have swum in the Colorado Basin for at least six million years, is the Moby Dick of the area's fish. Commonly called a Colorado salmon because of its migratory behavior, this minnow family member reached lengths of up to five or six feet and weighed eighty pounds. It was found throughout the basin, from the brackish estuary in Mexico to the mountains of Colorado. Remains have turned up in Anasazi ruins, while pioneer accounts and photos depict a popular fish with anglers, for both sport and food. The abundance of the pikeminnow was indicated by its use as fertilizer. The razorback sucker, nearly gone from the San Juan today, was also desired for food and supported commercial fisheries in the Colorado Basin. The razorback can grow three feet long and weigh thirteen pounds. Under normal conditions, an adult can live thirty or forty years.[68] Conditions along the San Juan and in the Colorado Basin, however, have been anything but normal during the past forty years. Only a small reproducing population of pikeminnows currently live in the San Juan. Razorback suckers are even rarer.

There have been many assaults on these native fish. Navajo and Glen Canyon Dams have been a major factor. Together with diversion dams in New Mexico, Navajo has altered stream flows, interrupting breeding habits. Glen Canyon has prevented migration upstream and down. Competition from introduced fish has also been a problem for pikeminnows and razorbacks. Moreover, pikeminnows lost a major prey, the bonytail chub *(Gila elegans)*, when that native fish disappeared from the San Juan. If these factors were not enough, pikeminnows and razorbacks declined because of various pollutants introduced directly and indirectly by people. Another crucial factor was poisoning. In 1961 the Fish and Wildlife Service, the same agency that is spending millions of dollars today to restore native fish, treated western waterways with rotenone. The chemical eradicated trash fish like carp and improved habitat for the popular sport fish, rainbow trout. Rotenone, however, dealt a crippling blow to all native fish.[69]

Finally, native fish have suffered from petrochemical pollution. The San Juan basin supports major oil-producing areas in Farmington and in Aneth. One kind of oil pollutant, polycyclic aromatic hydrocarbons (PAHs), are, according to a recent Fish and Wildlife Service study, "among the most potent carcinogens known to exist."[70] San Juan River fish are exposed to high levels of at least three PAH compounds. Fish absorb pollutants at an extremely high rate anyway, but the pikeminnow, a top-level predator, may be exposed to contaminants both directly in the water and indirectly through contaminated prey. PAHs enter the river from unlined waste pits next to pumping oil wells, as well as coal-fired plants, refineries, vehicular and heating emissions, motorized boats, and industrial sewage. These pollutants may be the most dire threat to pikeminnows and razorbacks. The Fish and Wildlife Service, however, has recently taken steps to reverse these trends.

The pikeminnow originally made the endangered species list in 1967. The first recovery plan, now largely scrapped, started in 1978. The razorback sucker joined the threatened list

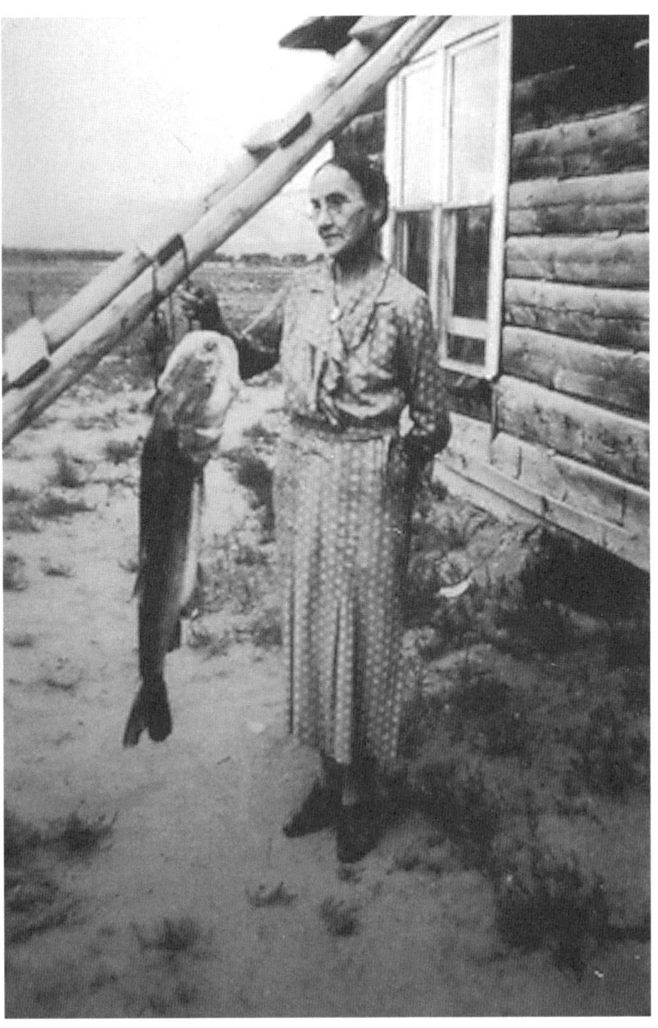

Florence Barnes stands next to a pikeminnow caught in Lily Park, Colorado, on the Green River during the 1930s (Upper Colorado River Endangered Fish Recovery Program, U.S. Fish and Wildlife Service)

in 1978 and the endangered one in 1991. Habitat recovery plans were based on cooperation among at least eleven federal and state agencies: BLM, NPS, the Bureau of Reclamation, the BIA, Utah Division of Wildlife Resources, New Mexico Department of Fish and Game, Arizona Department of Fish and Game, Colorado Division of Wildlife, Nevada Department of Wildlife, California Department of Fish and Game, and the U.S. Fish and Wildlife Service, the lead agency. These plans designated the San Juan as "critical habitat," according to ESA requirements. Thus, managers at Navajo Dam have adjusted water releases to mimic historic flows. For example, high late-spring and summer levels offered native fish calm backwater nurseries for their spawn, while very-low winter water stabilized habitat for growing fish and mature adults.[71]

The other principal action involves controlling PAHs and other petrochemical pollutants. Currently federal agencies, partly because of Navajo protests, are checking all wells in the San Juan basin for unlined waste pits. They clean out these pits and line them so that petrochemical pollutants do not wash down gullies and into the river nor seep into aquifers. The Fish and Wildlife Service is also monitoring native fish for other industrial and agricultural pollutants. Finally, using a fifteen-year budget of fifty-three million dollars for the whole basin, the service is employing radiotelemetry to monitor and analyze migration patterns of pikeminnows and razorbacks along the San Juan. It is considering augmenting existing populations (fifty to sixty reproducing pikeminnows and many fewer razorbacks) by stocking. The goal is to have self-sustaining populations of

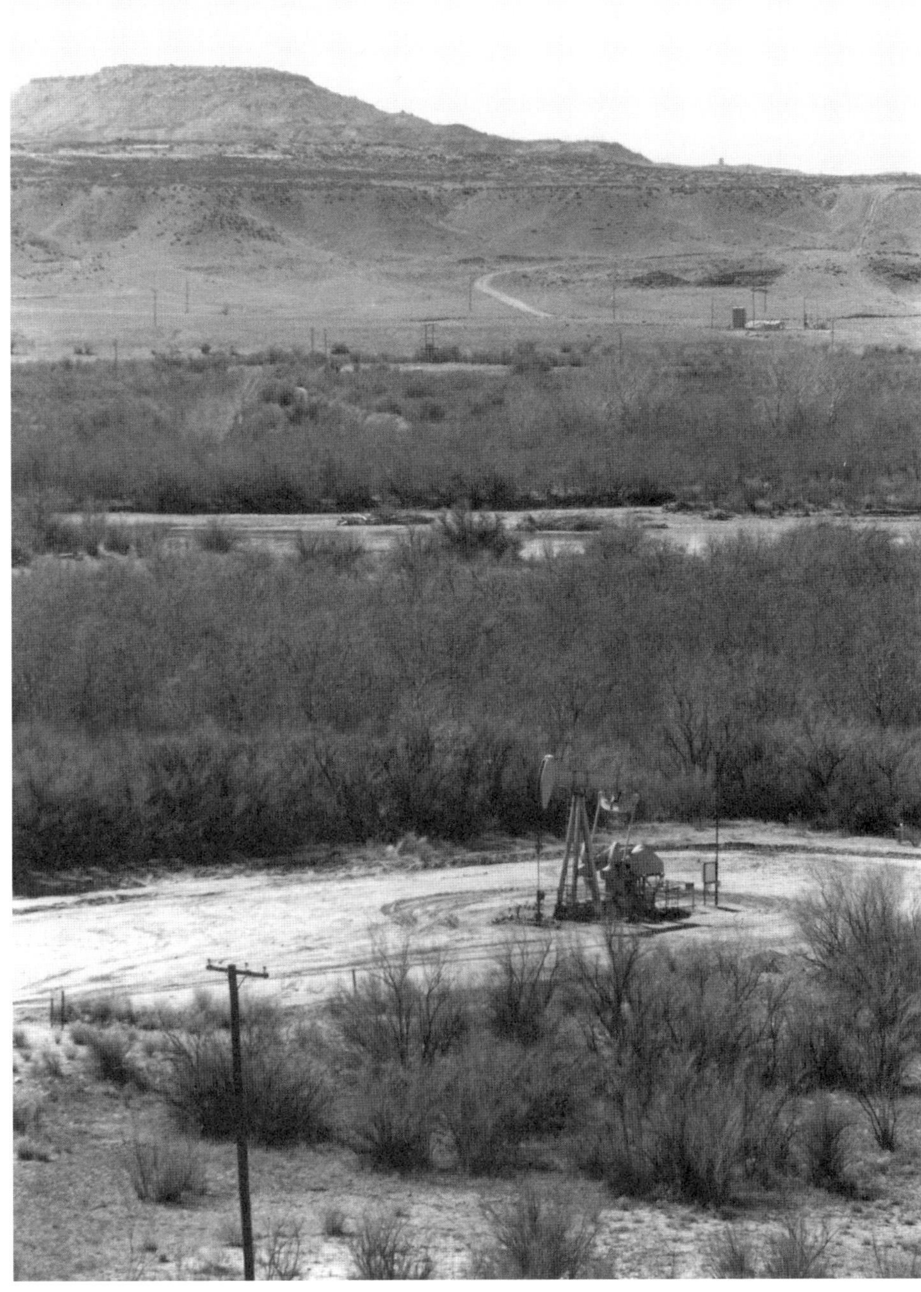

Oil wells near the San Juan like this one at Aneth have threatened native fish (James M. Aton photo)

these fish by 2006.[72] Cooperating agencies hope they can achieve the same success with native fish as they have with the peregrine falcon. The obstacles seem even more numerous and formidable.

An endangered species whose recovery faces many more challenges is the California condor *(Gymnogyps californicus)*, which made headlines during 1996 in San Juan County. Nearly exterminated, it has been the focus of a twenty-five-million-dollar recovery operation in California since 1987. The recovery team has planted this most-endangered species on BLM land in the Vermillion Cliffs, near the

Utah-Arizona border. The cliffs sit north of Grand Canyon National Park and west of GCNRA and the San Juan River. Federal officials expect the condor to take to the area and then migrate to other parts of canyon country to reestablish its range for the last 1.5 million years, including the Lower San Juan River. But San Juan County filed suit to stop reintroduction of this endangered species because of perceived threats to the area's ranching economy. This legal challenge failed.[73]

One animal that never attained endangered status but has topped the list of government restoration efforts in the San Juan region is the desert bighorn sheep *(Ovis canadensis nelsoni)*. Wildlife biologists estimate that the pre–Anglo-American agricultural West had numbers ranging from 1,500,000 to 2,000,000 bighorn sheep. Estimates now reckon the population at 40,000 to 50,000. The enemy is we. One biologist who has studied the south San Juan herd in the Red Canyon area near Glen Canyon says that bighorn sheep cannot tolerate mining, overhunting, domestic sheep, and cattle. Perhaps the greatest factor in their demise has been diseases from nonnative sheep, notably pneumonia and sinusitis.[74]

Historic, anecdotal evidence indicates many sightings of bighorn sheep in the San Juan canyons. As late as 1921, Trimble Expedition boatman Bert Loper recorded in his journal that the explorers saw "a bunch of mountain sheep" in the thirty-mile stretch between the Honaker Trail and Oljeto Wash.[75] Other twentieth-century travelers verified this. Today, however, all that remains of a once-thriving herd in the canyons is a small group at the Raplee Anticline between Chinle Wash and Mexican Hat, south of the river. This herd is managed by the Navajo Tribe and probably numbers around sixty animals. Fortunately for the herd, no Navajos run sheep in the area, and tribal wildlife biologists continue to work and talk with locals, not only to keep out domestics but also to learn about bighorn behavior and movement. At Eight Foot Rapid, a number of popular river-running camps have been closed because bighorns use that canyon for river access.[76]

While a few bighorns have been seen around the Goosenecks and the San Juan arm of Lake Powell, distinct herds no longer exist in this former range. The BLM, the Utah Department of Wildlife Resources, and the NPS, which manage bighorn herds in the Red Canyon area and Canyonlands National Park, have been cooperating to transplant, reestablish, monitor, and protect bighorn herds in San Juan County and elsewhere in Utah. One goal is to restore a herd north of the San Juan River, west of Mexican Hat. Wildlife biologists, however, are puzzled about why the south San Juan herd has not moved into its former range around Wilson Mesa, near the former confluence of the San Juan and Colorado. Nonetheless, the transplant program, begun in 1975, has fared well.

Interestingly, one of the ways that different state and federal agencies fund their expensive relocation programs is to auction off usually one hunting permit per herd at the annual Foundation for North American Wild Sheep banquet. The permits, which currently bring in fifty-to-sixty-thousand dollars apiece, allow the winner to hunt a mature ram. This money has been a good source of revenue for the different agencies and has also given biologists an opportunity to dissect dead animals and study them. Although cautious at this point, wildlife biologists believe that the future bodes well for bighorn sheep along the canyons of the San Juan.[77]

The federal presence on the San Juan has produced positive and negative effects on the river's environmental health. On the one hand, federally constructed dams and approved industrial, agricultural, and mineral development have profoundly damaged plants, animals, soil, water, and air along the San Juan. It is hard to overestimate just how deleterious dams have been. On the other hand, environmental laws have directed federal agencies to bring the lands they administer into compliance with recognized principles. Laws reflect national sentiment, and poll after poll suggests that Americans want environmental protection. Some residents of San Juan County and those who live along the river have often opposed environmental regulations.

Nevertheless, federal presence along the San Juan will continue to be omnipresent. It is not unreasonable to predict that within the next decade or two, the public will demand that the

San Juan be studied with the same scope and depth as the recent and massive Glen Canyon Environmental Studies (GCES), which commenced in 1982. Even though a final environmental impact statement was issued in 1995, individual scientific reports are still forthcoming. The GCES constitute the most extensive analysis of a river ecosystem in American history.[78] The San Juan and other major western rivers could well receive the same attention in the not-too-distant future. The SJRIP was a step in that direction. Either way, however, the San Juan River is and will continue to be a "federal river" well into the future, with all the legal, ethical, environmental, and social complexities that implies.

9 San Juan of the Imagination: *Local and National Values*

This book has focused primarily on the riparian landscape that people found along the San Juan and what they did with it. Clovis hunters stalked mammoths and mastodons and perhaps killed them to extinction. Indians, from the Clovis down to contemporary Utes, Paiutes, and Navajos, gathered ricegrass, hunted bighorn sheep, and later planted corn. Spanish and Anglo explorers and settlers introduced European-based agriculture and domestic animals. Later, Americans developed highly sophisticated technology to control water in the San Juan basin. The ripple effects of that technology—dams—are still being discovered, felt, and analyzed.

Underlying the physical adaptations are the values that shaped the day-to-day decisions people made as they lived in the San Juan area and used its resources. A particular group's cultural values will always influence the way they interact with a landscape's plants and animals. The first two chapters discussed the values of Indians in the Lower San Juan. Because Euro-Americans have had the greatest impact on the San Juan landscape, we have spent more time discussing it, but because more is known about their values, we haven't talked about them. This chapter, however, will show the ways various Anglo-American mythologies have tried to illuminate and so have affected the San Juan.

No single story has dominated. American mythologies have evolved from national as well as local trends and events. That local-national dichotomy of values, in fact, has often surfaced as groups and individuals contended for the river's resources. More often than not, conflict rather than concord has been the theme. Local writers who lived along the river and knew it from making a living often exhibited a proprietary attitude toward the land that excluded other views. National writers and artists sometimes condescended toward local Anglos. Both usually ignored Indian values and experience, although national writers sometimes paid lip service to them. Signs of change abound.

The Montana writer William Kittridge writes about "living in a story" and finding the right story to live in: the mythology that defines us and the way we interact with others and the landscape. These stories, images, songs, buildings, dances, even billboard advertisements say, "This is who we are, and this is what we believe." A mythology, Kittridge writes, "can be understood as a story that contains a set of implicit instructions from a society to its members, telling them what is valuable and how to conduct themselves if they are to preserve the things they value."[1] Since people along the San Juan have told diverse stories, their conduct toward the landscape has varied greatly. Although narratives have tended to break down along the local-national fault, a new kind of narrative may be evolving which combines parts of the two Anglo traditions, as well as the Indian experience. The San Juan area, one must remember, has never had the national significance of the Grand Canyon. Therefore, it has never attracted the great artists or writers. Nonetheless, in many ways, the story of the San Juan may be maturing.

Writers like Albert R. Lyman, Kumen Jones, and many current San Juan County residents who descended from Mormons have represented local ideas. On the other hand, novelists and nature writers such as Wallace Stegner, Ann

Zwinger, and Edward Abbey and photographers like William H. Jackson, Ansel Adams, and Alfred Bailey reflected the influence of national ideas about landscape as they depicted the San Juan in word and image. Similarly, best-selling novelists Louis L'Amour and Tony Hillerman also expressed national values to varying degrees. More importantly, their popularity exposed the San Juan area to a national audience as never before. This happened in the late 1980s.

Decades earlier, however, river runners like Norman Nevills and Kenny Ross moved to Mexican Hat and Bluff, respectively, and translated national ideas into local terms in the way they advertised their river trips. There is another kind of blend in the turn-of-the-century photographs of Charles Goodman and especially in the current writing of Ellen Meloy and Ann Weila Walka. Like Stegner, Zwinger, and Abbey, they came to the San Juan influenced by larger social trends; unlike at least Stegner and Zwinger, they stayed (Abbey lived for a time in nearby Moab). In reshaping local mythologies, they drew not only from national movements but pioneer ideas and Indian values. Whether local, national, or hybrid, all these artists have influenced the way thousands of others have thought about, interacted with, and tried to control the San Juan landscape.

When historians and literary critics of the American West examine pioneer recollections, writings, and diaries, they note a common theme: denigrate nature and exaggerate its hazards to emphasize the magnitude of pioneer accomplishments. This kind of narrative not only lionized the heroism of those who created order out of what appeared to be chaos but also sanctified their "blood bond" with the land. Historian Richard White believes this kind of pioneer mythos announces, "We created whatever is good in this place."[2] In Utah's San Juan country, such sacred-bond creation stories began with the prolific writings of Albert R. Lyman, an early settler of both Bluff and Blanding.[3] Lyman clearly exemplifies White's thesis. He created what Mormon scholar Charles S. Peterson has called "the San Juan mystique," the belief that the Hole-in-the-Rock settlers forged something unique and precious on the frontier.[4] This belief, in turn, fostered a proprietary attitude by the Mormons toward the land and its resources.

In three narratives about the settlement of San Juan County, "History of San Juan County, 1879–1917" (1918), "Fort on the Firing Line" (1948–49), and *Indians and Outlaws* (1962), Lyman worked with similar material, often even using the same wording. *Indians and Outlaws,* his most polished version of San Juan's founding, also contains his strongest metaphors regarding the river and local Indians. Although he began writing these narratives in the late 1910s and '20s, a generation after the 1880 founding of Bluff, Lyman and many in his initial audience had lived through pioneering events. He reflected the basic values of San Juan settlers regarding Bluff, the river, and local Indians. As we have seen, the river became the colonists' foe throughout their early years as they attempted to farm in Bluff.[5]

Indians and Outlaws characterized the Bluff settlement in the familiar Mormon terms of a "mission." In the literal sense, Lyman and his fellow Mormons came to the San Juan to convert what he often called "savage" Indians to the LDS faith. But in a broader sense, he saw the mission as the purveyor of Euro-American ideas of order, private property, and civilization to people (Indians and outlaw Texas cattlemen) who threatened to unleash a plague of evil and disorder on what he considered an untamed landscape. Thus, when Lyman described the San Juan River, he used the same kind of metaphors and language—"grim monster," "evil," "wild," "abominable," "ravages," "ruthless"—as he did when talking about Paiutes, Navajos, and outlaw cattlemen.

Lyman's first version of the story, the unpublished "History of San Juan County, 1879–1917," characterized the river in adversarial terms but also with wistful affection; he often called it "the old river." Over the course of his writing career, however, his figures of speech became harsher. His later writing is powerful and engaging because of his colorful language, but his metaphors also reveal a writer who viewed Bluff's settlement in the black-and-white terms of a struggle between good and evil. He saw the Mormon mission taming the wild and savage elements of "the seething triangle" between the San Juan and Colorado Rivers in the same way that he depicted taming the natural forces of the San Juan.[6] Interestingly, as he

Albert R. Lyman, 1880–1973, was born right after the Hole-in-the-Rock group settled Bluff. He grew up on the frontier, and his writings helped shape "the San Juan mystique." (San Juan Historical Commission)

moved further away in time from the very real struggles that he and fellow pioneers faced with the San Juan, his writing began to emphasize conflict with nature and Indians more. It was as if those elements became magnified in his mind.

Since *Indians and Outlaws* is one of a number of published versions of San Juan County's founding, that harsher vision has prevailed, especially in the imaginations of Mormon descendants of early settlers. In speaking of nature and the river in such demonic terms, Lyman was writing out of a nineteenth-century Mormon tradition that adopted its idioms and ideas from the Old Testament, says Charles Peterson. Nineteenth-century Romanticism did not shape the imaginations of Mormons like Lyman.[7] But the grip of his vision on subsequent local writers, as well as the general Mormon populace in San Juan County, has been remarkably strong. Examples of his influence are apparent in works such as Andrew Jenson's "History of San Juan Stake"; Cornelia Adams Perkins, Marian Gardner Nielson's, and Lenora Butt Jones's *Saga of San Juan;* and Norma Perkins Young's *Anchored Lariats on the San Juan Frontier.*[8] Jenson, for example, idealized the Bluff community's efforts to construct an irrigation ditch, saying, "in this wonderful colony . . . there remained a splendid element of invincibility. . . . that invincible spirit clinched its jaws tighter, and attacked the Bluff ditch with angry force." Or listen to Marian Gardner Nielson describe the heroism of the Hole-in-the-Rock group in a poem, ". . . dedicated pioneers . . . young zealots with heads high . . . confident in their manhood and the integrity of their quest."[9] In addition to these local antiquarians, *Blue Mountain Shadows,* a journal of San Juan County history, often publishes articles by locals whose ideas and spirit are clearly directed by Lyman. It is the spirit of triumphalism.

If Lyman's vision greatly influenced local perceptions of the San Juan, his greatest legacy is ultimately his scholarship. From his youth he manifested an interest in writing and the history of the San Juan colony. The collection of written and oral materials on San Juan County was his life's work. Both his published and unpublished writings are a treasure trove of information on the area for historians. Nonetheless, Albert R. Lyman's recording of the settlement of the San Juan, depiction of the river, and vision of the county's history have contributed in no small way to the embattled attitude county residents often hold for "outsiders." (Many forget that in 1880 Mormons were outsiders.) Sometimes these outsiders have been eastern do-gooders like the Indian Rights Association. More recently, they have taken the shape of the federal government and environmentalists. Both groups reflect national values, something Mormon pioneers have often viewed as poisonous.

In contrast to these local writers are a score of writers and photographers who did not grow up or work in the San Juan region and brought a decidedly different perspective to the river landscape. Their aesthetics, values, and recommendations for land use sometimes sharply disagreed with those of Lyman and the Bluff pioneers. These national artists lacked day-to-day experience of working the land along the river. Unlike the Grand Canyon, the San Juan has not yet found its great poet, its Clarence E. Dutton or John Wesley Powell. Nor has it inspired a great painter like Thomas Moran, W. H. Holmes, or Gunnar Widforss. Nonetheless, the writers and artists who have interpreted the San Juan have profoundly affected the way Americans have looked at the river. At the same time, these artists reflected a changing attitude toward the environment in the nation as a whole. That sea change began with Romanticism.

The first and foremost Romantic artists to visit the San Juan were photographers. Their images made the area familiar to many Americans, as well as international audiences. One example from the earth sciences will suffice. From the late nineteenth century to the present, photographers from Charles Goodman to Ansel Adams have been drawn to the spectacular view of the river winding back on itself—the Goosenecks—just downriver from Mexican Hat. Published photographs of the Goosenecks eventually caught geologists' attention. To them the view was more than beautiful; it dramatized the geologic principle of the entrenched meander. Thus, images of the Goosenecks now show up in geology texts as frequently as any other single landscape feature in the United States.[10] In 1962 Utah enshrined this view as Goosenecks of the San Juan State Park by sectioning off a ten-acre parcel of land on the ledge overlooking the river. More than fifty thousand visitors a year peer over the edge and contemplate this geological and aesthetic wonder.[11]

But beyond textbooks and scientific illustration, photographic images of the San Juan began seeping into the American consciousness as early as 1875. That year, one of the greatest western photographers, William H. Jackson, visited the San Juan while working for the Hayden Survey. He started a procession of photographers that have included the lesser-known but increasingly recognized Charles Goodman, as well as such luminaries as Ansel Adams and Timothy O'Sullivan.

William Henry Jackson grew up with the frontier movement, and his classic images of Yellowstone, the Colorado Rockies, western railroads, and other scenes helped shape the story of the western experience. Indeed it is hard for a late-twentieth-century viewer to think of the frontier experience without subconsciously calling up one of his photographs. In his own time, he defined the West as much as any single artist for an eastern public hungry for frontier images. Most of his work was sold as stereographs, and virtually every parlor in America owned a device to view them. Jackson's photographs were among the most popular landscape stereographs circulating at the time.[12] He was the right artist in the right place at the right time. Like his contemporaries—Timothy O' Sullivan, Carlton Watkins, Jack Hillers, Charles Savage, and Andrew Russell—Jackson became a kind of point man for American culture. He was, as Jackson scholar Peter B. Hale says, "raised on the nature-worshiping milk of American Romanticism and the bread of American democratic acquisitiveness."[13]

Jackson and his photographs embodied many of the contradictory ways Americans looked at nature in the post–Civil War era. On the one hand, they marveled at the sublime and awesome landscapes of the West. Simultaneously they celebrated the conquest and acquisition of these grand scenes as they were absorbed into the nation. Thus, in what might seem to today's viewers as a contradiction, Manifest Destiny and nature appreciation coexisted in the same images.

Ferdinand V. Hayden moved his survey to the San Juan mining district in southwestern Colorado, then west into southeastern Utah. One of his motivations was discovering and publicizing

Charles Goodman took the first photograph of the Goosenecks of the San Juan in 1895. It is the most famous view of the entire river. (Manuscripts Division, Marriott Library, University of Utah)

more Anasazi ruins like the ones found in the Mesa Verde area.[14] Survey member Jackson photographed many of the ruins north of the San Juan River, as well as Casa del Echo Ruin (also known as Sixteen-Window Ruin) across from Bluff. He also pointed his camera at unique geologic features. Like his contemporaries, Jackson was drawn to anthropomorphic qualities in the landscape. His famous 1873 image of a cross formed by snow in the cracks of a mountain in the Rockies, "Mountain of the Holy Cross," is a good example.[15] His images of the river, however, say much about American culture's views of nature at the time.

Jackson took all his photographs of the San Juan just below the mouth of Chinle Wash, which he called Rio de Chelly—the headwaters of Chinle. At that point, the river enters its first canyon, so Jackson conveniently set up his wet-plate photography outfit there. He took at least three images from either shore. All were entitled "Canyon of the San Juan," all appeared as stereographs, and all carried the inscription: "A few miles below the mouth of the Rio de Chelly immense great walls of dark brown sandstone hem the river closely in, and which grow in height and crowd still closer upon the river until they accumulate in the great canyon of the Colorado." Jackson's words suggested the remoteness and grandeur of the San Juan canyons. There was a serenely smooth river in the foreground or center of the photograph. In one a small figure sat in the lower right-hand corner, perfectly positioned on a rock which angled into the water.

This common technique of western landscape photographers derived from the Luminist painters like Martin Johnson Heade, John F. Kensett, and Fitz Hugh Lane. Luminism describes a group of mid-nineteenth-century, East Coast painters whose landscapes tried to capture the subtle effects of light. Often their canvases were organized horizontally and displayed calm water. In contrast to some of the large, grand, operatic paintings of artists like Frederick Church and Thomas Moran, Luminist landscapes were small and quiet, inviting the onlooker to transcendental contemplation. Often they contained small figures who sat or stood, watching the light. These figures instructed viewers to contemplate the serene landscape in front of them.[16] Behind Jackson's figure

The great western photographer, William H. Jackson, made the first images of the San Juan below Four Corners in 1875. Here Jackson was working in the Luminist tradition, depicting a quiet contemplation of nature. (Photo, #538, U.S. Geological Survey)

loomed dark, shadowy, almost-engulfing cliffs and tangled vegetation. In the foreground, by contrast, flowed placid waters, framed by massive layers of sedimentary cliffs.

Although many, if not most, of Jackson's photographs of San Juan country celebrated the triumph of American civilization, these images of the river asked the viewer to meditate on nature's stillness. In doing so, they joined the paintings of Lane, Heade, and other Luminists who extolled the quietly feminine sublime. Ultimately, they expressed some of the values of the national culture which eventually led, as Alfred Runte has shown, to the creation of national parks and then wilderness areas.[17] These developments have often clashed with the values of both San Juan County Indians and Mormons, who feel they have a deeper attachment to the land because they live there.

In addition to photographs, people in the world at large have learned about the San Juan through commercial river trips. River companies, originating with Norman Nevills of Mexican Hat in the 1930s, drew their ideas from photographers and writers promoting scenery and the wilderness experience. They created an advertising package based on those values and in turn encouraged other image makers and wordsmiths to follow and promote their businesses. Nevills set the trend.

Born in California and the college-educated son of an oil prospector, Nevills took the first commercial passengers down the San Juan from 1935 until his tragic death in his airplane in 1949. He introduced a whole generation of outsiders to the San Juan. At his death, Nevills himself and at least some others considered the Mexican Hat resident to be "the world's number-one fast water man." With his wife Doris, who often accompanied him on trips and provided the glue that held expeditions together, Nevills was a one-man publicist for the San Juan and river running in general. He was a short, brawny, theatrical, athletic figure who delighted passengers with all manner of performances and stories about the river, natural history, and

Norman Nevills billed himself as "the world's number one fastwater man." He introduced hundreds of paying customers to the San Juan country and was a larger-than-life character. (Manuscripts Division, Marriott Library, University of Utah)

legends. Few men have influenced public perception of the San Juan more because he was so successful in attracting well-educated, relatively well-healed clients to this remote corner of America. He especially sought out writers and photographers.[18]

In looking at trips conducted by Nevills and later businesses like Wild Rivers (now the oldest San Juan river-running company), one must consider the very delicate and slippery interplay between the preconceptions of tourist writers like Alfred Bailey, Ernie Pyle, Wallace Stegner, Tony Hillerman, and Edward Abbey and the expectations created by the river companies' advertisements. Moreover, river tourism was part of a larger boom in western tourism from the 1920s on, expanding in importance after World War II. Western tourism capitalized on what it thought visitors wanted to see. This is certainly true in the way Nevills once advertised and Wild Rivers and others promote now. In one sense these river-running companies script the way tourists encounter the river.

It remains to be seen if tourism and river running in San Juan country will become what Hal Rothman describes as a "devil's bargain": Success attracts so many people that the cultural and environmental amenities that make a place special are destroyed.[19] So far the footprints of river runners, strictly controlled as they are, have been fairly light along the river, especially compared to dams and extractive industries. A few well-worn paths and stepped-on plants hardly compare to the devastation wrought by dams. As previously noted, a recent study of campsites showed relatively little impact by river runners.

Norman Nevills hardly had to worry about too many people despoiling the San Juan. He did not have much money for promotion early in his career. Instead, he relied on word of mouth, voluminous personal letters, films, and articles by national magazine writers. Wild Rivers spends much more on advertising than Nevills ever dreamed but still gets most of its passengers from personal recommendations. The themes of their ads, however, have been similar from the 1930s to the present: scenery, wilderness, adventure, history, and education. In the post–Earth Day era, wilderness, nature, and especially education get more attention in promotional ads.[20] Popular mystery novelist Tony Hillerman summarizes Wild Rivers's philosophy through Joe

Leaphorm, a character in *A Thief of Time,* who says,

> This is Wild Rivers Expeditions out of Bluff. More into selling education. Take you down with a geologist to study the formations and the fossils, or with an anthropologist to look at the Anasazi ruins up the canyons, or maybe with a biologist to get you into the lizards and leeches and bats. . . . Older people go. More money. Not a bunch of overaged adolescents hoping to get scared shitless going down the rapids.[21]

Most San Juan River companies (as well as river-running outfits throughout the West) try to combine profit with environmental education. Few, if any, are getting rich.

Inviting writers to publicize his trips was something Nevills pioneered and mastered. Ernie Pyle, Alfred Bailey, and Wallace Stegner typified those who floated the San Juan with him and wrote articles about their experience.[22] They all pictured the West as an exotic landscape, part of the frontier legacy. Except for Stegner, who grew up on a Saskatchewan ranch, they did not possess working experience of the western landscape. But they all emphasized exotic scenery, adventure, and colorful locals. Bailey's *National Geographic* article in particular reminded the magazine's large readership that the San Juan landscape represented the last part of a vanishing American legacy.[23]

Wallace Stegner gave that theme full play in his piece, which first appeared in the prestigious *Atlantic Monthly.* At the beginning of his article, Stegner twice mentioned being at "the end of the world," pointing out that the San Juan country "is the heart of the last great wilderness." In his conclusion, he even considered using the river as a hideout from the coming Armageddon (he was writing at the beginning of the Cold War and nuclear-arms race). He then dismissed the notion, citing evidence that the outside world was already creeping in. He ended the article lamenting, "This is the way things were when the world was young; we had better enjoy them while we can."[24]

Stegner sounded a common theme of national environmental groups like the Sierra Club and the Wilderness Society, following Frederick Jackson Turner's argument. To some extent, they still voice it: Only a small portion of the wild American frontier remains, a little of the legacy that made America great.[25] While Stegner strongly advocated wilderness, his piece represented the first serious look at the San Juan in terms of the national discussion about wilderness preservation that began after World War II. Stegner used the same argument for "the geography of hope" in his famous Wilderness Letter: Modern life makes us a little crazy; we need to flee to those wild places where American values were shaped, or, as he put it, "the challenge against which our character as a people was formed." He later talked about wilderness as "a means of reassuring ourselves of our sanity as creatures."[26] The basic theme of wilderness preservation for spiritual renewal has formed the backbone of most discussions of the issue, along the San Juan or elsewhere. Certainly arguments have grown more sophisticated and less anthropocentric, but Stegner was the first writer to sound the theme in relation to the San Juan. He did not talk specifically about threats by dams, but he soon learned of them because of his involvement with the Echo Park controversy.

Although Stegner and other authors wrote about the river in national magazines like *National Geographic* and *Atlantic Monthly,* most Americans did not think much about the San Juan. In fact, it was excluded from the *Rivers of America* series that was just being published, an omission which Stegner lamented in his article. The San Juan was too far away from most developed, urban areas for many people other than a few writers, tourists, readers, and photographers to think about. Not until river running exploded as a tourist activity with the threat to the Grand Canyon did the San Juan become a must-do trip on anyone's agenda. All of these articles, however, paved the way for the next generation of writers, who came armed with knowledge, a sense of advocacy, and a different vision of the relationship between people and landscape. Most were tourists, but like Nevills and Ross, some eventually settled in the area and became locals.

After World War II, the environmental movement began to change its colors. The movement became more populist, political, and scientific. Science especially influenced nature writers and landscape artists. In a technical sense, it manifested itself in the way writers spent more time learning the basic ecology behind the landscapes they were writing about.

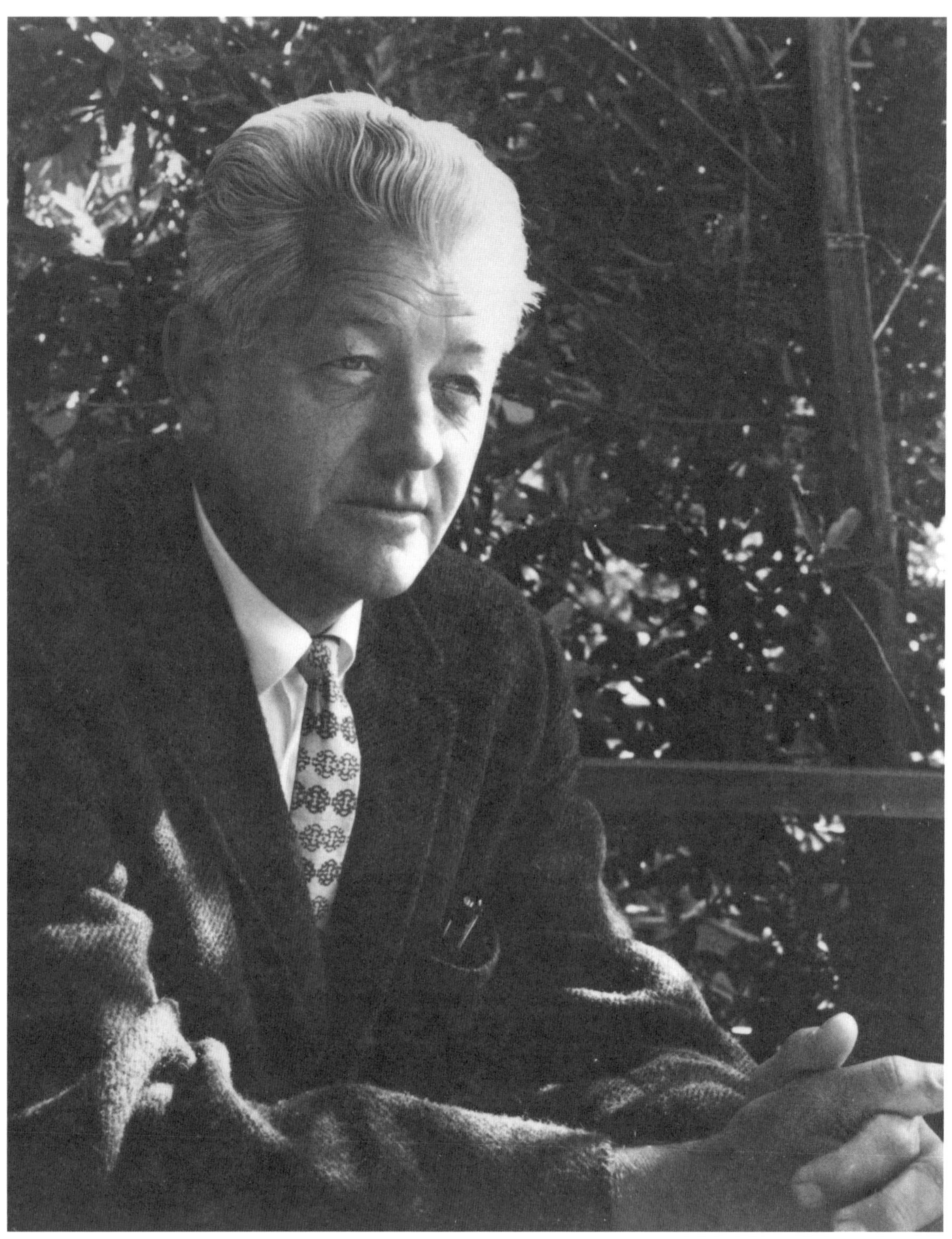

Wallace Stegner, dean of western writers, wrote about the canyon country in his novels, biographies, essays, and historical works. He floated the San Juan with Nevills in 1948. (Manuscripts Division, Marriott Library, University of Utah)

This was not entirely new. Thoreau and many successive nature writers knew the land in a scientific way. This trend, however, became the norm for writers of the postwar period.

The science of ecology also affected the values of the environmental movement and its writers. Ecology emphasizes humans as part of the sea of life rather than sitting on the throne, directing the flow. Ecological studies also reported that the planet was in big trouble because of human agency. Many nature writers took these reports as a call to arms. Nature writing not only became more urgently political, but its arguments for nature preservation were less human centered and more life centered. Wallace Stegner's Stanford student, Edward Abbey, typified these writers.

A number of activists have argued to protect the San Juan, but no one has been more influential on the national scene than Abbey. As someone who spanned the era of dams—old enough to have floated Glen Canyon before it was flooded and young enough to have opposed it—Abbey was in a unique position as a writer to espouse activism. He wrote on the cusp of the sixties in its politically charged atmosphere. His pugnacious, yet eloquent, writings on behalf of the Southwest moved a generation of baby boomers to radical environmental defense. Earth First!, for example, took its inspiration directly from his work. Ever the populist democrat, Abbey thought government had stopped listening to people and become the tool of big business. In the spirit of the Boston Tea Party, Abbey and Earth First! advocated monkey wrenching, or ecotage, as a truly patriotic defense of the American landscape against the dark forces of capitalist totalitarianism. Moreover, Abbey and his fellow activists supported a different set of values on people and the planet. Rather than arguing for wilderness as safety valve, refuge, and spiritual resource, Abbey took the life-centered position that nature has rights, too, and a basic core of democracy exists there. With that baseline, he could claim that "the wilderness idea needs no defense—only more defenders."[27]

When he wrote about the San Juan in *Down the River,* Abbey adopted the same humorous, ironic tone that characterized his two more-famous books, *Desert Solitaire* (1968) and *The Monkey Wrench Gang* (1975). Although he sprinkled his San Juan essay with plenty of nature appreciation, historical information, and camp humor, he saved his most pungent comments for those who would divert a river from its natural course: "Like many rivers these days the San Juan is . . . condemned by industrial agriculture to expire in a thousand irrigation ditches. . . . the rivers are too penned and domesticated and diverted through manifold ingenious ways . . . into the bottomless gut of the ever-expanding economy." He concluded his essay by attacking his favorite target, Glen Canyon Dam, lamenting that rather than flowing to meet the Colorado as it had for millions of years, the San Juan now expired into Lake Powell, "better known as Lake Foul, or Government Sump, or the Gangrene Lagoon or Glen Canyon National Recreation Slum."[28] Often sarcastic, Abbey aimed his barbs at wilderness despoilers and represented a new way of thinking about the San Juan River. Many writers have followed in Abbey's polemical tradition. Of special significance for the San Juan was Paul W. Rea, who wrote about the river in his book, *Canyon Interludes: Between White Water and Red Rock.*[29]

At about the same time Abbey was blasting wilderness wasters, his contemporary (and friend) Ann Zwinger was extending an older nature-writing tradition in San Juan country with her book, *Wind in the Rock* (1978). If Abbey represented a newer trend, the political nature writer, Zwinger updated the nineteenth-century tradition: the nature writer as naturalist. Both approaches derive from Thoreau, someone each wrote about.[30] Zwinger set her book in the side-canyon drainages north of the San Juan—John's Canyon, Slickhorn, Grand Gulch, Whirlwind Draw, and Steer Gulch. Her nonfiction essays combined science, history, poetic prose, and art. Trained as an artist, Zwinger schooled herself to become a respected naturalist. As she traveled by both foot and horseback down these canyons toward the river, she looked for stories in the plants, animals, and human artifacts she found. These stories might consist in the formation of a particular geological stratum, the replenishment of underground aquifers, the life cycle of Mormon tea, the exploration of the area by botanist Alice

Eastwood, or Anasazi lifeways on Cedar Mesa. Zwinger researched all aspects of human and natural history. Although she wrote personal narratives, her book is heavily footnoted; the notes are almost as interesting as her text.

Zwinger's method of writing about the country paralleled her nearly sixty-five pencil sketches of plants, animals, and artifacts: Focus narrowly on some feature in the landscape, study it, elicit its spirit, then convey it in poetic, yet measured, prose. She preferred small, private epiphanies to grand rhapsodies. After walking down John's Canyon to a cliff overlooking the river, she wrote,

> This brutal, dry, thorny landscape with the minimal river below is beautiful! . . . I have earned the uneasy euphoria of edging closer and closer to the drop of the cliff until I sit with my feet part way over, peering down only into time and water. . . . The exhilaration is worth every bit of the discomfort and the duress. . . . Perhaps once in a while everyone needs a little glory after lunch.[31]

Zwinger's voice was quiet—unlike her friend Abbey's—but in her own subdued way, her ideas reinforced a deep ecological consciousness of people and landscape. Hers was an apolitical voice for the San Juan landscape. Yet, like many writers who came to the San Juan, Zwinger did not reside there. Although as knowledgeable, perhaps even more so, than any native, she, like most of her readers, was an adventurer-tourist who could retire comfortably to her home on Colorado's front range.

The same can be said about two popular, contemporary fiction writers, Louis L'Amour and Tony Hillerman. Neither, however, lived too far from the Lower San Juan. L'Amour called Durango, Colorado, home while Hillerman still resides in Albuquerque. Both wrote about the San Juan River, L'Amour in *Haunted Mesa* (1987) and Hillerman in *A Thief of Time* (1988), and both popularized the area. One of the best-selling authors in American history, if not internationally, L'Amour was noted for his western novels pitting tough, cowboy heroes against the forces of evil and the rigors of the rugged western landscape. In *Haunted Mesa,* he departed from his usual nineteenth-century cowboy setting. In the late twentieth century, the story's hero, a writer named Mike Raglan, journeys to the San Juan to find a friend who has disappeared somewhere around No Man's Mesa.[32] The plot involves some bizarre twists, including people moving back and forth between this world and another.

Besides depicting the San Juan as one of the wildest, least-inhabited regions in the nation (which it is), L'Amour made some interesting comments about water development. In what he calls the Third World, descendants of the Anasazi live in a very carefully balanced, conservationist environment. They utilize every drop of water, recycle everything they have to nourish the soil, and only cut trees that are dead or dying. Although they use water and other natural resources intensely, they live, at least in L'Amour's mind, in a kind of perfect balance with nature.[33] In doing so, they suggest his model for civilization in an arid landscape.

L'Amour hinted at some of the environmental issues that have always confronted people along the San Juan—how to live in a dry land—but he largely used the landscape as a backdrop for his protagonists to struggle against evil. Nevertheless, the very fact that he wrote about the San Juan at all acquainted a large group of readers with the country and its most divisive issue—water use.

Mystery writer Hillerman is an Anglo who sets his detective novels on the Navajo Reservation and whose heroes are two Navajo policemen, Jim Chee and Joe Leaphorn. He often writes about the area near the San Juan. But his 1988 novel, *A Thief of Time,* is set on and around the river between Bluff and Chinle Wash (which he calls Many Ruins Canyon). As we saw earlier, Hillerman has Leaphorn comment on responsible river running. Near the climax of the novel, the writer sends Leaphorn on a midnight kayak trip downriver in search of a possible murder victim, a woman archaeologist. Leaphorn's wife, Emma, has recently died, and he is still grieving. Although he is considering retirement, Leaphorn finds that searching for this missing woman restores meaning to his life. Floating the river at night, he has a mystical experience when he sees a snowy egret along the bank. Like egrets, Leaphorn believes that he mated for life with Emma.[34] By the time he pulls his kayak to shore at the mouth of Many Ruins Canyon, he has regained his equilibrium and solves the crime. In this Hillerman novel,

Mystery writer Tony Hillerman visits Walter E. Mendenhall's camp about one hundred years after the gold rush that brought Mendenhall and other miners to the San Juan. Hillerman made numerous trips with Wild Rivers Expeditions to research for his book, *A Thief of Time.* (San Juan Historical Commission)

Leaphorn's San Juan trip functions like Stegner's "geography of hope": The wild river restores the detective's sanity.

Besides identifying the river as a place to restore one's spirit, Hillerman, like L'Amour, emphasized the San Juan's remoteness. Many Ruins Canyon is so far away from civilization that a schizophrenic can hide out undetected for twenty years, archaeologists can still uncover great treasures, and murder can occur unnoticed in broad daylight. Hillerman and L'Amour followed the theme of many tourist writers in emphasizing the value of the San Juan's rugged wilderness. At the same time, they introduced the area to a large readership, many of them not outdoor types. At least one Boston-based travel company now offers four tours yearly of Hillerman Country, which includes the San Juan.[35]

Except for Norman Nevills and Kenny Ross, most writers and artists who depicted the San Juan were, like Zwinger, Hillerman, and L'Amour, outsiders. And since the San Juan includes so much federal land, national values have often shaped management decisions and laws. This situation has often rankled San Juan residents. On the other hand, some image makers of the San Juan came with aesthetic and environmental ideas shaped by larger currents but decided to stay. They in turn forged a local-national hybrid mythos of the San Juan. Photographer Charles Goodman is the most important.

If William Henry Jackson was the preeminent photographer of the West, Charles Goodman was a unique chronicler of San Juan country. Goodman's images, as much as Albert R. Lyman's words, helped fashion the mythos of settling the San Juan frontier, but they also created an aesthetic that fostered landscape preservation. His work appears throughout this book. Although not much is known about Goodman personally, his photographs of San Juan gold and oil mining are archetypal images. The New York–born Goodman followed Colorado's mining booms in the 1880s from Pueblo to Aspen, Montrose, Creede, and eventually Mancos in the extreme southwest. By 1893, however, he had moved west into Utah, following the short-lived gold rush on the San Juan. Settling in Bluff, Goodman remained there until he died in 1912. Ironically, even though he lived nearly two decades in the Mormon village, at his death a local antiquarian described him as a "transient."[36] Whether transient or transplanted, Goodman clearly took to the San Juan and supported himself photographing landscapes, mining scenes, pioneer life, and Indians. Many of his photographs ended up as stereographs, sold to locals and miners passing through the country. Goodman was a skilled artist who handled his equipment well (he probably used the dry-plate method), framed shots artistically, and clearly loved his subject matter, whether human or natural.

Like Jackson's photographs, Goodman's work reconciles the seeming contradictions between awe of nature and celebration of technological triumph. He found many occasions to point his camera at the exploits of miners and

Technology spanning the wilderness. E. L. Goodridge's first oil well, 1908. (Charles Goodman photo, Manuscripts Division, Marriott Library, University of Utah)

village builders at Bluff, many of which appear in earlier chapters of this book. Although his close-up of the gold placer at Raplee's camp is probably his most-reproduced photograph exalting technology (see page 117), his image of E. L. Goodridge's first oil well near Mexican Hat in 1908 is one of his best and most evocative. With the spires of Alhambra in Monument Valley barely registering on the perfectly flat horizon, Goodridge's triangular oil rig frames and dominates the whole scene—land and sky. The men clustered in the picture seem to extend from the steel wheels, frame, and tank of the pumping apparatus. The dominant triangle of technology almost euphorically exults in its triumph in this remote desert landscape.

When it came to capturing pioneer life in Bluff, Goodman reveled in everyday activities. Probably few of these images were commissioned. One of Goodman's favorite views of Bluff came from Twin Rocks, looking down at the town and the river flowing south. In many ways these Twin Rocks images not only reflected but helped shape the idea of the village as a monument to pioneer commitment. Charles S. Peterson wrote,

> As a landscape form, the village marked Mormon Country generally, but in combination with desert wilderness and scenic wonders it became a special insignia of Southern Utah, a form on the land that highlighted the already luminous landscape. It was in effect a human verification of a land naturally unique but now doubly set apart to become a scenic and cultural resource to the nation.[37]

What Goodman celebrated was a "middle landscape," a harmony between pastoral and wild.[38] Looking at the same San Juan canyon that W. H. Jackson peered down, Goodman depicts a wild, awesome scene. But perched on the brink of this magnificent wilderness to define it sits a garden of human order. According to Goodman's aesthetic, a blend of the two formed the ideal western scene.

Another Goodman photograph of the Raplee Anticline placer mine and camp pushes this idea even further. The shadows on the right rake across the river bottom, nicely matching the lines in the sedimentary layers on the left. Raplee's placer mine, dimly figured in the center of the image, extends naturally from the low

Charles Goodman made many images of A. R. Raplee's camp on the San Juan in 1893 and 1894. This is his best. (Manuscripts Division, Marriott Library, University of Utah)

cliffs, growing out into the water to meet the approaching shadows. Such photographs suggest that people can live in balance with this rugged river landscape. They can blend in and become part of the land.

Blending in is a philosophy shared by two late twentieth-century immigrants, Ellen Meloy and Ann Weiler Walka. Both women resided at least part-time in Bluff, came to the area for the river itself, and were experienced river guides. Both were baby boomers who cut their teeth on Earth Day and the postwar environmental movement. Both writers were well schooled in the natural and cultural history of the river, and both sought to portray a new way of living along a river, inhabiting a landscape on its terms. Like Albert R. Lyman, each writer imagined creating a sacred bond with the landscape, but through accommodation rather than conflict.

Meloy, whose *Raven's Exile* so fully evoked the Green River in Desolation and Grey Canyons, wrote with scholarly attention, humor, and elegance about the San Juan. In various pieces about the river that flowed in front of her house, she said that after the better part of ten seasons, "I find home in motion itself, in a meandering ribbon of bright water and a bed under a cottonwood tree or on a slender crescent of sand at the river's edge."[39] Meloy wanted to achieve a Keatsian "negative capability," where the river and San Juan landscape spoke through her. She did not imagine herself in conflict with the landscape but tried to find a way to merge her life and consciousness with the river's.[40]

If Albert R. Lyman spoke for the frontier San Juan, Ann Weiler Walka wrote for the post–World War II green generation. In *Waterlines,* a book that is primarily poetry, she sought to explore "the mysterious border between a river's canyon and the terrain of my imagination."[41] Thus, rather than trying to grasp the river and twist it to her uses, Walka, like Meloy, let it seize her imagination, transform her consciousness, and speak through her. Walka did not have a literary imperialist's spirit, come to pluck off some scenery and relate a little history and local color like travel writers of an earlier generation. Her poems emanated

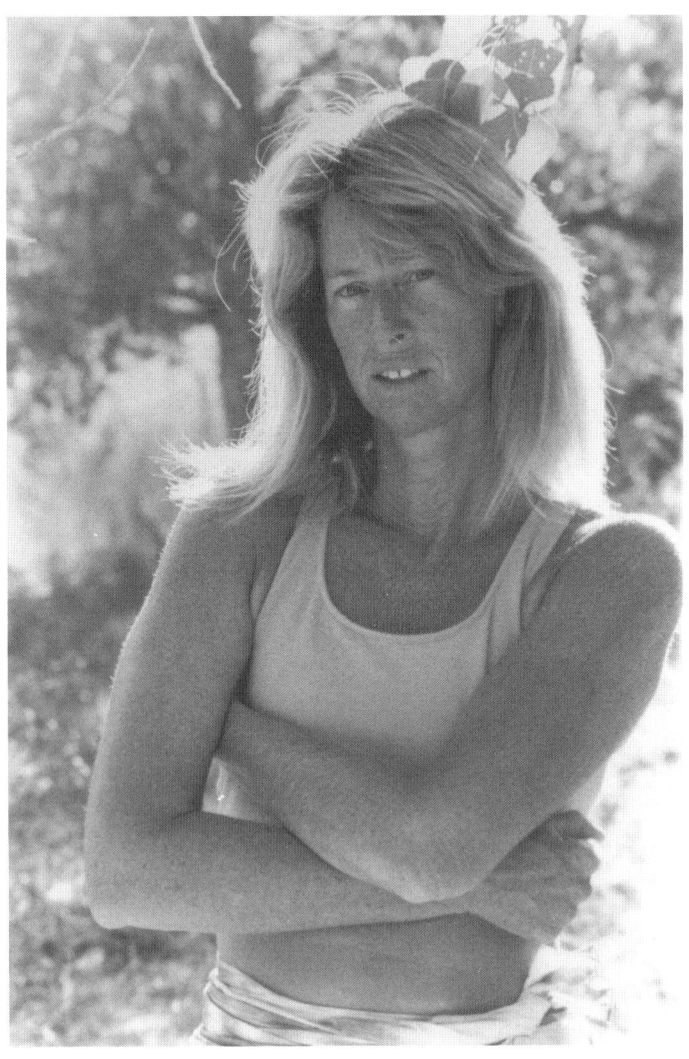

Writer Ellen Meloy lives along the banks of the San Juan (Mark Meloy)

from a deep and long-lived attachment to the river landscape. She knew, for example, where the cliff swallows emigrate from, why the Mormons had to climb San Juan Hill, where the bighorn sheep live, how cobblestone bars formed on the benches above the river, and how an Anasazi potter made her bowls. She was aware of and respected all the traditions that had grown up around the San Juan—Indian, Mormon, and non-Mormon Anglo—and tried to incorporate these values in her work.

Like Meloy and Zwinger, Walka celebrated the San Juan's cultural and biological richness, imagined a life attuned to its natural patterns, and quietly argued for its value. As she wrote in her preface,

> I've come here again and again until the river feels like home. I've applied myself to learning the names and histories and relationships of the locals—the rocks and river channels, plants and animals. I've tracked down stories of two-leggeds, natives and newcomers, settlers and adventurers, and imagined the ways they changed this place and been changed by it. I've come to know how the ground feels under my feet and what it smells like when it rains and where I can find good shade.[42]

Meloy and Walka wrote from the same set of environmental values as the hard-hitting Edward Abbey, but in a quieter way. Moreover, these women put values learned from the larger culture into practice in a local setting. The Navajos thought of the San Juan as a male river because of its raging, sediment-choked floods and northern origins. It is interesting that the current poets of the San Juan are women. Perhaps it represents a kind of Navajo union or meeting of the two different spirits.

Although the current generation of writers partially reflects the philosophy and ethics of the Utes and Navajos about the sacred, there are significant differences. For traditional Indians, the San Juan landscape is a specific, god-inspired and god-inhabited world. Various landscape features do not just stand for something sacred; the feature *is* a holy being. Thus, for the Utes, the water baby who pulls people underwater and drowns them is not a metaphor for the river's dangerous currents; it is a real being. Writers like Abbey, Rea, Zwinger, Meloy, and Walka, however, have an indistinct sense of nature's sacredness, nor have they attempted to develop a cosmology around features of the landscape the way Indians have. These writers are clearly searching for something greater than themselves in the natural world, but they acknowledge the interplay of their own consciousness and nature. For traditional Indians, on the other hand, nature's spirit is real and tangible. It may be too much to expect these nature writers eventually to combine traditions and create a unique San Juan religion.

Besides nature writers and photographers, another kind of voice speaking for the landscape emerges from two periodicals published in San Juan County, *Blue Mountain Shadows* and the *Canyon Echo.* The differences between these journals say much about the conflict over environmental issues that has surfaced over the years, especially when local values clashed with national ones. The older *Blue Mountain Shadows* originates in Blanding and advertises itself as "the magazine of San Juan County history." Therefore, it covers more than just the river area, although in 1993 it devoted half an issue to the San Juan.[43] This journal appears biannually and often contains scholarly, footnoted articles. It cleaves more to the local pioneer tradition initiated by Albert R. Lyman.

The *Canyon Echo,* a monthly journal from the town of Bluff, published a variety of news about the area between 1993 and 1997, but every issue contained numerous articles about the river, everything from reports of water flows to current scientific studies. It twice devoted entire issues to river-related articles, once in June 1995 and again in June 1997.[44] The *Canyon Echo* envisioned itself as a voice for the natural and cultural landscape linked by the San Juan River, and its editorial slant was generally "green" and multicultural. Its values, in fact, often conflicted with the more conservative *Blue Mountain Shadows.* In many ways, these periodicals defined the ongoing conflict of values over the river landscape. Whatever the views and talents of San Juan journalists, nature writers, photographers, or artists, however, the river has yet to find, like the Grand Canyon, its great artistic interpreter.

To say that the San Juan is a contested landscape may be stating the obvious. Most public lands in the West are. Throw in competing, conflicting cultural values—from Mormons, Indians, the federal government, and environmentalists—and it is sometimes hard to hear above the din what artists and writers have to say. But their voices have been heard. Moreover, as the world becomes smaller, no one system of values—pioneer, Indian, or national—will prevail. Perhaps a synthesis is occurring as people decide how to live in this landscape. Perhaps the future will bring threats not from mining and overgrazing but from too much love. Like other relatively unpopulated regions of the West, the San Juan may be close to being overphotographed. You can't help but wonder if one more published photograph of this redrock wilderness will help or hurt the landscape. Would it even be possible to limit the words and images about the San Juan the way the Bureau of Land Management restricts river travel—in the name of preservation?[45] Writers and photographers publicized the San Juan country to the world. Some influenced the laws and policies that led to land preservation. Will these artists find a new way to imagine the San Juan country?

Perhaps the new artists will look back to some of Charles Goodman's images for an updated, middle-landscape approach. This view may teach us that the San Juan is not a pristine wilderness, but it is also not a raked-over, spewed-out landscape where the human touch has been harsh and brutal. Perhaps, like Ellen Meloy, they will find a way to imagine how to live with the changes human beings have brought, yet somehow manage this naturalized, dam-controlled landscape along ecological lines. The San Juan is still a place of extraordinary beauty and natural diversity, a place where plants, animals, and natural processes can evolve as they have for eons. It is a place where people have lived and can continue, though not in too great a number. These artists may show us that the San Juan is a place where a future is possible which revolves neither around the always-false and deceptive image of pristine nature nor human monuments. Perhaps they will reveal how humans with computer models which regulate river flow can fit into a sacred river landscape.

EPILOGUE VISIONS: *Flowing from the Sunrise or a Water Spigot?*

Since the first Earth Day, 22 April 1970, predicting the planet's future has become almost an obsession for environmental prognosticators. While very few today envision a rosy scenario for the planet if the world continues on its present course, some positive developments have occurred since 1970. The 1997 Kyoto Global-Warming Treaty and population-control programs throughout the world are two examples of progress.[1] Nevertheless, hopeful or despairing, all predictions have one theme in common: Take better care of the environment, or not much of the planet will be worth inhabiting.

Predicting the San Juan River's environmental future is about as easy as peering through its sediment-choked waters and seeing the bottom. What makes it difficult is that the Lower San Juan is a small part of a larger physiographic province, the Colorado Plateau. The plateau, in turn, is but a section of the Rocky Mountains. The Rockies form a slice of North America, and so on. In the global scheme, the San Juan does not amount to much. In some ways, no matter what people along the river do to manage their land or local pollution, they can never control the effects of an event like the Krakatau El Niño of 1884 or global warming. Even if they agitate to reduce air pollution from, say, the Navajo Generating Station at Page, Arizona, they will have a much harder time counteracting what blows up from Las Vegas and Los Angeles. Thus, it is hard to predict how such exogenous factors will shape the San Juan's future. What happens along the river, however, lies within the grasp of area residents and government agencies. In these limited terms, predicting the environmental future of the San Juan is possible.

The most demanding task historians face is entering into the spirit of the past and, with the aid of good sources, recreating a narrative that approaches truth. Sources keep the historian from straying. Predicting the future allows more freedom, although it demands an equal sense of responsibility. Predictions must follow established patterns. Some of the patterns are clear: the role of the federal government and the Indianness of the San Juan. These will remain constant. The future of other aspects of the San Juan, like local control, irrigation, grazing, recreation, or larger problems like cooperation between constituencies, is not so obvious. Still, using caution as well as a little chutzpah, gazing into the crystal ball can be profitable.

Let's consider the federal government first, since it has played an increasingly dominant role during the last hundred years. Its main efforts have focused on water development, but it also directs activities along the river like grazing, mining, recreation and, since the 1970s, environmental protection. Numerous factors complicate its work, especially the conflicting missions of various agencies. Additionally, these missions have often clashed with local concerns, both Anglo and Indian. Nevertheless, the federal government's role and its various land agencies ensure that land use will remain primary during its second century of hegemony over the San Juan. And what will that next hundred years of federal jurisdiction look like?

In the future, the hand on the water spigot, the Bureau of Reclamation at Navajo Dam, will no longer operate by itself. Others will help turn the handle. Increasing population, greater demands from water users, and environmental concerns will force the bureau to deliver water

The Glen Canyon Institute and many other environmentalists would like to dismantle Glen Canyon Dam. For them it is a hated symbol of everything that is wrong with western water politics. (Bureau of Reclamation, Upper Colorado Region)

differently from the past. Environmental concerns are especially important. The Glen Canyon Environmental Studies (GCES) have changed the way the bureau operates the dam and releases water into the Grand Canyon. Power demands no longer predominate. The Grand Canyon Protection Act of 1992 mandates that the downstream consequences of water flow on beaches, birds, archaeological ruins, and vegetation must be considered by the bureau before releasing water.[2] Eventually the bureau will be forced to operate Navajo and other dams on western rivers equally conscientiously. The San Juan River Recovery Implementation Program (SJRIP) for native fish has already caused a change on the San Juan. The bureau even stated in 1987 that it was retooling to manage dams in an environmentally friendly way, and its website mission statement for the upper Colorado Basin now asserts it will "manage, develop, and protect water and related resources in an environmentally and economically sound manner."[3] Environmentalists will continue to pressure the bureau to live up to these promises.

In addition, some environmentalists are pushing to dismantle certain dams, specifically, Glen Canyon. In the summer of 1997, the Sierra Club and the newly formed Glen Canyon Institute announced a campaign to discuss terminating Glen Canyon Dam. They argue that the dam is a colossal economic and ecological mistake. Founded by GCES scientists and other environmentalists, the institute believes that recovery plans for native fish in the Colorado system are well intentioned, Band-Aid approaches doomed to failure. These environmentalists also claim that modifying dam operations won't really help. Sediment buildup in Lake Powell will create greater health and environmental problems because it contains heavy metals. They also point out that every five years the total amount of oil leaking into Lake Powell from boat engines equals an Exxon *Valdez* spill. Finally, they cite the fact that heavily subsidized Glen Canyon Dam only provides 3 percent of the power for the Four Corners states.[4]

Although current Utah politicians—a very conservative crowd—scoff, the idea drew support from archconservative, former Arizona senator and 1964 Republican presidential candidate Barry Goldwater before his death in 1998.[5] Draining Lake Powell would not only resuscitate a drowned part of the San Juan below Clay Hills Crossing but would offer researchers a unique opportunity to participate in the restoration of an injured riparian system. The plan to dismantle Glen Canyon Dam will acquire impetus if a sound alternative source of energy is developed during the new century. Still, the odds are against anything so radical occurring within the next twenty-five years.

Navajo Dam seems more solid politically. Controlling floods alone makes it especially valuable to downstream residents. An allied dam, though, the proposed Animas–La Plata (A-LP) near Durango, Colorado, appears doomed at this point. The longer that wrangling over A-LP continues, the higher the dam's price tag will rise, and the less likely it is to be built. In this case, time is on the environmentalists' side.

Besides pressure from environmentalists, the bureau must increasingly operate within the confines of environmental laws passed between 1964 and the present. The Endangered Species Act required the bureau and six other federal agencies to develop the SJRIP, primarily for the Colorado pikeminnow. Although some supporters of the program doubt it will succeed—too little, too late, they say—the fish study has greatly increased knowledge about the San Juan riparian system.[6] Oil pollution from wells near Aneth provides just one example. Understanding the effects of polycyclic aromatic hydrocarbons (PAHs), saline, and other chemical contaminants on fish has led to remedial efforts to keep these pollutants out of water systems. All living beings along the river will benefit. More of these kinds of controls will result from applying environmental laws. Further, environmental legislation requires more cooperation among agencies and between them and locals. The pikeminnow may never recover in the San Juan basin the way the peregrine falcon has throughout the West, but both political and natural systems will profit in the long run from plans like the SJRIP.

If the peregrine falcon is a success story of the Endangered Species Act, a San Juan resident that may follow in its wake is the desert bighorn sheep. At this point, observers are cautiously optimistic that the bighorns will reestablish parts of their traditional range along the San Juan. The much-heralded California condor, however, is at the forefront of the recovery plan. Recently three condors from the experimental program, based in House Rock Valley near Lees Ferry, flew 250 miles up the Colorado River to Grand Mesa in Colorado, fulfilling hopes of scientists that these birds would extend their range.[7] They may once again soar above the canyons of the Lower San Juan.

While some animals are coming back, traditional extractive practices are becoming "endangered." Oil drilling, ranching, and farming along the San Juan—always precarious occupations—are diminishing in importance, following a long-term trend in the West. Some local people blame the demise of these industries on environmental regulation and the public agencies which administer it. This is sometimes the immediate case, as with the Comb Wash decision mentioned in chapter 4, but the real forces that affect these occupations are state, national, and even global market economies.

Ranching, for example, will probably always exist along the river, but not the way it does today. Some locals assert that ranching is a substantial part of the economy. In fact, it is small and getting smaller. Economist Thomas Michael Power's studies show that the loss of all grazing on federal lands would have little overall effect on local economies in the West. Those economies have expanded in this century precisely because they diversified and moved away from agriculture and mineral extraction. Power maintains that keeping agriculture in the economic mix is a good thing, but he says it is wishful thinking to argue that western economies are based on grazing or farming. That idea runs counter to all evidence.

Ranching, farming, and mining will decline along the San Juan, although the first two will fade more slowly in Indian than in Anglo communities. Mineral extraction is predicted to comprise less than six-tenths of 1 percent of Utah's economy by the year 2010.[8] Oil production, however, will continue into the near future. What ranching and farming remain will become more environmentally sensitive in terms of native species, water use, riparian habitat, and herd rotation. Additionally some farming and ranching operations may survive with the help of the land-trust, open-space movement.

Recreation, on the other hand, is becoming more popular everywhere on public lands in the West. River running is the main event on the San Juan, and the demand for permits continues to outstrip their availability. Nothing will change that. The Bureau of Land Management (BLM), however, will face more pressure from both commercial and private boaters for access. Ironically these same boaters will want their river journey to be a kind of wilderness adventure. They will want to experience the same thrill as an E. L. Goodridge, for example, first running the river to look for gold. It is hard to feel that thrill when boats are lined up bow to stern as they sometimes are at Slickhorn Canyon.

The BLM's job of balancing access, solitude, and protection will get stickier than a goathead patch. The agency's job is further complicated by the fact that the Navajo Nation is now beginning to assert jurisdiction over its part of the river—theoretically the entire southern half—and is issuing permits as well. Currently, one Navajo river-running company, Bighorn Tours, has a permit from the BLM to conduct day trips from Sand Island to Mexican Hat.[9] In the future, look for more Navajo river companies and political presence on San Juan issues. No matter the source of the pressure, more public input and participation can democratize the process and guide the agency through the thorns of recreation management. Again, cooperation, participation, and coordination are necessary to maintain the golden egg of recreation. Things could go the other way as well. Recreators sometimes love a landscape to death.

Small towns along the river must look more and more to land planning. As populations grow, tourism increases, and outsiders move in, towns like Bluff will be hard pressed to avoid "the Moab syndrome." This is Bluff's term for unplanned growth, which it perceives in nearby Moab during the last decade, a boom hastened by mountain biking. Bluff, however, is beginning to take steps to control growth, preserve open space and cultural sites, and still keep the town thriving but rural in atmosphere.[10] It is a delicate balancing act that has often divided the community. Dialogue and conversation are the keys, just as with the larger issues along the San Juan. Bluff is still small enough to make that happen.

Affecting open space and the wilderness experience, both on the river and around it, is the intention of various wilderness bills pending in Congress. No wilderness bill will pass as long as Utah's congressional delegation remains conservative and solidly Republican. Although poll after poll demonstrates that the majority of Utahns favors wilderness (ranging from 70 to 80 percent), the issue is still too contentious in the Beehive State.[11] Whichever proposal eventually passes will contain substantial sections of wilderness on or near the river. When wilderness designation comes, it will further restrict extractive industries while protecting plants, animals, and cultural sites. At the same time, it will increase the appeal of recreational activities along the river—boating, backpacking, and hunting. That will mean more recreators for the BLM, the Navajo Nation, and the Park Service to deal with. Meanwhile, wild and scenic river designation seems less likely to happen. If it does, it will only occur sometime after the wilderness issue is decided in Utah and Washington.

One plant that needs no protection, it appears, is tamarisk. Various federal land managers, the Navajo Nation, and local groups like ranchers will continue to look for ways to control, reduce, even eliminate this hardy tree. All will fail. Given enough time, however, natural processes may begin controlling tamarisk to some extent. As long as Navajo Dam stands, however, tamarisk will dominate San Juan River beach vegetation.

If some native plants have had a difficult time remaining rooted by the San Juan, the native inhabitants have fared much better. As the novelist Frank Waters once wrote about the San Juan-Four Corners area, "This is Indian country."[12] It will stay that way for the next hundred years and beyond. The Navajos, Utes, and Paiutes are deeply planted in San Juan country and will not be leaving. Their populations are expanding. More than two thousand Navajos, for example, live near the San Juan in the Aneth Chapter alone.[13] What will change for these tribes is the extent to which they control their own fate because of land-use decisions along the river. Indian influence on issues like endangered species, irrigation, extractive industries, and recreation will continue to expand. As the People become more adept at manipulating the political and legal processes governing natural resources, their power will become a force to be reckoned with. Navajos, Utes, as well as Hopis and Zunis, had considerable input in the Grand Gulch Plateau Cultural and Recreational Area Management Plan, the BLM's recent program to manage cultural and natural resources on Cedar Mesa.[14] The Navajos, because of their numbers and geographical proximity to the San Juan, will have far more influence than the Utes and Paiutes in this and future land-planning issues.

But as San Juan Indian tribes become more adept at operating in the dominant culture, their sacred relationship with the river corridor will evolve. Some already say that little of the sacred tribal relationship with the river still exists; only older Indians are aware of it, and younger ones are as materialistic as most Americans. Nevertheless, it will be interesting and instructive to watch tribal leaders and members blend "government speak"—NEPA, endangered species, launch permits, and PAHs—with traditional beliefs and see how this fusion influences land decisions.

Speaking of the river in sacred terms is not the sole province of Indians. More and more Anglo-Americans will look at the San Juan the same way as an Ann Zwinger, Ansel Adams, Wallace Stegner, or Ann Weila Walka. More and more will be seized by a vision of the river landscape's power. Perhaps the artists among them will link their images of the land with Indian ones, meld the secular with the sacred, and imagine a new, yet old, way to live in this very-difficult landscape. Maybe these Anglo and Indian artists, armed with science, local intelligence, and imagination, will help others create the future environmental history of San Juan. There is no such thing as returning to pure nature. We need to get beyond our guilt at having supposedly destroyed paradise. That old saw should be retired. We do need, however, to find ways to talk about computer models and petroglyphs, PAHs and sacred space, cows and water babies in the same breath. Dealing with a river and all its inhabitants—human and nonhuman—is an extraordinarily messy, complicated process. Still, it would be nice to think that the San Juan's future will lean a little less toward the River as Water Spigot and a little more toward the River Flowing from the Sunrise.

Notes

Introduction

1. Frank Waters, *The Colorado* (New York: Rinehart and Co., 1946), 10. Waters claims that the San Juan is the largest tributary of the Colorado, which is not correct; the Green River is. But even this statement is complicated by the fact that geologists now consider the Green the main stem river where it meets the Colorado in Canyonlands.
2. Adrian N. Hansen, "The Endangered Species Act and Extinction of Reserved Indian Water Rights on the San Juan River, "*Arizona Law Review* 37 (1995): 1305–44.
3. Richard White, *The Organic Machine: The Remaking of the Columbia River* (New York: Hill and Wang, 1995), 109–13.
4. Just a few of the scores of book that emphasize the Colorado are David Lavendar, *Colorado River Country* (New York: E. P. Dutton, 1982); Philip Fradkin, *A River No More: The Colorado River and the West* (Tucson: University of Arizona Press, 1984); David Lavendar, *River Runners of the Grand Canyon* (Grand Canyon, Ariz.: Grand Canyon Natural History Association, 1985); Marc Reisner, *Cadillac Desert : The American West and Its Disappearing Water* (New York: Viking, 1986); Russell Martin, *A Story That Stands Like a Dam: Glen Canyon and the Struggle for the Soul of the West* (New York: Henry Holt, 1989); Loren D. Potter and Charles L. Drake, *Lake Powell: Virgin Flow to Dynamo* (Albuquerque: University of New Mexico Press, 1989); Roy Webb, *Call of the Colorado* (Moscow: University of Idaho Press, 1994); Donald L. Baars and Rex C. Buchanan, *The Canyon Revisited: A Rephotography of the Grand Canyon, 1923/1991* (Salt Lake City: University of Utah Press, 1994); Louise Teal, *Breaking into the Current: Boatwomen of the Grand Canyon* (Tucson: University of Arizona Press, 1994); Stephen J. Pyne, *How the Canyon Became Grand: A Short History* (New York: Viking, 1998); Jared Farmer, *Glen Canyon Damned: Inventing Lake Powell and the Canyon Country* (Tucson: University of Arizona Press, 1999); P. T. Reilly, *Lee's Ferry: From Mormon Crossing to National Park,* ed. Robert H. Webb (Logan: Utah State University Press, 1999).
5. Patricia Nelson Limerick, *The Legacy of Conquest: The Unbroken Past of the American West* (New York: Norton, 1987), 27.
6. For a pioneering study of the Gila River, see Henry F. Dobyns, *From Fire to Flood: Historic Human Destruction of Sonoran Riverine Oases,* Ballena Press Anthropological Papers, no. 20, ed. Lowell John Bean and Thomas C. Blackburn (Socorro, N. Mex.: Ballena Press, 1981). A more popular, less scholarly treatment of the same subject is Gregory McNamee's *Gila: The Life and Death of an American River* (New York: Orion Books, 1994). An excellent environmental history of a nonsouthwestern river is White's *The Organic Machine.* Environmental history, as an academic field, has matured greatly in the two decades since Dobyns wrote his book. For summaries of the field's development, see Donald Worster, "History as Natural History: An Essay on Theory and Method," *Pacific Historical Review* 53 (1984): 1–19; Richard White, "Historiographic Essay, American Environmental History: The Development of a New Field," *Pacific Historical Review* 54 (1985): 297–335; and Donald Worster et al., "A Roundtable: Environmental History," *Journal of American History* 76, no. 4 (March 1990): 1087–1147.
7. Charles. S. Peterson, *Water Rights on the Little Colorado River–First Draft* (Missoula, Mont.: History Research Associates, 1986), 89.
8. Geologist and river guide Donald L. Baars divides the river into three canyon segments: Chinle Wash to Mexican Hat, Mexican Hat to Clay Hills Crossing, and Clay Hills to the old Colorado confluence. See Donald L. Baars, ed., *Geology of the Canyons of the San Juan River*

(Durango, Colo.: Four Corners Geological Society, 1974), 1–4.

9. Information on San Juan geology comes primarily from Baars, *Geology of the Canyons of the San Juan River,* and Gene Stevenson, consulting geologist, interviews by James Aton, 16 May 1996 and 29 October 1998. For background, see also Herbert E. Gregory, *The San Juan Country: A Geographic and Geologic Reconnaissance of Southeastern Utah,* U.S. Geological Survey Professional Paper 188 (Washington, D.C.: GPO, 1938); Donald L. Baars, *Redrock Country: The Geologic History of the Colorado Plateau* (Garden City, N.Y.: Doubleday/Natural History Press, 1971); and Donald L. Baars, *Navajo Country: A Geological and Natural History of the Four Corners Region* (Albuquerque: University of New Mexico Press, 1995).

10. Charles G. Oviatt, "Late Quaternary Geomorphic Changes along the San Juan River and Its Tributaries Near Bluff, Utah," in *Contributions to Quaternary Geology of the Colorado Plateau,* Special Studies 64, ed. G. E. Christenson et al. (Salt Lake City: Utah Geological and Mineral Survey, 1985), 32–47.

11. Kendall R. Thompson, *Characteristics of Suspended Sediment in the San Juan River Near Bluff, Utah,* Water Resources Investigation Report no. 82-4104 (Salt Lake City: U.S. Geological Survey, 1982), 2, 12.

12. Stevenson, interview, 29 October 1998.

13. For a map detailing the geomorphic provinces of the San Juan, see Baars, *Redrock Country,* 95. For a discussion of the geomorphic provinces in Utah, see William Lee Stokes, *Geology of Utah* (Salt Lake City: Utah Museum of Natural History, 1986), 235–37.

14. Paul B. Holden and William Masslich, *San Juan River Recovery Implementation Program: Summary Report, 1991-1996* (Logan, Utah: Bio-West, 1997), 5–8.

15. For a good survey of the San Juan Paiutes, see Pamela A. Bunte and Robert J. Franklin, *From the Sands to the Mountain: Change and Persistence in a Southern Paiute Community* (Lincoln: University of Nebraska Press, 1987).

16. See Robert S. McPherson, *The Northern Navajo Frontier, 1860–1900: Expansion through Adversity* (Albuquerque: University of New Mexico Press, 1988).

17. David E. Miller, *Hole-in-the-Rock: An Epic in the Colonization of the Great American West,* 2d ed. (Salt Lake City: University of Utah Press, 1966), 5–9.

18. For a summary of these affairs, see Floyd O'Neil, ed., *Southern Ute: A Tribal History* (Ignacio, Colo.: Southern Ute Tribe, 1972) and Gregory C. Thompson, "The Unwanted Utes: The Southern Utes in Southeastern Utah," *Utah Historical Quarterly* 49 (Spring 1981): 189–203.

19. Stevenson, interview, 16 May 1996.

20. Mark W. T. Harvey, *Symbol of Wilderness: Echo Park and the American Conservation Movement* (Albuquerque: University of New Mexico Press, 1994), xi–xviii.

21. Donald Worster, "Appendix: Doing Environmental History," in *The Ends of the Earth: Perspectives on Modern Environmental History,* ed. Donald Worster (New York: Cambridge University Press, 1988), 289–307.

22. George B. Chittenden, "Report of George B. Chittenden, Topographer of the San Juan Division, 1875," in F. V. Hayden, *Ninth Annual Report of the United States Geological and Geographical Survey for the Year 1875* (Washington, D.C.: GPO, 1877), 361.

23. Wallace Stegner, "San Juan and Glen Canyon," in *The Sound of Mountain Water: The Changing American West* (New York: E. P. Dutton, 1980), 120.

24. Worster, "Appendix," 293.

Chapter 1

1. *Anasazi* is a Navajo word meaning "enemy ancestors." Archaeologists adopted it as the official designation of this prehistoric culture after the distinguished archaeologist Alfred V. Kidder introduced it in 1936, even though further work has shown that Navajos entered the area after the collapse of the Anasazi lifeway and had no contact with them. Two groups of Anasazi descendants, the Hopi and Zuni, currently refer to their ancestors as Hisatsinom and Enote:que, respectively. The Hopi word means "people of long ago," while the Zuni word means "our ancestors." Hopis and Zunis resent that a Navajo word has become the official name of what they consider as the origin of their culture. Another problem arises from using a single word for a people whose various Pueblo descendants more often referred to themselves by clan names. Unfortunately, we are probably stuck with the Navajo name, although a few reference works now use Ancestral Puebloans. We employ the more familiar term, *Anasazi,* just as we also mostly use the more common term, *Indian,* rather than *Native American.*

   The evolving lifestyles of the group are given these names and dates: Basketmaker II (early), 1500 B.C–A.D. 50.; Basketmaker II (late), A.D. 50–500; Basketmaker III, A.D. 500–750; Pueblo I, A.D. 750–900; Pueblo II, A.D. 900–1150; Pueblo III, A.D. 1150–1350. Within the periods, certain names identify specific cultural and geographic divisions: Kayenta, Mesa Verde, Chaco, and Virgin River (or Western) Anasazi. At different times, the Anasazi in the San Juan exhibited influences from the Chaco, Kayenta, and Mesa Verde cultures.

2. Lisa Nelson, *Ice Age Mammals of the Colorado Plateau* (Flagstaff: Northern Arizona University Press, 1990), 21; Julio L. Betancourt, "Late Quaternary Plant Zonation and Climate in Southeastern Utah," *The Great Basin Naturalist* 44 (1984): 7; R. Dale Guthrie, "Mosaics, Allelochemics and Nutrients: An Ecological Theory of Late Pleistocene Megafauna Extinctions," in *Quaternary Extinctions: A Prehistoric Revolution,* ed. Paul S. Martin and Richard G. Klein (Tucson: University of Arizona Press, 1984), 282.
3. Betancourt, "Late Quaternary Plant Zonation," 21; Owen Davis et al., "The Pleistocene Dung Blanket of Bechan Cave, Utah," in *Contributions to Quaternary Vertebrate Paleontology,* Special Publication of Carnegie Museum of Natural History, no. 8, ed. Hugh H. Genoways and Mary R. Dawson (Pittsburgh: Carnegie Museum, 1984), 281.
4. Paul S. Martin, "Who or What Destroyed Our Mammoths?: (A Bedtime Story for the Visitors at the Hot Springs, South Dakota, Mammoth Site)," in *Megafauna and Man: Discovery of America,* Scientific Papers, vol. I, ed. Larry D. Agenbroad, Jim I. Mead, and Lisa W. Nelson, (Hot Springs, S. Dak.: The Mammoth Site of Hot Springs, South Dakota, 1990), 109.
5. For a summary of the debate, see Eliot Marshall, "Clovis Counterrevolution," *Science* 249 (1990): 738–41.
6. Nelson, *Ice Age Mammals,* 19.
7. William E. Davis. "The Lime Ridge Clovis Site," *Utah Archaeology* 2, no. 1 (1989): 66–76; William E. Davis and Gary M. Brown, "The Lime Ridge Clovis Site," *Current Research in the Pleistocene* 3 (1986): 1–3.
8. Larry D. Agenbroad, "Clovis People: The Human Factor in the Pleistocene Megafauna Extinction Equation," in *Ice-Age Origins,* Ethnography Monograph, no. 12, ed. Ronald C. Carlisle (Pittsburgh: University of Pittsburgh Press, 1988), 64–65; Brian M. Fagan, *The Great Journey: The Peopling of North America* (New York: Thomas and Hudson, 1987), 179; William W. Dunmire and Gail D. Tierney, *Wild Plants and Native Peoples of the Four Corners* (Santa Fe: Museum of New Mexico Press, 1997), 13; and Nelson, *Ice Age Mammals,* 4–16. Dunmire's and Tierney's book is the best single source on plant use by native peoples in the Four Corners area.
9. Nelson, *Ice Age Mammals,* 4–5.
10. Owen Davis et al., "The Pleistocene Dung Blanket," 281; Owen Davis et al., "Riparian Plants Were a Major Component of the Diet of Mammoths of Southern Utah," *Current Research in the Pleistocene* 2 (1985): 81.
11. Gary Haynes, *Mammoth, Mastodonts, and Elephants: Biology, Behavior, and the Fossil Record* (Cambridge: Cambridge University Press, 1991), 312; Agenbroad, "Clovis People," 64.
12. William Davis, "Lime Ridge Clovis Site," 66; Davis and Brown, "Lime Ridge Clovis Site," 3. A mammoth bone found in nearby Butler Wash is now on display in a museum in Blanding. See Winston Hurst, "The Prehistoric Peoples of San Juan County, Utah," in *San Juan County, Utah: People, Resources, and History,* ed. Allan Kent Powell (Salt Lake City: Utah State Historical Society, 1982), 23.
13. Jeffrey J. Saunders, "Immonence, Configuration, and the Discovery of America's Past," in *Megafauna and Man,* 142.
14. Haynes, *Mammoth, Mastadonts and Elephants,* 296.
15. T. R. Van Devender and W. G. Spaulding, "Development of Vegetation and Climate in the Southwestern United States," *Science* 204 (1979): 701.
16. Paul S. Martin, "The Discovery of America," *Science* 179 (1973): 973.
17. Martin, "Who or What," 116.
18. Russell W. Graham and Ernest L. Lundelius, Jr., "Coevolutionary Disequilibrium and Pleistocene Extinctions," in *Quaternary Extinctions,* 243–44.
19. Guthrie, "Mossaics, Allelochemics and Nutrients," 282–83.
20. Haynes, *Mammoth, Mastadonts and Elephants,* 317; Bjorn Kurten and Elaine Anderson, *Pleistocene Animals of North America* (New York: Columbia University Press, 1980), 363.
21. See Robert Brightman, *Grateful Prey: Rock Cree Human-Animal Relationships* (Los Angeles: University of California Press, 1993); also Robert S. McPherson, *Navajo Land, Navajo Culture: Persistence and Change in Southeastern Utah* (Norman: University of Oklahoma Press, forthcoming).
22. For a discussion of buffalo in Utah, especially northern Utah, see Karen D. Lupo, "The Historical Occurrence and Demise of Bison in Northern Utah," *Utah Historical Quarterly* 64 (Spring 1996): 168–80.
23. Winston B. Hurst, consulting archaeologist, interview by James Aton, 6 June 1994, Blanding, Utah.
24. Alan R. Schroedl, "Archaic of the Northern Colorado Plateau" (Ph.D. diss., University of Utah, 1976), 11.

25. See Alexander J. Lindsay, Jr., et al., *Survey and Excavations North and East of Navajo Mountain, Utah, 1959–1962*, Museum of Northern Arizona Bulletin, no. 45, Glen Canyon Series no. 8 (Flagstaff: Museum of Northern Arizona, 1968), 119; Jesse D. Jennings, "Summary and Conclusions," in *Sudden Shelter,* University of Utah Anthropological Papers (UUAP), no. 103, ed. Jesse D. Jennings (Salt Lake City: University of Utah Press, 1980), 197–202; Dale Davidson et al., "San Juan County Almost 8000 Years Ago: Ongoing Excavations at Old Man Cave," *Blue Mountain Shadows* 13 (1994): 7–12.
26. Timothy M. Kearns, "Aceramic Sites and the Archaic Occupation along the Middle San Juan River, Southeast Utah," paper presented at the 54th Meeting of the Society for American Archaeology, Las Vegas, Nevada, 1990, part of Sampling, Time, and Population along the San Juan River.
27. Jennings, "Summary and Conclusions," 199.
28. Janette M. Elyea and Patrick Hogan, "Regional Interaction: The Archaic Adaptation," in *Economy and Interaction along the Lower Chaco River,* ed. Patrick Hogan and Joseph C. Winter (Albuquerque: Maxwell Museum, 1983), 393.
29. Ibid., 400–401.
30. Nancy J. Coulam and Peggy R. Barnett, "Paleoethnobotanical Analysis," in *Sudden Shelter,* 191.
31. Ibid., 187, 191.
32. Dr. James E. Bowns, Southern Utah University range ecologist, interview by James Aton, Spring 1996, Cedar City, Utah; Dunmire and Tierney, *Wild Plants and Native Peoples,* 196.
33. Coulam and Barnett, "Paleoethnobotanical Analysis," 188–90.
34. Elyea and Hogan, "Regional Interaction," 396.
35. Winston B. Hurst, "Regional Cultural History," in "U-262 Report," 1992, p. 19, Abajo Archeology, Bluff, Utah.
36. See Alan R. Schroedl, *Kayenta Anasazi Archeology and Navajo Ethnohistory on the Northwestern Shonto Plateau: The N-16 Project* (Salt Lake City: P-III Associates, 1989), 808; Karen R. Adams, "Seeds and Large Plant Remains," in *Basketmaker Settlement and Subsistence along the San Juan River, Utah,* ed. Robert B. Neilly (Salt Lake City: Utah Division of State History, 1982), 340–82; Linda S. Cordell, *Prehistory of the Southwest* (San Diego: Academic Press, 1984), 31; and William E. Davis, "Summary and Evaluation of Results," in "U-262 Report," 6.
37. William D. Lipe, "The Basketmaker Period in the Four Corners Area," in *Anasazi Basketmaker: Papers from the 1900 Wetherill-Grand Gulch Symposium,* BLM Cultural Resource Series, no. 24, ed. Virginia M. Atkins (Salt Lake City: Department of the Interior, Bureau of Land Management, 1993), 7; Jesse D. Jennings, *Glen Canyon: A Summary,* UUAP, no. 81 (Salt Lake City: University of Utah Press, 1966), 21.
38. Linda J. Scott, "Pollen Analysis," in *Dolores Archaeological Program: Anasazi Communities at Dolores: Grass Mesa Village,* ed. William D. Lipe et al. (Denver: Department of the Interior Bureau of Land Management, 1988), 1210; Kenneth L. Peterson et al., "Implications of Anasazi Impact on the Landscape," in *Dolores Archaeological Program: Supporting Studies: Settlement and Environment,* ed. Kenneth L. Peterson and Janet D. Orcutt (Denver: Department of the Interior, Bureau of Land Management, 1987), 154; Steven D. Emslie, "Faunal Remains," in *Basketmaker Settlement and Subsistence,* 422–43.
39. Winston B. Hurst, "The Mysterious Telluride Blanket," *Blue Mountain Shadows* 13 (1994): 68–69.
40. Hurst, interview.
41. Alex Patterson, *A Field Guide to Rock Art Symbols of the Greater Southwest* (Boulder, Colo.: Johnson Books, 1992), 182.
42. Alexander J. Lindsay, Jr., "The Beaver Creek Agricultural Community on the San Juan River, Utah," *American Antiquity* 27 (1961): 181–84; Lindsay, et al., *Survey and Excavations,* 136–46.
43. R. G. Matson, William D. Lipe, and William R. Haase III, "Adaptational Continuities and Occupational Discontinuities: The Cedar Mesa Anasazi," *Journal of Field Archaeology* 15 (1980): 258; Dunmire and Tierney, *Wild Plants and Native Peoples,* 66.
44. Julio L. Betancourt and Thomas R. Van Devender, "Holocene Vegetation in Chaco Canyon," *Science* 214 (1981): 658.
45. Geoffrey W. Spaulding, *Paleoecological Investigations at the Coombs Site (42 GA 34) Megafauna and Man* (Las Vegas: Dames and Moore, 1994), 11; Timothy A. Kohler and Meredith H. Matthews, "Long-Term Anasazi Land Use and Forest Reduction: A Case Study from Southwest Colorado," *American Antiquity* 53, no. 3 (1988): 537–64; Peterson et al., "Implications of Anasazi Impact," 150.
46. Jack Oviatt, "Environmental Setting and Geomorphic History," in *Basketmaker Settlement and Subsistence,* 61.
47. Peterson et al., "Implications of Anasazi Impact," 158.
48. Hurst, "Regional Cultural History," 81.

49. See J. Richard Ambler and Mark Q. Sutton, "The Anasazi Abandonment of the San Juan Drainage and the Numic Expansion," *North American Anthropologist* 10 (1989): 39–57 and Steven A. Leblanc, *Prehistoric Warfare in the American Southwest* (Salt Lake City: University of Utah Press, 1999), 309, 312. Although not necessarily connected at this point with abandonment theories, there is increasing evidence of intracultural violence and cannibalism among Chaco Anasazi and in Chaco outliers, two of which are along the San Juan—one near Bluff at St. Christopher's Mission and one near Navajo Mountain, now under Lake Powell. Noted archaeologists Christy and Jacqueline Turner believe that warrior cultists from Mexico, dedicated to the Tezcatlipoca-Xipe-Totec complex, with its emphasis on human sacrifice, infiltrated the Chaco area around A.D. 900. The Turners think that these thugs terrorized the Chaco population into adopting the hierarchical and bloody system practiced in Meso-America, where they achieved their objectives through warfare, human sacrifice, and cannibalism. See Christy G. Turner II and Jacqueline A. Turner, *Man Corn: Cannibalism and Violence in the Prehistoric American Southwest* (Salt Lake City: University of Utah Press, 1999).
50. Robert S. McPherson, *Sacred Land, Sacred View* (Provo, Utah: Brigham Young University Press, 1992).
51. Jerold G. Widdison, ed., *Anasazi: Why Did They Leave? Where Did They Go?* (Albuquerque: Southwest Cultural Heritage Association, 1991), 26, 35.
52. Ibid., 37–40.
53. Christy G. Turner II, "Revised Dating for Early Rock Art of the Glen Canyon Region," *American Antiquity* 36 (1971): 469.
54. Christy G. Turner II, *Petroglyphs of the Glen Canyon Region,* Museum of Northern Arizona Bulletin, no. 48 (Flagstaff: Museum of Northern Arizona, 1963), 29.
55. Turner, "Revised Dating," 469–70.
56. See D. L. Schwartz et al., "Split-Twig Figurines in the Grand Canyon, *American Antiquity* 23 (1958): 273; Robert C. Euler, "The Canyon Dwellers, "*American West* 4 (1967): 23; A. P. Olson, "Split-Twig Figurines from NA5607, Northern Arizona," *Plateau* 38 (1966): 63; and Alan R. Schroedl, "The Grand Canyon Figurine Complex," *American Antiquity* 42 (1977): 254–65.
57. Turner, *Petroglyphs of Glen Canyon;* Turner, "Revised Dating," 469–71; Polly Schaafsma, *Indian Rock Art of the Southwest* (Santa Fe: School of American Research, 1980), 72–75.
58. Schaafsma, *Indian Rock Art,* 114–19.
59. Ibid., 135; Nancy H. Olsen, *Hovenweep Rock Art: An Anasazi Visual Communication System,* Occasional Paper, no. 14 (Los Angeles: UCLA Institute of Archaeology, 1985), 134–37.
60. Schaafsma, *Indian Rock Art,* 153.
61. Ibid., 136, 140.
62. Klaus F. Wellman, "Kokopelli of Indian Paleology: Humpbacked Rain Priest, Hunting Magician, and Don Juan of the Old Southwest," *JAMA* 212 (1970): 1678–82; Schaafsma, *Indian Rock Art,* 141.
63. Winston B. Hurst and Joe Pachak, *Spirit Windows: Native American Rock Art of Southeastern Utah* (Blanding, Utah: Edge of the Cedars Museum, 1992), 8, 20; Schaafsma, *Indian Rock Art,* 134.

Chapter 2

1. For a discussion of this prehistoric period of Numic-speaking people, see Alan D. Reed, "Ute Cultural Chronology," in *An Archaeology of the Eastern Ute: A Symposium,* Occasional Papers, no. 1, ed. Paul R. Nickens (Denver: Colorado Council of Professional Archaeology, 1988), 80–81; C. S. Fowler and D. D. Fowler, "The Southern Paiute: A.D. 1400–1776," in *The Protohistoric Period in the North American Southwest, A.D. 1350–1700,* Archaeological Research Papers, no. 24, ed. D. R. Wilcox and W. B. Masse (Tempe: Arizona State University Press, 1981), 129–62.
2. Isabel T. Kelly and Catherine S. Fowler, "Southern Paiute," in *Handbook of North American Indians,* vol. 11, ed. Warren D'Azevedo (Washington, D.C.: Smithsonian Institution, 1986), 368, 396.
3. For the most complete history of the San Juan Paiutes, see Pamela A. Bunte and Robert J. Franklin, *From the Sands to the Mountain: Changes and Persistence in a Southern Paiute Community* (Lincoln: University of Nebraska Press, 1987).
4. For further discussion of the history of the Utes and Paiutes living along the Lower San Juan River, see Robert S. McPherson, *A History of San Juan County: In the Palm of Time,* Utah Centennial County History Series (Salt Lake City: Utah State Historical Society, 1995).
5. Edward Dutchie, Sr., interview by Robert McPherson, 7 May 1996, transcript in possession of author.
6. E. L. Hewitt, "Field Notes 1906–09" (Edge of the Cedars Museum, Blanding, Utah, unpublished typescript, photocopy on file); Billy Mike, interview by Aldean Ketchum and Robert McPherson, 13 October 1993, transcript in possession of author; Dutchie, interview; Stella Eyetoo, interview by Aldean Ketchum and Robert McPherson, 21 December 1994, transcript in possession of

author; Chester Cantsee, Sr., interview by Aldean Ketchum and Robert McPherson, 6 September 1994, transcript in possession of author; John W. Van Cott, *Utah Place Names* (Salt Lake City: University of Utah Press, 1990), 264.

7. Harold Lindsay Amoss, Jr., "Ute Mountain Utes" (Ph.D. diss., University of California, 1951), 90.
8. Frank Silvey, "Information on Indians," 26 September, 1936, pp. 1–2, Utah State Historical Society, Salt Lake City. While Silvey's work addresses this topic specifically, many of the sites mentioned derive from several historical documents that span a hundred years.
9. Amoss, "Ute Mountain Utes," 55–61; Ralph V. Chamberlain, "Some Plant Names of the Ute Indians," *American Anthropologist* 2, no. 1 (January–March, 1909): 27–37.
10. Amoss, "Ute Mountain Utes," 49–51.
11. Fred A. Conetah, *A History of the Northern Utes,* ed. Kathryn L. McKay and Floyd A. O'Neil (Salt Lake City: Uintah Ouray Ute Tribe, 1982), 9; Florence Hawley et al., "Culture Process and Change in Ute Adaptation—Part I," *El Palacio* (October 1950): 325; and Anne M. Smith, *Ethnography of the Northern Utes* (Albuquerque: Museum of New Mexico Press, 1974), 61–62, 64.
12. Marvin Kaufman Opler, *The Southern Ute of Colorado* (New York: D. Appleton-Century, 1940): 137–38, 140.
13. Amoss, "Ute Mountain Utes," 37–38; Chamberlain, "Some Plant Names," 27–40; James Jefferson, Robert W. Delaney, and Gregory Thompson, *The Southern Utes: A Tribal History* (Ignacio, Colo.: Southern Ute Tribe, 1972), 72–73.
14. Terry Knight, spiritual leader of the Ute Mountain Utes, interview by Mary Jane Yazzie and Robert McPherson, 19 December 1994, tape in possession of author.
15. "The Ute Indians of Southwestern Colorado," compiled by Helen Sloan Daniels, unpublished manuscript, 1941.
16. Dutchie, interview.
17. Jefferson, Delaney, and Thompson, *Southern Utes,* 74–75; Dutchie, interview; Anne M. Smith, *Ute Tales* (Salt Lake City: University of Utah Press, 1992), 110.
18. Smith, *Ethnography of Northern Utes,* 39, 109–13; Jefferson, Delaney, and Thompson, *Southern Utes,* 74–75.
19. Dutchie, interview.
20. Mary Jay, interview by Robert McPherson, 22 February 1991, San Juan County Historical Commission (SJCH), Blanding, Utah; Isabelle Lee, interview by Robert McPherson, 13 February 1991, SJCH; Vernon O. Mayes and Barbara Lacy, *Nanise': A Navajo Herbal* (Tsaile, Ariz.: Navajo Community College, 1989), 29–30, 87, 106.
21. Lee, interview: Ben Whitehorse, interview by Baxter Benally and Robert McPherson, 30 January 1991, SJCH.
22. Cyrus Begay, interview by Baxter Benally and Robert McPherson, 14 May 1991, transcript in possession of author.
23. Florence Begay, interview by Robert McPherson, 30 January 1991, transcript in possession of author.
24. Lee, interview; Margaret Weston, interview by Robert McPherson, 13 February 1991; Monument Valley and Navajo Mountain residents, interview by Jim Dandy, n. d., manuscript in possession of author; Mary Bitsili, interview by Aubrey Williams and Maxwell Yazzie, 19 January 1961, p. 3, Doris Duke 710, Special Collections, Marriott Library, University of Utah, Salt Lake City.
25. Walter E. Mendenhall testimony, *United States v Utah* (1931), Colorado River Bed Case, October 1929, p. 1616, Utah State Historical Society, Salt Lake City; also Robert S. McPherson, *Sacred Land, Sacred View* (Provo, Utah: Brigham Young University Press, 1992), 90–91.
26. Dozens of witnesses during the Colorado River Bed Case testified that boats and makeshift rafts were rarely, if ever, used by Navajos on the river. Once the trading post was introduced, boats were utilized and even operated by Navajos, but an inherent fear of traveling on the water persisted.
27. John Wetherill testimony, *United States v Utah* (1931), Colorado River Bed Case, September 1929, pp. 1327, 1333, Utah State Historical Society, Salt Lake City.
28. John Norton, interview by Baxter Benally and Robert McPherson, 16 January 1991, SJCH; Ray Hunt, interview by Robert McPherson, 21 January 1991, SJCH.
29. Robert W. Young and William Morgan, *Navajo Historical Selections* (Lawrence, Kans.: Bureau of Indian Affairs, 1954), 38; Charles Kelly, "Aneth," Charles Kelly Papers, Special Collections, Marriott Library, University of Utah, Salt Lake City; Whitehorse, interview.
30. Billy Smiley, interview by Robert McPherson, 14 January 1991, SJHC; Margaret Weston, interview; Jerry Begay, interview by Baxter Benally and Robert McPherson, 16 January 1991, SJHC; Norton, interview; Lee, interview; Jane Silas, interview by Robert McPherson, 27 February 1991.
31. Lee, interview.

32. Cyrus Begay, interview.
33. Franc Johnson Newcomb, *Navaho Folk Tales* (Albuquerque: University of New Mexico Press, 1967), 23–32.
34. Klara Bonsack Kelley and Harris Francis, *Navajo Sacred Places* (Indianapolis: Indiana University Press, 1994), 37.
35. Newcomb, *Navajo Folktales,* 31–32; Washington Matthews, *Navaho Legends* (reprint, Salt Lake City: University of Utah Press, 1994), 63–64.
36. Matthews, *Navajo Legends,* 73–74, 77, 212; Gladys Reichard, *Navaho Religion—A Study of Symbolism* (Princeton: Princeton University Press, 1963), 490; Franciscan Fathers, *An Ethnologic Dictionary of the Navajo Language* (Saint Michaels, Ariz.: Saint Michaels Press, 1910), 156–57, 507.
37. Aileen O'Bryan, *Navaho Indian Myths* (New York: Dover Publications, 1993), 109.
38. Gerald Hausman, *The Gift of the Gila Monster* (New York: Simon and Schuster, 1993), 101–7.
39. Matthews, *Navajo Legends,* 160–70.
40. Personal knowledge; Pat Seltzer, principal of Monument Valley High School and community member, telephone conversation with Robert McPherson, 12 June 1996.
41. Washington Matthews, *Navaho Legends* (New York: Houghton, Mifflin and Company, 1897), 211; Charlie Blueeyes, interview by Robert McPherson, 28 August 1988, SJHC; Ernest Nelson, Long Salt, and Karl Luckert as quoted in Karl Luckert, *Navajo Mountain and Rainbow Bridge Religion* (Flagstaff: Museum of Northern Arizona, 1977), 24, 40, 113, 117.
42. Charlie Blueeyes, interview by Robert McPherson, 7 June 1988, transcript in possession of author.
43. Tallis Holiday, interview by Robert McPherson, 3 November 1987; Fred Yazzie, interview by Robert McPherson, 5 November 1987; transcripts in possession of author.
44. Ada Black, interview by Robert McPherson, 11 October 1991; transcript in possession of author.
45. For a more detailed study of the way concepts of power, prayers, and protection are tied to geographical forms, see McPherson, *Sacred Land, Sacred View.*
46. Luckert, *Navajo Mountain and Rainbow Bridge Religion,* 40, 44–45, 94, 103, 112.
47. Ibid., 117.
48. Mendenhall testimony, 1616.
49. Jared Farmer, *Glen Canyon Dammed: Inventing Lake Powell and the Canyon Country* (Tucson: University of Arizona Press, 1999), 160.
50. Department of the Interior, National Park Service, *Glen Canyon National Recreation Area Annual Report,* 1999. (San Juan County Tourist Information Bureau, Monticello, Utah, on file).
51. Farmer, *Glenn Canyon Damned,* 168–69.
52. Luckert, *Navajo Mountain and Rainbow Bridge Religion,* 45, 141–42.
53. Chris Smith and Elizabeth Manning, "The Sacred and Profane Collide in the West," *High Country News,* 26 May 1997, pp. 1, 8–12.
54. Tom Dougi, interview by Robert McPherson, 8 April 1992, transcript in possession of author.

Chapter 3

1. William H. Goetzmann, *New Men, New Lands: America and the Second Great Age of Discovery* (New York: Viking, 1986), 3–8.
2. Stephen J. Pyne, *How the Canyon Became Grand: A Short History* (New York: Viking, 1998), 7–8, 22.
3. Iris H. W. Engstrand, *Spanish Scientists in the New World: The Eighteenth-Century Expeditions* (Seattle: University of Washington Press, 1981), 3, 11.
4. Clell G. Jacobs, "The Phantom Pathfinder: Juan Maria Antonio de Rivera and His Expedition," *Utah Historical Quarterly* 60 (Summer 1992): 201. See also Austin Nelson Leiby, "Borderland Pathfinders: The Diaries of Juan Maria Antonio de Rivera," (Ph.D. diss., Northern Arizona University, 1985). For a general survey of Spanish exploration, see Joseph P. Sanchez, *Explorers, Traders, and Slavers: Forging the Old Spanish Trail, 1678–1850* (Salt Lake City: University of Utah Press, 1997).
5. Jacobs, "Phantom Pathfinder," 201; see also Donald C. Cutter, "Prelude to a Pageant in the Wilderness," *Western Historical Quarterly* 8 (January 1977): 6.
6. Engstrand, *Spanish Scientists,* xi.
7. Leiby, "Borderland Pathfinders," 130–31.
8. Walter Briggs, *Without Noise of Arms: The 1776 Dominguez-Escalante Search for a Route from Santa Fe to Monterey* (Flagstaff, Ariz.: Northland Press, 1976), 189–90.
9. For an excellent discussion of this shift, see Keith Thomas, *Man and the Natural World: A History of the Modern Sensibility* (New York: Pantheon, 1983), 254–69.
10. Fray Angelico Chavez and Ted Warner, eds., *The Dominguez-Escalante Journal* (Provo, Utah: Brigham Young University Press, 1976), 78, 93. Perhaps because the padres were in the missionary business, they were much less tolerant of native ways than Rivera. See pp. 90–91 of Chavez and Warner for an example of the fathers' cultural intolerance.
11. Leiby, "Borderland Pathfinders," 105, 190.

12. Joseph J. Hill, "Spanish and Mexican Exploration and Trade Northwest from New Mexico into the Great Basin, 1765–1853," *Utah Historical Quarterly* 3 (January 1930): 16. Also see Sondra Jones, *The Trial of Don Pedro Leon Lujan: The Attack against Indian Slavery and Mexican Traders in Utah* (Salt Lake City: University of Utah Press, 1999), and Sondra Jones, "'Redeeming' the Indian: The Enslavement of Indian Children in New Mexico and Utah," *Utah Historical Quarterly* 67 (Summer 1999): 220–41.
13. David M. Brugge, "Vizcarra's Navajo Campaign of 1823," *Arizona and the West* 6 (1964): 223–44; Leroy R. Hafen and Ann Hafen, *Old Spanish Trail: Santa Fe to Los Angeles* (Glendale, Calif.: Arthur H. Clark, 1965), 154–69.
14. David J. Weber, *The Taos Trappers: The Fur Trade in the Far Southwest, 1540–1846* (Norman: University of Oklahoma Press, 1968), 8–23, 67, 79. See also Iris Higbee Wilson, "William Wolfskill," in *The Mountain Men and the Fur Trade of the Far West,* vol. 2, ed. Leroy R. Hafen (Glendale, Calif.: Arthur H. Clark, 1965), 352.
15. Alfred Glen Humphreys, "Thomas L. (Peg-Leg) Smith," and Forbes Parkhill, "Antoine Leroux," in *The Mountain Men and the Fur Trade of the Far West,* vol. 5, ed. Leroy R. Hafen (Glendale, Calif.: Arthur H. Clark, 1966), 312 and 174.
16. Clifton Kroeber, ed., "The Route of James O. Pattie on the Colorado in 1826: A Reappraisal by A. L. Kroeber," *Arizona and the West* 4 (Summer 1964): 130–31; James O. Pattie, *The Personal Narrative of James O. Pattie,* ed. William H. Goetzmann (Philadelphia: J. B. Lippincott, 1962), 90.
17. Stephan D. Durrant and Nowlan K. Dean, "Mammals of Glen Canyon," in *Ecological Studies of the Flora and Fauna in Glen Canyon,* University of Utah Anthropological Papers (UUAP), no. 40, ed. Angus Woodbury et al. (Salt Lake City: University of Utah Press, 1959), 87.
18. David J. Wishart, *The Fur Trade of the American West, 1807–1840: A Geographical Synthesis* (Lincoln: University of Nebraska Press, 1979), 31–33.
19. Weber, *Taos Trappers,* 224–25; William deBuys, *Enchantment and Exploitation: The Life and Hard Times of a New Mexico Mountain Range* (Albuquerque: University of New Mexico Press, 1988), 98.
20. Linda Richmond, river ranger, Bureau of Land Management, San Juan Resource Area, interview by James Aton, 9 July 1997.
21. Henry F. Dobyns, *From Fire to Flood: Historic Human Destruction of Sonoran Riverine Oases,* Ballena Press Anthropological Papers, no. 20, ed. Lowell John Bean and Thomas C. Blackburn (Socorro, N. Mex.: Ballena Press, 1981), 105–16.
22. Robert M. Utley, *A Life Wild and Perilous: Mountain Men and the Paths to the Pacific* (New York: Henry Holt, 1997); Goetzmann, *New Men, New Lands,* 128, 148–49.
23. W. D. Huntington, "Interesting Account of a Trip to the Navajos and of the Ancient Ruins in that Region," *Deseret News,* 28 December 1854, p. 3.
24. Alfred N. Billings, "Account Book and Diary of Alfred N. Billings," 15–16; and Ethan Pettit, "Diary of Ethan Pettit, 1855–1881," Utah State Historical Society, Salt Lake City.
25. In 1853, five years before Macomb, Major Henry L. Kendrick, commander of the military post at Fort Defiance, New Mexico, led a military expedition into the Four Corners area and down the San Juan to the mouth of Chinle Wash. He was pursuing Navajos who had stolen and killed. His short, four-page report commented on the barrenness of the river and the difficulty of grazing horses. He also noted that cottonwoods were the only wood for building, coal was present, there were many signs of sheep, and little corn was raised along the river. See Major Henry L. Kendrick, "Report to Maj. Wm. A. Nichols," 15 August 1853, Old Military Records, National Archives, Washington, D.C.
26. J. N. Macomb, *Report of the Exploring Expedition from Santa Fe, New Mexico, to the Junction of the Grand and Green Rivers of the Great Colorado of the West in 1859* (Washington, D.C.: GPO, 1876). See also F. A. Barnes, *Hiking the Historic Route of the 1859 Macomb Expedition* (Moab, Utah: Canyon Country Publications, 1989).
27. William H. Goetzmann, *Army Exploration in the American West, 1803–1863* (New Haven: Yale University Press 1959; reprint, Lincoln: University of Nebraska Press, 1979), 375.
28. Macomb, *Report of the Exploring Expedition,* 6.
29. Ibid., 103.
30. Ibid., 104.
31. Pyne, *How the Canyon became Grand,* 48–49.
32. Ibid., 6, 89.
33. Macomb, *Report of the Exploring Expedition,* 109.
34. Richard A. Bartlett, *Great Surveys of the West* (Norman: University of Oklahoma Press, 1962), 20. The standard biography of Hayden's life and career is Mike Foster, *Strange Genius: The Life of Ferdinand Vandeveer Hayden* (Niwot, Colo.: Roberts Rinehart Publishers, 1994).
35. See F. V. Hayden, *Ninth Annual Report of the United States Geological and Geographical Survey for*

the Year 1875 (Washington, D.C.: GPO, 1877), 12–15, 361; W. H. Jackson, "Report on the Ruins Examined in 1875 and 1877," in F. V. Hayden, *Tenth Annual Report of the United States Biological and Geological Survey for the Year 1875* (Washington, D.C.: GPO, 1878), 411-431.

36. Hayden, *Ninth Annual Report,* 361; Jackson, "Report on the Ruins," 412.
37. Biographical information on Alice Eastwood comes primarily from Carol Green Wilson, *Alice Eastwood's Wonderland: The Adventures of a Botanist* (San Francisco: California Academy of Sciences, 1955); Alice Eastwood, "Memoirs," Alice Eastwood Archives, California Academy of Sciences, San Francisco; Maurine S. Fletcher, ed., *The Wetherills of the Mesa Verde: Autobiography of Benjamin Alfred Wetherill* (1977; reprint, Lincoln: University of Nebraska Press, 1987); and James Thomas Howell, "'I Remember When I Think . . .'" *Leaflet of Western Botany* 2 (August 1954): 153–64.
38. Alice Eastwood, "Report on a Collection of Plants from San Juan County, in Southeastern Utah," *California Academy of Sciences Proceedings* 6 (August 1896): 272–76; "General Notes of a Trip through Southeastern Utah," *Zoe* 3 (January 1893): 360.
39. Eastwood, "Notes on the Cliff Dwellers," *Zoe* 3 (January 1893): 375–76.
40. Eastwood, "Report on a Collection of Plants," 279.
41. Eastwood, "Memoirs,"
42. Virginia M. Atkins, ed., *Anasazi Basketmaker: Papers from the 1990 Wetherill-Grand Gulch Symposium,* BLM Cultural Resource Series, no. 24 (Salt Lake City: Department of the Interior, Bureau of Land Management, 1993).
43. See Helen Sloan Daniels, *Adventures with the Anasazi of Falls Creek,* Occasional Papers of the Center for Southwest Studies, no. 3 (Durango, Colo.: Fort Lewis College, 1954) for a copy of Graham's diary. See also Fred M. Blackburn and Victoria M. Atkins, "Handwriting on the Wall: Applying Inscriptions to Reconstruct Historic Archaeological Expeditions," in *Anasazi Basketmaker,* 41-100; and Warren K. Moorhead, "The Great McLoyd Collections," part 12 of "In Search of a Lost Race," *Illustrated American,* 12 August 1892, 24.
44. Moorhead, "In Search of a Lost Race." 177.
45. Ibid., 409–10.
46. T. Mitchell Prudden, "An Elder Brother to the Cliff Dwellers," *Harper's New Monthly Magazine* 95 (1897): 55–62; "The Prehistoric Ruins of the San Juan Watershed," *American Anthropologist* 5 (1903): 224–88; *On the Great American Plateau* (New York: G. P. Putnam's, 1907); and *Biographical Sketches and Letters of T. Mitchell Prudden, M.D.* (New Haven: Yale University Press, 1927).
47. Prudden, *On the Great American Plateau,* 94.
48. Neil M. Judd, "The Discovery of Rainbow Bridge," in *The Discovery of Rainbow Bridge,* Bulletin 1 (Tucson: Cummings Publication Council, 1959), 8–13; Byron R. Cummings, "Field Notes of 1909," Arizona State Museum, University of Arizona, Tucson; and Neil M. Judd, *Men Met along the Trail: Adventures in Archaeology* (Norman: University of Oklahoma Press, 1968), 3–45.
49. See Frank McNitt, *Richard Wetherill: Anasazi: Pioneer Explorer of Southwestern Ruins,* rev. ed. (Albuquerque: University of New Mexico Press, 1966); Fletcher, *The Wetherills of the Mesa Verde;* and Francis Gillmor and Louisa Wade Wetherill, *Traders to the Navajos* (Albuquerque: University of New Mexico Press, 1952).
50. Warren G. Moorhead, "Across the Desert," *Illustrated American,* 16 July 1892, 410.
51. Herbert E. Gregory, "Scientific Explorations in Southern Utah," *American Journal of Science* 243 (October 1945): 527–49; "The Navajo Country," *American Biographical Society Bulletin* 8 (1915): 561–672; *The Navajo Country: A Geographic and Hydrographic Reconnaissance of Parts of Arizona, New Mexico, and Utah,* U.S. Geological Survey Water Supply Paper 380 (Washington, D.C.: GPO, 1916); *Geology of the Navajo Country: A Reconnaissance of Parts of Arizona, New Mexico, and Utah,* U. S. Geological Survey Professional Paper 93 (Washington, D.C.: GPO, 1917); and *The San Juan Country: A Geographic and Geologic Reconnaissance of Southeastern Utah,* U.S. Geological Survey Professional Paper 188 (Washington, D.C.: GPO, 1938).
52. See Gary Topping, "Herbert E. Gregory: Humanistic Geologist," in *Glen Canyon and the San Juan Country* (Moscow: University of Idaho Press, 1997), 209–29.
53. Gregory, *The Navajo Country,* 9.
54. Gregory, *The San Juan Country,* 9, 26–27, 35, 102; Herbert E. Gregory, "Field Notes," Book 8 (1913), 8 June 1913, U.S. Geological Survey Archives, Denver.
55. Gregory, *The Navajo Country,* 3.
56. Wetherill appears in nearly every chapter of Gary Topping's *Glen Canyon* and the *San Juan Country.* Until we have a full scholarly biography of him, Topping's book is the best portrait of the famous guide and trader.

57. William A. Myers, *Iron Men and Copper Wires: A Centennial History of the Southern California Edison Company* (Glendale, Calif.: Trans-Anglo Books, 1983), 179.
58. For accounts of the expedition, see Hugh D. Miser, *The San Juan Canyon, Southeastern Utah: A Geographic and Hydrographic Reconnaissance,* U.S. Geological Survey Water Supply Paper 538 (Washington, D.C.: GPO, 1924); Richard E. Westwood, *Rough Water Man: Elwyn Blake's Colorado River Expeditions* (Reno: University of Nevada Press, 1992); H. Elwyn Blake, "Boating the Wild Rivers," Marston Collection, Huntington Library, San Marino, California; and Bert Loper, "U.S.G.S. Survey of the San Juan River, 1921," p. 58, Marston Collection, Huntington Library, San Marino, California.
59. Hugh D. Miser to Heber Christensen, 16 March 1925, Marston Collection, Huntington Library, San Marino, California.
60. Hugh D. Miser, "Field Records," 1921, Accession no. 4124-A, U.S. Geological Survey Archives, Denver. Slickhorn is one of the few places north of the river between Mexican Hat and Clay Hills Crossing-Paiute Farms where stock can reach the river. In the fall of 1990 on a San Juan River trip, one of the authors found two horses which had wandered down to the river at Slickhorn. They had literally stripped all the vegetation in the bottomland at the mouth of Slickhorn and two miles along the river to Government Rapid. The animals were later removed, and the native vegetation of Indian ricegrass, Mormon tea, and blackbrush has returned.
61. Miser, "Field Records;" Bert Loper, "This Is to Be an Attempt to Describe My Second San Juan Voyage," p. 8, Marston Collection, Huntington Library, San Marino, California; Miser, *The San Juan Canyon,* 32–33, 60.
62. Miser, *The San Juan Canyon,* 57.
63. Hugh D. Miser testimony, *United States v Utah* (1931), Colorado River Bed Case, November 1929, Utah State Historical Society, Salt Lake City.
64. For a good summary of the RBMVE's work, see Andrew L. Christenson, "The Last of the Great Expeditions: The Rainbow Bridge/Monument Valley Expeditions 1933–1938," *Plateau* 58 (1987):1-32.
65. Ibid., 27–28; see also Lyndon Lane Hargrave, *Report on Archaeological Reconnaissance in the Rainbow Plateau Area of Northern Arizona and Southern Utah* (Berkeley: University of California Press, 1935).
66. Christenson, "Last of the Great Expeditions," 27; Angus M. Woodbury and Henry N. Russell, Jr., *Birds of the Navajo Country,* University of Utah Bulletin, Biological Series, no. 9. 1: 35 (Salt Lake City: University of Utah Press, 1945).
67. Officially known as the Upper Colorado River Basin Archaeological Salvage Project, the survey not only modeled new procedures for salvage operations but also spawned the University of Utah Anthropological Papers (UUAP), which continue to publish important findings in monographs. Information for this section comes from Jesse D. Jennings, *Glen Canyon: A Summary,* UUAP, no. 81 (Salt Lake City: University of Utah Press, 1966); Jesse D. Jennings and Floyd W. Sharrock, "The Glen Canyon: A Multi-Discipline Project," *Utah Historical Quarterly* 33 (Winter 1965): 34–50; C. Gregory Crampton, "Historical Archaeology on the Colorado River," in *The American West: An Appraisal,* ed. Robert G. Ferris (Santa Fe: Museum of New Mexico, 1963): 213–18; C. Gregory Crampton, *Outline History of the Glen Canyon,* UUAP, no. 42 (Salt Lake City: University of Utah Press, 1959); and Jesse D. Jennings, *Accidental Archaeologist: Memoirs of Jesse D. Jennings* (Salt Lake City: University of Utah Press, 1994).
68. Alexander J. Lindsay, Jr., et al., *Survey and Excavations North and East of Navajo Mountain, Utah, 1959-1962,* Museum of Northern Arizona Bulletin, no. 45 Glenn Canyon Series, no. 8 (Flagstaff: Museum of Northern Arizona, 1968); C. Gregory Crampton, *The San Juan Canyon Historical Sites,* UUAP, no. 70 (Salt Lake City: University of Utah, 1964). See also Crampton's *Standing Up Country: The Canyon Lands of Utah and Arizona* (1969; reprint, Salt Lake City: Peregrine Smith Books, 1983) and his *Ghosts of Glen Canyon: History beneath Lake Powell* (St. George, Utah: Publishers Place, 1986).
69. Jennings, *Glen Canyon,* 45.
70. See Angus Woodbury et al., *Preliminary Report on Biological Resources of the Glen Canyon Reservoir,* UUAP, no. 31 (Salt Lake City: University of Utah Press, 1958); Angus Woodbury, Stephen D. Durrant, and Seville Flowers, *Survey of Vegetation in the Glen Canyon Reservoir Basin,* UUAP, no. 36 (Salt Lake City: University of Utah Press, 1959); Angus Woodbury et al., eds., *Ecological Studies of the Flora and Fauna in Glen Canyon;* and Angus Woodbury, *Notes on the Human Ecology of Glen Canyon,* UUAP, no. 74 (Salt Lake City: University of Utah Press, 1965).
71. Woodbury, *Ecological Studies,* 153.

Chapter 4

1. David M. Brugge, "Navajo Use and Occupation of Lands North of the San Juan River in Present-day Utah to 1935," unpublished manuscript, 1966, 1–7.
2. See Robert S. McPherson, *The Northern Navajo Frontier, 1860–1900: Expansion through Adversity* (Albuquerque: University of New Mexico Press, 1988), 5–19.
3. Navajo Council to the Military in February or March 1866, quoted in J. Lee Correll, *Through White Men's Eyes—A Contribution to Navajo History,* vol. 6 (Window Rock, Ariz.: Navajo Heritage Center, 1979), 113.
4. Ralph Grey, interview by Fern Charley and Dean Sundberg, June 1974, *Navajo Stock Reduction Interviews,* compiled by the Utah State Historical Society and California State University at Fullerton (Fullerton, Calif.: California State University at Fullerton Press, 1984), 7.
5. Martha Nez, interview by Robert McPherson, 2 August 1988, transcript in possession of author.
6. Slim Benally, interview by Robert McPherson, 8 July 1988, transcript in possession of author.
7. Nez, interview.
8. Charlie Blueeyes, interview by Robert McPherson, 8 July 1988, transcript in possession of author.
9. Charles Hite to Galen Eastman, 17 April 1883, cited in Brugge, "Navajo Use and Occupation," 16.
10. Henry L. Mitchell et al., "Evaluation of Property Destroyed in Kane County, Utah," 24 December 1879, Record Group 75, Bureau of Indian Affairs, Consolidated Ute Agency Records, Federal Records Center, Denver.
11. Wade and Talbot to Stanley, 19 November 1883, quoted in Brugge, "Navajo Use and Occupation," 69, see also 17, 19; Park to McHintey, 22 August 1897, cited in Brugge, 69.
12. Brugge, "Navajo Use and Occupation," 117, 141, 142, 153, 159.
13. Garrick Bailey and Roberta Glenn Bailey, *A History of the Navajos—The Reservation Years* (Santa Fe: School of American Research Press, 1986), 41–42.
14. David E. Miller, *Hole-in-the-Rock: An Epic in the Colonization of the Great American West* (Salt Lake City: University of Utah Press, 1959), 53.
15. Francis A. Hammond to Editor, 8 January 1887, *Deseret News;* Francis A. Hammond to Editor, 23 November 1887, *Deseret News,* cited in "San Juan Stake History," LDS Church History Archives, Church of Jesus Christ of Latter-day Saints, Salt Lake City.
16. *Deseret News,* 28 September 1887, p. 9.
17. Charles S. Peterson, *Look to the Mountain: Southeastern Utah and the La Sal National Forest* (Provo, Utah: Brigham Young University Press, 1975), 93–94, 96, 100; Bryant L. Jensen, "An Historical Study of Bluff City, Utah, from 1878 to 1906" (master's thesis, Brigham Young University, 1966), 60–63; *Deseret News,* 28 September 1887, p. 9.
18. Francis A. Hammond, quoted in Daniel K. Muhlestein, "The Rise and Fall of the Cattle Companies in San Juan, 1880–1900," n. d., p. 43, Utah State Historical Society, Salt Lake City.
19. For an excellent overview of the cattle industry in this region, see Charles S. Peterson, "San Juan: A Hundred Years of Cattle, Sheep, and Dry Farms," in *San Juan County, Utah: People, Resources, and History,* ed., Allan Kent Powell (Salt Lake City: Utah State Historical Society, 1982), 171–203. For a historical and contemporary overview of cowboying in the deserts along the San Juan and Colorado, see Gary Topping, *Glen Canyon and the San Juan Country* (Moscow: University of Idaho Press, 1997), 145–62. For an understanding of Al Scorup's operations, see Neal Lambert, "Al Scorup: Cattleman of the Canyons," *Utah Historical Quarterly* 32, no. 3 (Summer 1964): 301–20.
20. The following analysis is based upon the work of Stanley W. Trimble and Alexandra C. Mendel, "The Cow as a Geomorphic Agent—A Critical Review," *Geomorphology* 13 (1995): 233–53.
21. Ibid., 243.
22. Robert D. Ohmart, Wayne O. Deason, and Constance Burke, "A Riparian Case History: The Colorado River," paper presented at symposium on Importance, Preservation, and Management of Riparian Habitat, Tucson, Arizona, 9 July 1977.
23. Senate, *The Western Range—Letter of Secretary of Agriculture,* 74th Cong., 2d sess., 24 April 1936, S. Doc. 199, 316–20.
24. Frank H. Hyde testimony, *United States v Utah* (1931), Colorado River Bed Case, September 1929, pp. 1208–09, Utah State Historical Society, Salt Lake City.
25. Ibid.
26. *Deseret News,* 17 August 1885, p. 15.
27. McPherson, *Northern Navajo Frontier,* 63–78.
28. D. M. Riordan to Commissioner of Indian Affairs, 31 December 1883, microfilm. Record Group 75, Letters Received by the Office of Indian Affairs 1881–1907, National Archives, Washington, D.C.
29. Ernest B. Hyde testimony, *United States v Utah* (1931), Colorado River Bed Case, October

1929, pp. 1691–92, Utah State Historical Society, Salt Lake City.

30. Ibid.
31. Ibid; Frank Hyde testimony.
32. Sand waves are created when the water on the surface of the river moves faster than that on the bottom. Sand waves on the San Juan in some cases can be ten feet high.
33. Martha Nez, interview by Robert McPherson, 10 August 1988, transcript in possession of author.
34. For a fuller account of Henry Mitchell, see McPherson, *Northern Navajo Frontier*, 39–50.
35. Frank McNitt, *The Indian Traders* (Albuquerque: University of New Mexico Press, 1962), 309; Fern D. Ellis, *Come Back to My Valley* (Cortez, Colo.: Cortez Printers, 1976), 188.
36. Walter Dyk, *A Navaho Autobiography* (New York: Viking Fund, 1947), 37.
37. Walter Dyk and Ruth Dyk, *Left Handed: A Navajo Autobiography* (New York: Columbia University Press, 1980), 370–71.
38. Dyk, *A Navaho Autobiography*, 87.
39. According to an interview on 11 June 1958 with Old Lady Sweetwater, "Ugly Trader" (unidentified) built a ferry with a cable at Aneth around 1923. One of the anchor cables is still visible at the post, but the system did not last long before it was washed down the river. Doris Duke Oral History 13. Special Collections, Marriott Library, University of Utah, Salt Lake City.
40. Ray Hunt, interview by Robert McPherson, 21 January 1991, transcript in possession of author; Dyk, *A Navaho Autobiography*, 141.
41. Anthony L. Klessert, "Inventory Report—BIA," 17 August 1989, Report no. BIA—NAO NTM 89-205, Navajo Archives, Navajo Nation, Window Rock, Arizona.
42. Kumen Jones, "Writings of Kumen Jones," p. 213, Special Collections, Harold B. Lee Library, Brigham Young University, Provo, Utah.
43. Platte D. Lyman, "Diary," pp. 292, 301, Special Collections, Harold B. Lee Library, Brigham Young University, Provo, Utah.
44. Albert R. Lyman, "History of San Juan County, 1879–1917," p. 36, Special Collections, Harold B. Lee Library, Brigham Young University, Provo, Utah.
45. Dyk, *A Navaho Autobiography*, 85.
46. Captain E. A. Sturgis to Adjutant, Fort Wingate, 12 August 1908, Record Group 75, Letters Received by the Office of Indian Affairs 1881–1907, National Archives, Washington, D.C.
47. Benally, interview; Mamie Howard, interview by Robert McPherson, 2 August 1988, tape in possession of author; Nez, interview, 10 August 1988.
48. Senate, *Navajo Indians in New Mexico and Arizona*, 52nd Cong., 1st sess., 4 August 1892, Ex. Doc. 156, 15.
49. For a more complete account of Antes, see Robert S. McPherson, "Howard Antes and the Navajo Faith Mission: Evangelist of Southeastern Utah," *Blue Mountain Shadows* 17 (Summer 1996): 14–24.
50. Antes to President Theodore Roosevelt, 18 April 1904, cited in Brugge, 107.
51. Antes to Secretary of the Interior, 14 November 1898, cited in Brugge, 73.
52. Ibid, 77.
53. Commissioner of Indian Affairs to Secretary of the Interior, 2 December 1898, cited in Brugge, 79.
54. Kate Perkins to Commissioner of Indian Affairs, 15 January 1900, cited in Brugge, 95.
55. Antes to Theodore Roosevelt, 18 April 1904, cited in Brugge, 107.
56. "Navaho Reservation, Utah—Cancellation of Lands Set Apart in Utah," Executive Order 324A, cited in Charles J. Kappler, *Indian Affairs—Laws and Treaties*, vol. 3 (Washington, D.C.: GPO, 1913), 690.
57. "Minutes of County Commission of San Juan County, Utah, from April 26, 1880 to March 5, 1900," p. 14, Recorder's Office, County Courthouse, Monticello, Utah.
58. Jones, "Writings of Kumen Jones," 213.
59. Lyman Hunter, "San Juan Remembered," p. 3, Special Collections, Marriott Library, University of Utah, Salt Lake City.
60. Warren K. Moorhead, "In Search of a Lost Race," *Illustrated American*, 16 July 1892, 411.
61. Francis F. Kane and Frank M. Riter, "A Further Report to the Indian Rights Association on the Proposed Removal of the Southern Utes," 20 January 1892, p. 13, Utah State Historical Society, Salt Lake City.
62. "Minutes of County Commission," 26-29.
63. John Meadows, interview by Robert McPherson, 30 May 1991, transcript in possession of author.
64. Ray Hunt, interview by Robert McPherson, 5 September 1991.
65. Richard E. Klinck, *Land of Room Enough and Time Enough* (Salt Lake City: Peregrine Smith Books, 1984), 47; Kappler, *Indian Affairs*, 407, 433, 491, 575.
66. "Funds for Indian Highway," *Mancos Times-Tribune*, 18 February 1916, p. 1; "Highway North of Shiprock to Be Completed to State Line This Summer," *Cortez Journal-Herald*, 27 February 1930, p. 1.

67. Ira Freeman, *A History of Montezuma County, Colorado* (Boulder, Colo.: Johnson Publishing, 1958), 209.
68. "Trade Relations," *Mancos Times-Tribune,* 25 April 1913, p. 7; "Mancos Best Trading Point," *Mancos Times-Tribune,* 9 July 1915, p. 1.
69. *Annual Report,* Bureau of Indian Affairs, 1930, Navajo Archives, Navajo Nation, Window Rock, Arizona.
70. For an excellent explanation of the ecological and economic impact of livestock reduction, see Richard White's *The Roots of Dependency: Subsistence, Environment, and Social Change among the Choctaws, Pawnees and Navajos.* (Lincoln: University of Nebraska Press, 1983), 212–323.
71. James Muhn and Hanson R. Stuart, *Opportunity and Challenge—The Story of BLM* (Washington, D.C.: Department of the Interior, Bureau of Land Management, 1988), 35–41.
72. *Annual Report,* Bureau of Indian Affairs, 1934, Navajo Archives, Navajo Nation, Window Rock, Arizona.
73. Ibid, 86.
74. Nedra Todich'ii'nii, interview by Fern Charley and Dean Sundberg, comps., 13 July 1972, *Navajo Stock Reduction Interviews,* 23; quotation from Ason Attakai, interview, cited in Ruth Roessel and Broderick H. Johnson, *Navajo Livestock Reduction: A National Disgrace* (Tsaile, Ariz.: Navajo Community College Press, 1974), 129.
75. Deneh Bitsilly, interview, *Navajo Stock Reduction Interviews,* 132; Ernest Nelson, interview, *Navajo Stock Reduction Interviews,* 159; Hite Chee, interview, *Navajo Stock Reduction Interviews,* 7–8; Betty Canyon, interview by Robert McPherson, 10 September 1991, transcript in possession of author; Todich'ii'nii, interview, 24; Guy Cly, interview by Robert McPherson, 7 August 1991, transcript in possession of author.
76. Chee, interview, 13.
77. *National Wildlife Federation, Southern Utah Wilderness Alliance, and Joseph Feller v Bureau of Land Management,* 20 December 1993, pp. 4, 33–36 (Bureau of Land Management Library, Monticello, Utah, on file).
78. The information concerning the current status of the north side of the San Juan River corridor was provided by Nick Sandberg and Paul Curtis, range managers, Bureau of Land Management, San Juan Resource Area, Monticello, Utah, 9 November 1999.
79. The information concerning the grazing status of Navajo-controlled lands on both the north and south sides of the San Juan River was provided by Steven Deeter, rangeland management specialist, Natural Resources Conservation Service, Monticello, Utah, 12 November 1999.

Chapter 5

1. Robert J. Foster, *General Geology* (Columbus, Ohio: Merrill Publishing, 1988), 124.
2. W. W. Hill, *The Agricultural and Hunting Methods of the Navaho Indians* (New Haven: Yale University Press, 1938), 13–14; R. Clayton Brough, Dale L. Jones, and Dale J. Stevens, *Utah's Comprehensive Weather Almanac* (Salt Lake City: Publisher's Press, 1987), 278.
3. Hill, 25.
4. "Bluff City Notes," *Grand Valley Times,* 23 September 1904, p. 1.
5. Walter E. Mendenhall testimony, *United States v Utah* (1931), Colorado River Bed Case, October 1929, p. 1603; John Wetherill testimony, *United States v Utah* (1931), Colorado River Bed Case, September 1929, p. 1331; Bert Loper testimony, *United States v Utah* (1931), Colorado River Bed Case, September 1929, p. 972, Utah State Historical Society, Salt Lake City.
6. Hill, 26–27, 30.
7. T. J. Morgan, "The Navajo Situation," in *Report of the Commissioner of Indian Affairs,* New Mexico Superintendency (Washington, D.C.: GPO, 1892), 125–26.
8. Odon Gurovitz, "Survey of the Navajo Reservation," 17 November 1892, cited in David M. Brugge, "Navajo Use and Occupation of Lands North of the San Juan River in Present-day Utah," unpublished manuscript, 1966, 47.
9. C. E. Vandever to Commissioner of Indian Affairs, 4 March1890, cited in Brugge, 44.
10. E. H. Plummer to Commissioner of Indian Affairs, 29 December 1893, cited in Brugge, 55.
11. Constant Williams to Commissioner of Indian Affairs, 11 December 1894, cited in Brugge, 64.
12. Platte D. Lyman, "Journal," pp. 34, 37, Special Collections, Harold B. Lee Library, Brigham Young University, Provo, Utah; quotation from Albert R. Lyman, "History of San Juan County, 1879–1917," pp. 23, 32, Special Collections, Harold B. Lee Library, Brigham Young University, Provo, Utah.
13. William Adams to T. J. Jones of Parowan, *Deseret Weekly,* 26 June 1882, p. 5.
14. "Water Index," in "San Juan Stake History," [1881], LDS Church History Archives, Church of Jesus Christ of Latter-day Saints, Salt Lake City.
15. Frank H. Hyde testimony, *United States v Utah* (1931), Colorado River Bed Case, September

1929, p. 1214, Utah State Historical Society, Salt Lake City.

16. "Water Records Index," in "San Juan Stake History."
17. Ibid.
18. Albert Lyman, 28.
19. John Morgan to the Editor, *Deseret Weekly* [1883], cited in Bryant L. Jensen, "An Historical Study of Bluff City, Utah, from 1878 to 1906" (master's thesis, Brigham Young University, 1966), 54.
20. Albert Lyman, 28, 31, 34.
21. Ibid., 42, 50, 58.
22. Constant Williams to Commissioner of Indian Affairs, 3 February 1895, cited in Brugge; Mary Eldridge to Commissioner of Indian Affairs, n. d., cited in Brugge.
23. W. A. Jones, "Irrigation," in *Report of the Commissioner of Indian Affairs* (Washington, D.C.: GPO,1897), 29–30.
24. "Annual Report of the Department of Indian Affairs," in *Report of the Commissioner of Indian Affairs* (Washington, D.C.: GPO,1901), 65–66.
25. George W. Hayzlett to Commissioner of Indian Affairs, 16 January 1903, cited in Brugge, 105.
26. George W. Hayzlett to Commissioner of Indian Affairs, 28 July 1903, cited in Brugge, 107; William T. Shelton to Commissioner of Indian Affairs, 28 December 1903, cited in Brugge, 105.
27. "Bluff City Notes," p. 1.
28. William T. Shelton to Commissioner of Indian Affairs, 30 April 1904, cited in Brugge, 110.
29. Walter Dyk, *A Navaho Autobiography* (New York: Viking Fund, 1947), 62.
30. William T. Shelton to Commissioner of Indian Affairs, 24 July 1905, cited in Brugge, 122; Harriet Peabody to Commissioner of Indian Affairs, 28 February 1905, cited in Brugge, 118; James M. Holley to William T. Shelton, 20 February 1905, cited in Brugge, 118.
31. Dyk, 86.
32. Ibid., 97–98.
33. William T. Shelton to Commissioner of Indian Affairs, 18 March 1907; William T. Shelton to Commissioner of Indian Affairs, 11 July 1906, J. Lee Correll Collection, Navajo Archives, Navajo Nation, Window Rock, Arizona.
34. William T. Shelton to Commissioner of Indian Affairs, 24 June 1912, J. Lee Correll Collection, Navajo Archives, Navajo Nation, Window Rock, Arizona.
35. Dyk, 101–3; William T. Shelton to James M. Holley, 2 October 1907, cited in Brugge.
36. William T. Shelton, "Report of Superintendent of San Juan School," in *Report of the Commissioner of Indian Affairs* (Washington, D.C.: GPO,1906), 280; Dyk, 127.
37. William T. Shelton, "Statements from Navajos," 1909, cited in Brugge, 142.
38. Dyk, 142.
39. Ibid., 145.
40. Ibid., 156–57.
41. William T. Shelton to Commissioner of Indian Affairs, 14 June 1912, cited in Brugge, 157.
42. Ibid.
43. Ray Hunt, interview by Robert McPherson, 21 January 1991; Robert Howell interview by Robert McPherson, 14 May 1991; Kay Howell, interview by Robert McPherson, 14 May 1991; Helen Redshaw, interview by Robert McPherson, 16 May 1991; all transcripts in possession of author.
44. Hunt, interview; Margaret Weston, interview by Robert McPherson, 13 February 1991, transcript in possession of author.
45. Hunt, interview; Jane Silas, interview by Robert McPherson, 27 January 1991, transcript in possession of author.
46. Harvey Oliver, interview by Baxter Benally and Robert McPherson, 7 May 1991, transcript in possession of author.
47. Ibid.
48. C. L. Christensen, "Tells of Strange Navajo Ceremonies," *The Times-Independent*, 9 February 1922, p. 3.
49. Evan W. Estep to Colonel Dorrington, 3 September 1922, Navajo Archives, Edge of the Cedars Museum, Blanding, Utah.
50. Herbert Redshaw to Evan W. Estep, 29 January 1921, cited in Brugge, 169.
51. Evan W. Estep to Commissioner of Indian Affairs, 9 February 1921, cited in Brugge, 170.
52. Evan W. Estep to Commissioner of Indian Affairs, 25 January 1923, Navajo Archives, Edge of the Cedars Museum, Blanding, Utah; Oliver, interview.
53. *Mancos Times*, 13 January 1899, p. 1.
54. "WIN Trainees Tap San Juan River," *San Juan Record*, 21 January 1971, p. 7; "UNDC, Navajos Build Biggest Irrigated Farm," *San Juan Record*, 3 October 1974, p. 10.
55. "Navajos Build Biggest Irrigate Farm," p. 10; Tully Lameman, previous director of the Utah Navajo Development Council's Division of Natural Resources, telephone conversation with Robert McPherson, 13 August 1996.
56. Lameman, telephone conversation.
57. "Overall Economic Development Plan 1992–1993," p. 18, Navajo Nation Division of Economic Development, Window Rock, Arizona.

58. Adrian N. Hansen, "The Endangered Species Act and Extinction of Reserved Indian Water Rights on the San Juan River," *Arizona Law Review* 37 (1995): 1305-44.

Chapter 6

1. Various accounts of the settlement of the San Juan include Cornelia Adams Perkins, Marian Gardner Nielson, and Lenora Butt Jones, *Saga of San Juan* (Monticello, Utah: San Juan Chapter, Daughters of Utah Pioneers, 1957); Faun McConkie Tanner, *The Far Country: A Regional History of Moab and La Sal* (Salt Lake City: Olympus Publishing, 1976); Charles S. Peterson, *Look to the Mountains—Southeastern Utah and the La Sal National Forest* (Provo: Brigham Young University Press, 1975); Lee Reay, *Incredible Passage through the Hole-in-the-Rock* (Salt Lake City: Publisher's Press, 1980); and C. Gregory Crampton, *Standing Up Country: The Canyon Lands of Utah and Arizona* (1969; reprint, Salt Lake City: Peregrine Smith Books, 1983). The best study of the trek to the San Juan is found in David E. Miller, *Hole-in-the-Rock: An Epic in the Colonization of the Great American West* (Salt Lake City: University of Utah Press, 1959).
2. Charles G. Oviatt, "Late Quaternary Geomorphic Changes along the San Juan River and Its Tributaries Near Bluff, Utah," in *Contributions to Quaternary Geology of the Colorado Plateau,* Special Studies 64, ed. G. E. Christenson et al. (Salt Lake City: Utah Geological and Mineral Survey, 1985), 33.
3. Erastus Snow to John Taylor, 6 September 1880, in "San Juan Stake History," LDS Church History Archives, Church of Jesus Christ of Latter-day Saints, Salt Lake City.
4. For a more detailed description of the Mormons' trip over the Hole-in-the-Rock Trail and settlement of the San Juan, see sources cited in Robert S. McPherson, *A History of San Juan County: In the Palm of Time,* Utah Centennial County History Series (Salt Lake City: Utah State Historical Society, 1995), 95–106.
5. Albert R. Lyman, "History of San Juan County, 1879–1917," p. 13, Special Collections, Harold B. Lee Library, Brigham Young University, Provo, Utah.
6. Andrew Jenson, no title, n. d., in "San Juan Stake History," p. 37.
7. Lyman "History," 15.
8. Ibid., 28.
9. Albert R. Lyman, "Journals," 25 March 1897, Special Collections, Harold B. Lee Library, Brigham Young University, Provo, Utah.
10. Lyman, "History," 50, 82, 95.
11. Platte D. Lyman, "Journal," p. 77, Special Collections, Harold B. Lee Library, Brigham Young University, Provo, Utah; Albert Lyman "History," 32.
12. John C. Kricher, "Needs of Weeds," *Natural History* 89, no. 12 (December 1980): 36–45.
13. Jens Nielsen to the Editor, *Deseret News,* 18 February 1885, p. 69.
14. Albert Lyman, "History," 64.
15. Ibid., 23; Platte D. Lyman, "Diary," pp. 279, 281, Special Collections, Harold B. Lee Library, Brigham Young University, Provo, Utah; for a description of pigweed and its effects, see Tom D. Whitson and Roy Riechenbeck, *Weeds and Poisonous Plants of Wyoming and Utah,* Cooperative Extension Series (Laramie: University of Wyoming, 1987), 32–33.
16. Albert Lyman, "History," 23.
17. Albert Lyman, "Journals," 14 December 1893.
18. Ernest B. Hyde testimony, *United States v Utah* (1931), Colorado River Bed Case, October 1929, pp. 1699–1700, Utah State Historical Society, Salt Lake City.
19. Otto J. Zahn testimony, *United States v Utah* (1931), Colorado River Bed Case, September 1929, p. 5058, Utah State Historical Society, Salt Lake City.
20. Francis A. Hammond to the Editor, *Deseret News,* 21 June 1885, p. 415.
21. "Utah Stream Water Quality," in *National Water Summary 1990–91,* U.S. Geological Survey Water Supply Paper 2400 (Washington, D.C.: GPO, 1991), 518.
22. See Ralf R. Woolley, *Cloudburst Floods in Utah, 1850–1938,* U.S. Geological Survey Paper 994 (Washington, D.C.: GPO, 1946),1–128; and Richard Hereford and R. H. Webb, "Historic Variation of Warm-Season Rainfall, Southern Colorado Plateau, Southwestern USA," in *Climate Change,* vol. 22 (Washington, D.C.: GPO, 1992), 235–56.
23. Richard Hereford, G. C. Jacoby, and V. A. S. McCord, *Geomorphic History of the Virgin River in Zion National Park Area, Southwestern Utah,* U.S. Geological Survey Open File Report 95-515 (Washington, D.C.: GPO, 1995), 1–75.
24. Richard White, *The Roots of Dependency: Subsistence, Environment, and Social Change among the Choctaws, Pawnees and Navajos* (Lincoln: University of Nebraska Press, 1983), 226–29.
25. Ibid.
26. "Utah Stream Water Quality," p. 521.
27. Kendall R. Thompson, *Characteristics of Suspended Sediment in the San Juan River Near Bluff, Utah,*

Water Resources Investigation Report no. 82-4104 (Salt Lake City: U.S. Geological Survey, 1982), 6.

28. Ibid., 5.
29. Gary C. Huber, "The Canyon of the San Juan River," in *Geology of the Canyons of the San Juan River,* ed. Donald L. Baars (Durango, Colo.: Four Corners Geological Society, 1974), 13.
30. "Utah Surface-Water Resources," in *National Water Summary 1985,* U.S. Geological Survey Water Supply Paper 2300 (Washington, D.C.: GPO, 1985), 456.
31. Rick Valdez, wildlife biologist, telephone conversation with Robert McPherson, 13 September 1996.
32. Kumen Jones testimony, *United States v Utah* (1931), Colorado River Bed Case, September 1929, p. 2333, Utah State Historical Society, Salt Lake City.
33. Platte Lyman, "Diary," 307.
34. Albert Lyman, "History," 38.
35. Samuel Rowley, "Autobiography," quoted in "San Juan Stake History," p. 6; "San Juan Utah," *San Juan Record,* 19 April 1951, p. 4.
36. Platte Lyman, "Journal," 73, 75–76; Albert Lyman, "History," 39–40.
37. For a more detailed analysis of the effects of Krakatau and El Nino on global weather patterns, see James M. Aton, "The River, the Ditch and the Volcano—Bluff, 1879–1884," *Blue Mountain Shadows* 12 (Summer 1993): 15–23.
38. Erastus Snow and Brigham Young to President Silas S. Smith, Platte D. Lyman, and the Saints, 5 September 1880. Platte D. Lyman Collection, Special Collections, Harold B. Lee Library, Brigham Young University, Provo, Utah.
39. Francis A. Hammond, "From Sunny San Juan," *Deseret News,* (Spring 1893), quoted in "San Juan Stake History," p. 539.
40. Albert R. Lyman, no title, 1896, cited in "San Juan Stake History."
41. Ibid.; Jens Nielsen testimony, *United States v Utah* (1931), Colorado River Bed Case, August 1929, p. 190, Utah State Historical Society, Salt Lake City.
42. Francis A. Hammond, "News from San Juan," *Deseret News,* 16 April 1897, p. 633.
43. Albert Lyman, "History," 95–96.
44. David M. Brugge, "Navajo Use and Occupation of Lands North of the San Juan River in Present-day Utah," unpublished manuscript, 97, 101.
45. Louisa Wetherill testimony, *United States v Utah* (1931), Colorado River Bed Case, October 1929, p. 1665, Utah State Historical Society, Salt Lake City.
46. "Bluff Items," *Montezuma Journal,* 14 October 1904, p. 4; "Locals and Personals," *Mancos Times-Tribune,* 11 March 1905, p. 4; no title, *Grand Valley Times,* 12 May 1905, p. 4.
47. William T. Shelton to Commissioner of Indian Affairs, 12 August 1905. J. Lee Correll Collection, Navajo Archives, Navajo Nation, Window Rock, Arizona.
48. Brugge, "Navajo Use and Occupation," 125, 128; no title, *Montezuma Journal,* 28 March 1907, p. 4.
49. William T. Shelton to Commissioner of Indian Affairs, 10 June 1907. J. Lee Correll Collection, Navajo Archives, Navajo Nation, Window Rock, Arizona.
50. "Bluff City," *Grand Valley Times,* 12 July 1907, p. 1; quotation from "Bluff City News," *Montezuma Journal,* 12 September 1907, p. 2.
51. "Bluff City Items," *Montezuma Journal,* 5 August 1909, p. 2.
52. Ibid., p. 6.
53. R. Clayton Brough, Dale L. Jones, and Dale J. Stevens, *Utah's Comprehensive Weather Almanac* (Salt Lake City: Publisher's Press, 1987), 290; Harold Muhlestein and Fay Muhlestein, *Monticello Journal—A History of Monticello until 1937* (Monticello, Utah: Authors, 1988), 105, 108.
54. William T. Shelton to Peter Pacquette, 9 October 1911; William T. Shelton to W. M. Peterson, 28 October 1911 J. Lee Correll Collection, Navajo Archives, Navajo Nation, Window Rock, Arizona; quotation from "Disastrous Flood," *Montezuma Journal,* 12 October,1911, p. 1.
55. Muhlestein, *Monticello Journal,* 105.
56. Zahn testimony, 2057.
57. Cord C. Bowen testimony, *United States v Utah* (1931), Colorado River Bed Case, September 1929, p. 205, Utah State Historical Society, Salt Lake City.
58. "Bluff Farmers Fight with S.C.S. Aid to Avert Disaster," *San Juan Record,* 7 August 1941, p. 1.
59. "Fight against Erosion at Bluff Progressing," *San Juan Record,* 8 July 1948, p. 1; "Bluff Streambank Project Approved," *San Juan Record,* 2 December 1948, p. 1.
60. Frank H. Hyde testimony, *United States v Utah* (1931), Colorado River Bed Case, September 1929, p. 1217, Utah State Historical Society, Salt Lake City.
61. A. L. Kroeger testimony, *United States v Utah* (1931), Colorado River Bed Case, September 1929, p. 1893, Utah State Historical Society, Salt Lake City.
62. Frank Hyde testimony, 1217.
63. Rose McKenney, Robert B. Jacobson, and Robert C. Wertheimer, "Woody Vegetation and

Channel Morphogenesis in Low-Gradient, Gravel-Bed Streams in the Ozark Plateaus, Missouri and Arkansas," in *Biogeomorphology, Terrestrial and Freshwater Aquatic Systems: Geomorphology,* vol. 13, ed. C. R. Hupp, W. R. Ostertramp, and A. D. Howard (Binghamton: State University of New York, 1995), 175–98.

64. Benjamin L. Everitt, Water Department, Division of Natural Resources, telephone conversation with Robert McPherson, 16 September 1996; D. E. Burkham, *Hydraulic Effects of Changes in Bottom-Land Vegetation on Three Major Floods, Gila River in Southeastern Arizona,* U.S. Geological Survey Professional Paper 655-J (Washington, D.C.: GPO, 1976), J-1–J-14.
65. Patty Fenner, Ward W. Brady, and David R. Patton, "Effects of Regulated Water Flows on Regeneration of Fremont Cottonwood," *Journal of Range Management* 38, no. 2 (March 1985): 135–38; Raymond M. Turner, *Quantitative and Historical Evidence of Vegetation Changes Along the Upper Gila River, Arizona,* U.S. Geological Survey Professional Paper 655-H (Washington, D.C.: GPO, 1974), H-1–H-19.
66. Jones testimony, 1168.
67. Frank Hyde testimony, 1213.
68. Ibid., 1213–14.
69. Christopher J. Hunter, *Better Trout Habitat—A Guide to Stream Restoration and Management* (Washington, D.C.: Island Press, 1991), 40–60.
70. Zahn testimony, 1385; John Wetherill testimony, *United States v Utah* (1931), Colorado River Bed Case, September 1929, p. 1327, Utah State Historical Society, Salt Lake City; Jones testimony, 168; quotation from Frank Hyde testimony, 1209.
71. Frank H. Karnell testimony, *United States v Utah* (1931), Colorado River Bed Case, November 1929, p. 1768, Utah State Historical Society, Salt Lake City.
72. Bert Loper testimony, *United States v Utah* (1931), Colorado River Bed Case, September 1929, p. 977, Utah State Historical Society, Salt Lake City.
73. "Governor Rampton Offers Aid to Flooded Bluff as Waters Recede," *San Juan Record,* 8 August 1968, p. 1.
74. Donna G. Anderson, "Trials and Triumphs: Bluff's Struggle with Water," *Blue Mountain Shadows* 2 (Fall 1988): 18–26.

Chapter 7

1. Bukkyo Dendo Kyokai, *The Teaching of Buddha* (Tokyo: Kosaido Printing Company, Ltd., 1966), 284–86.
2. For further information, see Robert S. McPherson, *The Northern Navajo Frontier, 1860–1900: Expansion through Adversity* (Albuquerque: University of New Mexico Press, 1988), 39–50.
3. C. Gregory Crampton, *Standing Up Country: The Canyon Lands of Utah and Arizona* (1969; reprint, Salt Lake City: Peregrine Smith Books, 1983), 123–28.
4. Herbert E. Gregory, *The San Juan Country—A Geographic and Geologic Reconnaissance of Southeastern Utah,* U.S. Geological Survey Professional Paper 188 (Washington, D.C.: GPO, 1938), 108.
5. "How to Rock the Placers," *Salt Lake Tribune,* 28 December 1892, p. 1.
6. C. Gregory Crampton, *Historical Sites in Glen Canyon Mouth of San Juan River to Lee's Ferry,* University of Utah Anthropological Papers, no. 46 (Salt Lake City: University of Utah Press, 1960), 74.
7. *Deseret News,* 13 December 1892, p. 5; "Up From San Juan," *Salt Lake Herald,* 29 December 1892, p. 6; "Many One-Ounce Nuggets,'" *Salt Lake Herald,* 31 December 1892, p. 1.
8. "No Place for Poor Men," *Salt Lake Herald,* 10 January 1893, p. 3.
9. "Senseless Stampede," *Salt Lake Herald,* 18 January 1893, p. 1; quotation from "San Juan Fake," *Salt Lake Tribune,* 27 January 1893, p. 7.
10. "Howard's Letter from Salina," *Salt Lake Tribune,* 30 January 1893, p. 3.
11. *Mancos Times,* 9 February 1894, p. 4; Bryant L. Jensen, "An Historical Study of Bluff City, Utah, from 1878 to 1906" (master's thesis, Brigham Young University, 1966), 77–78.
12. *Mancos Times,* 24 November 1893, p. 1.
13. *Mancos Times,* 24 November 1893, p. 4; Walter E. Mendenhall testimony, *United States v Utah* (1931), Colorado River Bed Case, October 1929, p. 1613, Utah Historical Society, Salt Lake City; "The San Juan Placers," *Mancos Times,* 3 August 1894, p. 4.
14. *Mancos Times,* 28 January 1898, p. 4; "Half Pint of Gold Nuggets," *Mancos Times-Tribune,* 21 October 1904, p. 1; "Report of Messrs Clay and Calhoun," *Mancos Times-Tribune,* 17 March 1906, p. 1.
15. "How Gold Nuggets Grow," *Montezuma Journal,* 28 August 1903, p. 2.
16. Mrs. E. J. (Billie) Yost to Otis Marston, 5 December 1955, unpublished letter. Yost was writing about the experiences of her father, William F. Williams, during the gold-rush era.

17. *Mancos Times,* 30 March 1894, p. 3; *Mancos Times,* 20 April 1894, p. 4.
18. *Mancos Times,* 6 July 1894, p. 1; *Mancos Times,* 17 August 1894, p. 3.
19. "San Juan Placer," *Mancos Times,* 21 September 1894, p. 1.
20. "San Juan Placer Mines," *Mancos Times-Tribune,* 7 February 1906, p. 1; "Rich Gold Placers on the San Juan River," *Denver Post,* 29 July 1907, p. 8; "Affairs Active in the Placer Region," *Mancos Times-Tribune,* 5 January 1912, p. 1; "Placer Mines Assured Success," *Mancos Times-Tribune,* 5 April 1912, p. 1.
21. Donald L. Baars, ed., *Geology of the Canyons of the San Juan River* (Durango, Colo.: Four Corners Geological Society, 1974), 79.
22. Bert Loper testimony, *United States v Utah* (1931), Colorado River Bed Case, September 1929, pp. 969–70, Utah State Historical Society, Salt Lake City.
23. Albert R. Lyman, "History of San Juan County, 1879–1917," Special Collections, Harold B. Lee Library, Brigham Young University, Provo, Utah; A. L. Raplee testimony, *United States v Utah* (1931), Colorado River Bed Case, September 1929, p. 196, Utah State Historical Society, Salt Lake City.
24. Otto J. Zahn testimony, *United States v Utah* (1931), Colorado River Bed Case, September 1929, pp. 867–68, Utah State Historical Society, Salt Lake City.
25. Frank H. Karnell testimony, *United States v Utah* (1931), Colorado River Bed Case, November 1929, pp. 1764–67, Utah State Historical Society, Salt Lake City.
26. Crampton, *Standing Up Country,* 131–41.
27. Charles Goodman, "History of the Oil Fields in San Juan County, Utah," *Salt Lake Mining Review,* 15 April 1910, pp. 17–18; for information concerning some of the personalities associated with the gold rush and oil industry in the Mexican Hat area, see Doris Valle, *Looking Back around the Hat—A History of Mexican Hat* (Mexican Hat, Utah: Author, 1986).
28. Herbert E. Gregory, "The San Juan Oil Field, San Juan County, Utah," in *Contributions to Economic Geology, 1909,* part 2, Bulletin no. 431 (Washington, D. C.: GPO, 1911), 11, 21–23.
29. "Mexican Hat Now a Village," *Grand Valley Times,* 2 December 1910, p. 1; Valle, *Looking Back,* 10.
30. "Oil Boom Continues," *Grand Valley Times,* 22 May 1908, p. 1.
31. "Monticello Happenings," *Grand Valley Times,* 8 January 1909, p. 5; "Happenings in San Juan," *Grand Valley Times,* 24 June 1910, p. 1; "Send Men to San Juan," *Grand Valley Times,* 22 July 1910, p. 1; "The New Bluff Gusher," *Grand Valley Times,* 1 July 1910, p. 1.
32. Frank H. Hyde testimony, *United States v Utah* (1931), Colorado River Bed Case, September 1929, p. 1211, Utah State Historical Society, Salt Lake City.
33. Gregory, *San Juan Country,* 111–13; "Government Report on Oil Resources of San Juan," *Grand Valley Times,* 10 May 1912, p. 1.
34. Curtis Palmer, oil field worker-driller for fifteen years near Mexican Hat, conversation with Robert McPherson, 25 July 1997.
35. Pamela A. Bunte and Robert J. Franklin, *From the Sands to the Mountain: Change and Persistence in a Southern Paiute Community* (Lincoln: University of Nebraska Press, 1987), 163–64.
36. House, *Amending the Act of March 1, 1933 . . . as an Addition to the Navajo Indian Reservation,* 90th Cong., 2d sess., 1968, S. Report 1324, 1–6.
37. Don Preston, ed., *A Symposium of the Oil and Gas Fields in Utah* (Salt Lake City: Intermountain Association of Petroleum Geologists, 1961), 8; Michael Rounds, "Indian Sovereignty Issue Concerns Operators," *Western Oil Reporter,* June 1978, p. 19; and Philip Reno, *Mother Earth, Father Sky, and Economic Development: Navajo Resources and Their Use* (Albuquerque: University of New Mexico Press, 1981), 127.
38. Robert W. Bernick, "Tribe Opens Bids Today on San Juan Oil Leases," *Salt Lake Tribune,* 1 November 1956, sec. C, p. 9; Robert W. Bernick, "Indian Land Buyers Term $27 Million 'Right Price,'" *Salt Lake Tribune,* 3 November 1956, sec C, p. 33.
39. Peter Iverson, *The Navajo Nation* (Albuquerque: University of New Mexico Press, 1981), 68.
40. Notes from the 1991 legislative audit of the Utah Navajo Development Council and the Utah Division of Indian Affairs, summary in possession of Robert S. McPherson.
41. Office of Program Development, *Navajo Nation Overall Economic Development Program* (Window Rock, Ariz.: Navajo Nation, 1980), 31; Garrick Bailey and Roberta Glenn Bailey, *A History of the Navajos—The Reservation Years* (Santa Fe: School of American Research Press, 1986), 237.
42. Reno, *Mother Earth, Father Sky,* 125, 127; Steve Marks, "Navajos Seize, Occupy Utah's Huge Aneth Field," *Western Oil Reporter,* April 1978, p. 23.
43. Frank Cole, *Well Spacing in the Aneth Reservoir* (Norman: University of Oklahoma Press, 1962),

vii; Harvey C. Moore, "Culture Change in a Navaho Community," in *American Historical Anthropology: Essays in Honor of Leslie Spier,* ed. Carroll L. Riley and Walter W. Taylor (Carbondale: Southern Illinois University Press, 1967), 127, 129–30; Clyde Benally, translator and spokesman during the Aneth conflict, telephone conversation with Robert McPherson, 30 April 1992.

44. Carlton Stowe, *Utah's Oil and Gas Industry: Past, Present, and Future* (Salt Lake City: Utah Engineering Experiment Station, University of Utah, 1979), 11, 51.
45. Program Development, *Navajo Economic Development Program,* 31.
46. "Bridge Dedication Opens New Way across San Juan," *San Juan Record,* 12 December 1958, p. 1.
47. John Norton, interview by Baxter Benally and Robert McPherson, 16 January 1991, transcript in possession of author.
48. Ibid.
49. John Knot Begay, interview by Baxter Benally and Robert McPherson, 7 May 1991, transcript in possession of author.
50. L. E. Spangler, D. L. Naftz, and Z. E. Peterman, *Hydrology, Chemical Quality, and Characterization of Salinity in the Navajo Aquifer in and Near the Greater Aneth Oil Field, San Juan County, Utah,* U.S. Geological Survey Report 96-4155 (Salt Lake City: U.S. Geological Survey, 1996), 86–87; "Aneth Technical Committee Fact Sheet," in *Using Geochemical Data to Identify Sources of Salinity to the Freshwater Navajo Aquifer in Southeastern Utah* (Salt Lake City: U.S. Geological Survey, 1995), 1–4.
51. Harvey Oliver, interview by Baxter Benally and Robert McPherson, 6 March 1991, transcript in possession of author; Jane Silas, interview by Baxter Benally and Robert McPherson, 27 February 1991, transcript in possession of author; Ben Whitehorse, interview by Baxter Benally and Robert McPherson, 30 January 1991, transcript in possession of author; Jerry Begay, interview by Baxter Benally and Robert McPherson, 16 January 1991, transcript in possession of author.
52. "Coalition for Navajo Liberation: Spreading Out from Shiprock," *Navajo Times,* 2 February 1978, sec. A, p. 2.
53. Ella Sakizzie, interview by Baxter Benally and Robert McPherson, 14 May 1991, transcript in possession of author.
54. "Inside Look at Aneth Occupation," *Navajo Times,* 13 April 1978, sec. A, p. 1; Benally, interview.
55. Jerry Begay, interview.
56. Oliver, interview; Margaret Weston, interview by Robert McPherson, 13 February 1991, transcript in possession of author; Sakizzie, interview.
57. "Agreement Reached with Texaco, U.S. Oil Companies," *Navajo Times,* 20 March 1997, p. 5.
58. "U.S. Accuses Texaco of Polluting Utah River," *Deseret News,* 28 March 1998; "EPA Files Lawsuit, Accuses Mobil of Polluting San Juan River," *Deseret News,* 31 March 1998.
59. Cyrus Begay, interview by Baxter Benally and Robert McPherson, 14 May 1991, transcript in possession of author.

Chapter 8

1. This provocative thesis has been worked out by Donald Worster in his *Rivers of Empire: Water, Aridity, and the Growth of the American West* (New York: Pantheon Books, 1985); *An Unsettled Country: Changing Landscapes of the American West* (Albuquerque: University of New Mexico Press, 1994); and in numerous articles. Also see Marc Reisner's *Cadillac Desert: The American West and Its Disappearing Water* (New York: Viking, 1986). The term *iron triangle* is explained in Daniel McCool, "Politics, Water, and Utah," in *Waters of Zion: The Politics of Water in Utah,* ed. Daniel C. McCool (Salt Lake City: University of Utah Press, 1995), 11.
2. See Donald J. Pisani's *To Reclaim a Divided West: Water, Law, and Public Policy, 1848–1902* (Albuquerque: University of New Mexico Press, 1992) for the best treatment of this thesis.
3. See Mark W. T. Harvey, *Symbol of Wilderness: Echo Park and the American Conservation Movement* (Albuquerque: University of New Mexico Press, 1994), 17; Samuel P. Hays, *Conservation and the Gospel of Efficiency: The Progressive Conservation Movement, 1890–1920* (Cambridge: Harvard University Press, 1959), 239–40; and Robert Follansbee, *Years of Increasing Cooperation, July l, 1919 to June 30, 1978,* vol. 2 of *A History of the Water Resources Branch of the United States Geological Survey,* U.S. Geological Survey Archives, Denver.
4. The definitive treatment of the Colorado River Compact is Norris Hundley, Jr., *Water and the West: The Colorado River Compact and the Politics of Water in the American West* (Berkeley: University of California Press, 1975).
5. E. C. LaRue, *Colorado River and Its Utilization,* U.S. Geological Survey Water Supply Paper 395 (Washington, D.C.: GPO, 1916); *Water Power and Flood Control of Colorado River below Green*

River, Utah, U.S. Geological Survey Water Supply Paper 556 (Washington, D.C.: GPO, 1925).

6. Russell Martin, *A Story That Stands Like a Dam: Glen Canyon and the Struggle for the Soul of the West* (New York: Henry Holt, 1989), 29.
7. The definitive study of Hoover Dam is Joseph E. Stevens's *Hoover Dam: An American Adventure* (Norman: University of Oklahoma Press, 1988).
8. "Utah's River Beds Battle Opens in S. L.," *Salt Lake Tribune,* 10 October 1929, pp. 1, 3. All testimony from the Colorado River Bed Case is housed at the Utah State Historical Society in Salt Lake City on microfilm.
9. See C. Gregory Crampton and Steven K. Madsen, "Boating on the Upper Colorado: A History of the Navigational Use of the Green, Colorado and San Juan Rivers and Their Major Tributaries," 1975, Utah State Historical Society, Salt Lake City.
10. Department of the Interior, Bureau of Reclamation, *The Colorado River: "A Natural Menace Becomes a National Resource," A Comprehensive Report on the Development of the Water Resources of the Colorado River Basin for Irrigation, Power Production, and Other Beneficial Uses in Arizona, California, Colorado, Nevada, New Mexico, Utah, and Wyoming* (Washington, D. C.: GPO, 1946), 25.
11. Ibid.
12. Ibid., 19.
13. Reisner, *Cadillac Desert,* 172.
14. Gary B. Weatherford and Phillip Nichols, *Legal-Political History of Water Resource Development in the Upper Colorado River Basin,* Lake Powell Research Bulletin, no. 4, part I (Los Angeles: Lake Powell Research Project, 1974), 8.
15. The best book so far that deals with CRSP is Harvey's *Symbol of Wilderness.* Also see Richard Allan Baker's *Conservation Politics: The Senate Career of Clinton P. Anderson* (Albuquerque: University of New Mexico Press, 1985), and Martin's *A Story That Stands Like a Dam.*
16. See Mark W. T. Harvey, "Echo Park, Glen Canyon, and the Postwar Wilderness Movement," *Pacific Historical Review* 60 (February 1991): 43–67 for a cogent analysis of the "great trade-off" myth.
17. Senate Subcommittee on Irrigation and Reclamation, *Colorado River Storage Project: Hearings on S.1555,* 83rd Cong., 2d sess., 1954, 662–69; House Subcommittee on Irrigation and Reclamation, *Colorado River Storage Project: Hearings on H. R. 270, H. R. 2836, H. R. 3383, H. R. 3384, and H. R. 4488,* 84th Cong., lst sess., 1955, 1141–42.
18. Senate Subcommittee on Irrigation and Reclamation, *Colorado River Storage Project: Hearings on S.500,* 84th Cong., lst sess., 1955, 679.
19. House Subcommittee on Irrigation and Reclamation, *Colorado River Storage Project: Hearings on H. R. 4449, H. R. 4443, and H. R. 4463,* 83rd Cong., 2d sess., 1954, 877; and House Subcommittee, *Colorado River Storage Project,* 1955, 1098.
20. For a complete summary of the entire Rainbow Bridge controversy, see Mark W. T. Harvey, "Defending the Park System: The Controversy over Rainbow Bridge," *New Mexico Historical Review* 73 (January 1998): 45–67. See also Jared Farmer, *Glen Canyon Damned: Inventing Lake Powell and the Canyon Country* (Tucson: University of Arizona Press, 1999), 137–71.
21. Chris Smith and Elizabeth Manning, "The Sacred and Profane Collide in the West," *High Country News,* 26 May 1997, pp. 1, 8–12; Farmer, *Glen Canyon Damned,* 168–91.
22. Baker, *Conservation Politics,* 67.
23. House Subcommittee, *Colorado River Storage Project,* 1954, 254.
24. Ira G. Clark, *Water in New Mexico: A History of Its Management and Use,* (Albuquerque: University of New Mexico Press, 1987), 507.
25. Dean E. Mann, "The Politics of Water Resource Development in the Upper Colorado River Basin," in *Legal-Political History of Water Resource Development in the Upper Colorado River Basin,* 36.
26. Mike Sims, senior production engineer, telephone conversation with James Aton, 22 October 1998, Farmington, New Mexico. The Farmington power facility at Navajo Dam generates up to fifteen megawatts of power. It is licensed under the Federal Energy Regulatory Commission (FERC).
27. See Jim Aton, "Lake Bluff?" *Canyon Echo* 4 (June 1997): 1, 14.
28. LaRue, *Colorado River and Its Utilization,* 213–14.
29. "Impounded Will Cover Bluff City," *Montezuma Journal,* 22 October 1914, p. 1.
30. Bureau of Reclamation, *The Colorado River,* 106, 145, 147, 216, 247.
31. Department of the Interior, Bureau of Reclamation, *Colorado River Storage Project and Participating Projects* (Salt Lake City: Author, 1949); *Colorado River Storage Project and Participating Projects* (Salt Lake City: Author, 1950).
32. A bureau report included in the House subcommittee hearings *(Colorado River Storage Project,* 1954, 105) indicated that Navajo Dam

would make the Bluff Dam less necessary; the high cost of building the dam also made it "infeasible under present procedures for project analysis."

33. Harvey, *Symbol of Wilderness,* 133–37; House Subcommittee, *Colorado River Storage Project,* 1954, 486; U. S. Grant, "The Dinosaur Dam Sites Are Not Needed," *Living Wilderness* 34 (Autumn 1950): 20.
34. Department of the Interior, Bureau of Reclamation, *San Juan Investigation: Utah and Colorado* (Washington, D.C.: GPO, 1969), 111–17.
35. See "A Review of Animas–La Plata: The West's Last Big Water Project," *High Country News Reports* (1996) for a summary of the project's history.
36. Angus Woodbury et al., eds., *Ecological Studies of the Flora and Fauna in Glen Canyon,* University of Utah Anthropological Papers, no. 40 (Salt Lake City: University of Utah Press, 1959), 102.
37. For a summary of the Lake Powell Research Project, see Lorin D. Potter and Charles L. Drake, *Lake Powell: Virgin Flow to Dynamo* (Albuquerque: University of New Mexico Press, 1989). Chapter IV compares the pre- and post-dam biota of the reservoir area.
38. Ibid, 149, 163–65, 224–25, 230.
39. For a history of tamarisk, see Jerome S. Horton, *Notes on the Introduction of Deciduous Tamarisk,* U.S. Forest Service Research Note RM-16 (Fort Collins, Colo.: Rocky Mountain Forest and Range Experiment Station, 1964); T. W. Robinson, *Introduction Spread and Areal Extent of Salt Cedar (Tamarix) in the Western United States,* U.S. Geological Survey Professional Paper 491-A (Washington, D.C.: GPO, 1965); and William L. Graf, "Fluvial Adjustments to the Spread of Tamarisk in the Colorado Plateau Region," *Geological Society of America Bulletin* 89 (October 1978): 1491–1501.
40. Angus Woodbury, Stephen D. Durrant, and Seville Flowers, *Survey of Vegetation in the Glen Canyon Reservoir Basin,* University of Utah Anthropological Papers, no. 36 (Salt Lake City: University of Utah Press, 1959), 22.
41. James L. Patterson and William P. Somers, "Part 9: Colorado Basin," in *Magnitude and Frequency of Floods in the United States,* U.S. Geological Survey Water Supply Paper 1683 (Washington, D.C.: GPO, 1966), 348–49.
42. Hugh D. Miser, *The San Juan Canyon, Southeastern Utah: A Geographic and Hydrographic Reconnaissance,* U.S. Geological Survey Water Supply Paper 538 (Washington, D.C.: GPO, 1924), 43–46; U.S. Geological Survey website, Washington, D.C., 15 July 1997, available at <h2o.usgs.gov/swr/>
43. Larry Stevens, "Scourge of the West: The Natural History of Tamarisk in the Grand Canyon," *The News: The Journal of Grand Canyon River Guides* [now *Boatman's Quarterly Review]* 6 (Summer 1993): 14–15; Potter and Drake, *Lake Powell,* 163.
44. Potter and Drake, *Lake Powell,* 165–66.
45. Steven W. Carothers and Bryan T. Brown, *The Colorado River through Grand Canyon: Natural History and Human Change,* vol. 2 (Tucson: University of Arizona Press, 1991), 149–50.
46. M. J. Kasprzyk and G. L. Bryant, *Results of Biological Investigations from the Lower Virgin River Vegetation Management Study* (Boulder City, Nev.: Department of the Interior, Bureau of Reclamation, 1989).
47. Stevens, "Scourge of the West," 15.
48. Carothers and Brown, *The Colorado River through Grand Canyon,* 188.
49. Potter and Drake, *Lake Powell,* 172.
50. See a series of three articles on tamarisk by Paul Larmer, "Tackling Tamarisk," "Fighting Exotics with Exotics," and "Killing Tamarisks Frees Water," *High Country News,* 25 May 1998. Sue Bellagamba, biologist and conservator, Matheson Slough, telephone conversation with James Aton, February 1996, Moab, Utah, said the Nature Conservancy has used garlon to clear small patches of tamarisk in the Matheson Slough. See also T. E. A. van Hylckama, *Weather and Evapotranspiration Studies in a Salt Cedar Thicket, Arizona* (Washington, D.C.: GPO, 1974); Josie Glausiusz, "Trees of Salt," *Discover* 17 (March 1996): 30–32; and Jim Woolf, "Experts May Sic Foreign Bugs on Pesky Trees," *Salt Lake Tribune,* 30 November 1996, sec A, pp. 1, 6.
51. For nearly daily coverage of the cop-killer saga, see *Salt Lake Tribune,* 6 June 1998 to 23 August 1998. McVean's apparent suicide was reported in Greg Burton, "Pile of Bones Believed to Be Cop Killer," *Salt Lake Tribune,* 2 November 1999, sec B, p. 1. Popular mystery writer Tony Hillerman referred frequently to these events in his 1999 novel, *Hunting Badger* (New York: HarperCollins).
52. Dave May, "Maligned or Malignant; Tamarisk: The Plant We Love to Hate," *Canyon Legacy* (Winter 1989): 3–7.
53. Eben Rose, "In Kinship with Tamarisk," *The News: The Journal of Grand Canyon River Guides* 6 (Winter 1992–93): 20–29.
54. Samuel P. Hays, *Beauty, Health and Permanence: Environmental Politics in the United States,*

1955–1985 (New York: Cambridge University Press, 1987), 3–13.

55. Besides the agencies listed, the U.S. Fish and Wildlife Service, Bureau of Reclamation, and Utah Department of Wildlife Resources play important roles in managing the San Juan River.
56. The Bureau of Land Management was created in 1946 by combining the U.S. Grazing Service, established in 1934 by the Taylor Grazing Act to administer federal grazing lands in the West, and the General Land Office. Grazing and mining were dominant BLM interests until the environmental laws of the 1960s and '70s forced it into adapting a multiple-use philosophy that includes recreational, ecological, and cultural uses. For an in-house history of the BLM, see James Muhn and Hanson R. Stuart, *Opportunity and Challenge: The Story of BLM* (Washington, D.C.: Department of the Interior, Bureau of Land Management, 1988).
57. The Utah Wilderness Coalition, a group of wilderness advocacy groups, supports H. R. 1500 (sponsored by New York Representative Maurice Hinkley), which originally proposed to protect 5.7 million acres of BLM and NPS land. In 1998 the coalition announced the results of a ten-year "Citizens' Inventory" of Utah's wildlands, pushing that total to 8.5 million acres. Utah's congressional delegation has put forth H. R. 1745, which would protect 1.8 million acres. See Utah Wilderness Coalition, *Wilderness at the Edge* (Salt Lake City: Author, 1990) for details of that organization's 5.7-million-acre proposal for BLM wilderness contained in H. R. 1500. See Southern Utah Wilderness Alliance's website, Salt Lake City, available at <www.suwa.org> for a delineation of the 8.5- million-acre proposal. For the BLM's 1.8-million-acre proposal, see Department of the Interior, Bureau of Land Management, *Utah Statewide Wilderness Study Report* (Salt Lake City: Author, 1991).
58. See Bureau of Land Management, *Utah State Wilderness Study Report.*
59. See two books by Tim Palmer: *Endangered Rivers and the Conservation Movement* (Berkeley: University of California Press, 1989) and *Wild and Scenic Rivers of America* (Washington, D.C.: Island Press, 1993.)
60. Department of the Interior, Bureau of Land Management, *Proposed Resource Management Plan and Final Environmental Impact Statement for the San Juan Resource Area, Moab District,* vol. 1 (Washington, D.C.: GPO, 1987), 280–83.
61. Phil Gezon, outdoor recreation planner, Bureau of Land Management, San Juan Resource Area, interview by James Aton, 18 January 1996, Monticello, Utah.
62. See Nancy Nelson, *Any Time, Any Place, Any River: The Nevills of Mexican Hat* (Flagstaff, Ariz.: Red Lake Books, 1991); Roy Webb, "'Never Was Anything So Heavenly': Nevills' Expeditions on the San Juan River," *Blue Mountain Shadows* 12 (Summer 1993): 35–50; and P. T. Reilly, "Norman Nevills: Whitewater Man of the West" *Utah Historical Quarterly* 55 (Spring 1987): 181–200.
63. Gezon, interview.
64. See Department of the Interior, Bureau of Land Management, "Supplement No. 6 to Memorandum of Understanding between National Park Service, Utah State Office and Bureau of Land Management, Utah State Office," 1 January 1979, Bureau of Land Management Library, Monticello, Utah; Jerry Ballard, "San Juan River Management Plan—Staff Report," 18 August 1979, Bureau of Land Management Library, Monticello, Utah.
65. Debra Jane Sholly, "A Study of Campsite Impacts on the San Juan River, Utah," (master's thesis, University of Montana, 1991), 75.
66. Linda Richmond, river ranger, Bureau of Land Management, San Juan Resource Area, interview by James Aton, 17 January 1996, Bluff, Utah; Gezon, interview.
67. Bob Woyewodzic, wildlife biologist, Bureau of Land Management, San Juan Resource Area, interview by James Aton, 17 January 1996, Monticello, Utah; Clive Pinnoch, wildlife specialist, Glen Canyon National Recreation Area, interview by James Aton, 16 January 1996, Page, Arizona.
68. Information on the history of pikeminnows and razorbacks comes from R. J. Behnke and D. E. Benson, *Endangered and Threatened Fishes of the Upper Colorado River Basin,* Cooperative Extension Service Bulletin 503A (Fort Collins, Colo.: Colorado State University, 1983); Department of the Interior, Fish and Wildlife Service, "Endangered And Threatened Wildlife and Plants: Determination of Critical Habitat for Four Colorado River Endangered Fishes: Final Rule," *Federal Register* 59, no. 54 (21 March 1994): 13374–400; Department of the Interior, Fish and Wildlife Service, "Memorandum: Biological Opinion on the Effects of Water Well Depletion," 8 December 1994, Fish and Wildlife Service Office, Denver; and Fred Quartarone, *Historical Accounts of Upper Colorado River Basin Endangered Fish* (Denver: Department of the Interior, Fish and Wildlife Service, 1995). These

fish have also been called white fish, white salmon, Colorado River salmon, Colorado white salmon, landlocked salmon, and silver salmon.

69. Department of the Interior, Fish and Wildlife Service, *Colorado River Squawfish Recovery Plan* (Denver: Author, 1991), 15–16; Behnke and Benson, *Endangered and Threatened Fishes,* 19–21; Quartarone, *Historical Accounts of Endangered Fish,* 34–48.

70. Department of the Interior, Fish and Wildlife Service, "Memorandum: Formal Section 7 Consultation, Biological Opinion for the Proposed Permitting of Oil and Gas Activities," 20 July 1993, Fish and Wildlife Service, Albuquerque, New Mexico (Bureau of Land Management Office, Monticello, Utah, photocopy); Fish and Wildlife Service, "Memorandum: Biological Opinion on the Effects," 5.

71. Fish and Wildlife Service, *Colorado River Squawfish Recovery Plan,* 32–34.

72. Ibid. Fish and Wildlife Service, "Memorandum: Biological Opinion for the Proposed Permitting"; Department of the Interior, Fish and Wildlife Service, "Staff Report on the December 13, 1994 Meeting between the U. S. Fish and Wildlife Service and Bureau of Land Management to Discuss Oil and Gas, Water Quality and Threatened and Endangered Fish Habitats of the San Juan River," 16 December 1994 (Bureau of Land Management Office, Monticello, Utah, photocopy); Bob Woyewodizic, telephone conversation with James Aton, 26 January 1996.

73. Todd Wilkinson, "Homecoming," *National Parks,* May–June 1996, 40–45; Jim Woolf, "Rare Birds of a Feather to Flock to New Home Near Utah Border," *Salt Lake Tribune,* 29 October 1996, sec. A, pp. 1, 11.

74. Lanny O. Wilson, *Distribution and Ecology of the Desert Bighorn Sheep in Southeastern Utah,* Publication no. 68-5 (Salt Lake City: Utah State Division of Fish and Game, 1968); Charles A. Irvine, *The Desert Bighorn Sheep of Southeastern Utah,* Publication no. 69-12 (Salt Lake City: Utah State Division of Fish and Game, 1969); William Wishart, "Bighorn Sheep," in *Big Game of North America,* ed. John L. Schmidt and Douglas L. Olbert (Harrisburg, Pa.: Stockpole Books, 1978), 161–71; Kathleen McCoy, wildlife biologist, Navajo Tribe, telephone conversation with James Aton, 1 February 1996; and Richard Manville, "The Origin and Relationships of American Wild Sheep," in *The Desert Bighorn: Its Life History, Ecology and Management,* ed. Gale Monson and Lowell Summer (Tucson: University of Arizona Press, 1980), 106.

75. Bert Loper, "This Is to Be an Attempt to Describe My Second San Juan Voyage," p. 7, Marston Collection, Huntington Library, San Marino, California.

76. McCoy, telephone conversation.

77. Woyewodzic, interview, 17 January 1996; McCoy, telephone conversation; Jim Karpowitz, regional wildlife manager, Utah Department of Wildlife Resources, telephone conversation with James Aton, February 1996; Pinnoch, interview.

78. See Department of the Interior, Bureau of Reclamation, *Operation of Glen Canyon Dam: Final Environmental Impact Statement* (Salt Lake City: Author, 1995). Individual scientific reports, which number nearly one hundred, are available from either Glen Canyon Environmental Studies, P. O. Box 22459, Flagstaff, AZ 86002-2459 or the Colorado River Studies Office, Bureau of Reclamation, 125 South State, Room 6107, Salt Lake City, UT 84138-1102. Also see George Sibley, "Glen Canyon: Using a Dam to Heal a River," *High Country News,* 22 July 1996, pp. 1, 8–12. Finally, for a projection of dam management, see Michael Collier, Robert H. Webb, and John C. Schmidt, *Dams and Rivers: A Primer on the Downstream Effects of Dams,* U.S. Geological Survey Circular 1126 (Washington, D.C.: GPO, 1996).

Chapter 9

1. William Kittridge, *Owning It All* (St. Paul: Graywolf Press, 1987), 62.

2. See Barre Toelken, "Folklore in the American West," in *A Literary History of the American West,* ed. J. Golden Taylor et al. (Fort Worth: Texas Christian University Press, 1987), 29–67; Richard White, *"It's Your Misfortune and None of My Own": A History of the American West* (Norman: University of Oklahoma Press, 1991), 613–32; and Clyde A. Milner, II, "The Shared Memory of Montana's Pioneers," *Montana* 37 (Winter 1987): 2–13.

3. Lyman's many published, as well as unpublished, writings include *Indians and Outlaws: Settling the San Juan Frontier* (1962; reprint, Salt Lake City: Publisher's Press, 1980); "Fort on the Firing Line," serialized in *Improvement Era* in 1948 and 1949; "History of San Juan County, 1879–1917," Utah State Historical Society, Salt Lake City; Special Collections, Harold B. Lee Library, Brigham Young University, Provo, Utah; and Bureau of Land Management Library, Monticello, Utah; "Journals" in Special Collections, Harold B. Lee Library, Brigham

Young University, Provo, Utah; *The Outlaw of Navajo Mountain* (Salt Lake City: Deseret Books, 1963); *The Trail of the Ancients* (Blanding, Utah: The Trail of the Ancients Association, 1972) and many other works of history, personal memoir, fiction, and biography.

4. Charles S. Peterson, *Look to the Mountain: Southeastern Utah and the La Sal National Forest* (Provo, Utah: Brigham Young University Press, 1975), 53.
5. Pioneer writers like Lyman adhered to collective values and remembered events of community history. Personal memories were sometimes altered to conform to group recollections. The writings of Lyman's lesser-known contemporary, Kumen Jones, are an example. See Milner, "The Shared Memory," 4. Also see Kumen Jones, "The Writings of Kumen Jones" and "San Juan Mission to the Indian," Joel E. Ricks Collection, Utah State Historical Society, Salt Lake City. To the contemporary reader, Jones comes across as a little less culturally biased than Lyman.
6. Lyman, *Indians and Outlaws,* 53, 73, 78; "Fort on the Firing Line," (November 1948): 689.
7. Charles S. Peterson, "'A Utah Moon': Perceptions of Southern Utah" (the Juanita Brooks lecture at Dixie College, St. George, Utah, 1984), 3.
8. Cornelia Adams Perkins, Marian Gardner Nielson, and Lenora Butt Jones, *Saga of the San Juan,* (Monticello, Utah: San Juan Chapter, Daughters of Utah Pioneers, 1957); Andrew Jenson, "History of San Juan Stake," Special Collections, Harold B. Lee Library, Brigham Young University, Provo, Utah; and Norma Perkins Young, *Anchored Lariats on the San Juan Frontier* (Provo, Utah: Community Press, 1985).
9. Jenson, "History of San Juan Stake," 90; Perkins, Nielson, and Jones, *Saga of the San Juan,* 37.
10. See, for example, the following college introductory geology textbooks: Harold L. Levin, *Contemporary Physical Geology,* 2d ed. (Philadelphia: Saunders College Publishing, 1986), 325; James S. Monroe and Reed Wicander, *Physical Geology: Exploring the Earth* (St. Paul: West Publishing, 1992), 478; and Carla W. Montgomery, *Physical Geology,* 2d ed. (Dubuque: Wm. C. Brown, 1988), 288.
11. Robert S. McPherson, *A History of San Juan County: In The Palm of Time,* Utah Centennial County History Series (Salt Lake City: Utah State Historical Society, 1995), 350.
12. Peter B. Hale, *William Henry Jackson and the Transformation of the American Landscape* (Philadelphia: Temple University Press, 1988), 124.
13. Ibid., 82.
14. For information on Hayden's life and work, see Mike Foster, *Strange Genius: The Life of Ferdinand Vandeveer Hayden* (Niwot, Colo.: Roberts Rinehart Publishers, 1994).
15. See Mark Klett et al., *Second View: The Rephotographic Survey Project* (Albuquerque: University of New Mexico Press, 1984), 29.
16. Hale, *William Henry Jackson,* 75. The standard discussion of luminism is Barbara Novak's *Nature and Culture: American Landscape Painting, 1825–1875* (New York: Oxford University Press, 1980), 28–29, 40, 192. For a fuller discussion of the evolution of the concept of the sublime from masculine violent cataclysm to gentle feminine nature, see Angela Miller, *The Empire of the Eye: Landscape Representation and American Cultural Politics, 1825–1875* (Ithaca, N. Y.: Cornell University Press, 1993), 249–52.
17. Alfred Runte, *National Parks: The American Experience,* 3rd ed. (Lincoln: University of Nebraska Press, 1997), 198.
18. See Nancy Nelson, *Any Time, Any Place, Any River: The Nevills of Mexican Hat* (Flagstaff, Ariz.: Red Lake Books, 1991); Roy Webb, "Never Was Anything So Heavenly: Nevills Expeditions on the San Juan River," *Blue Mountain Shadows* 12 (Summer 1993): 35–50; and P. T. Reilly, "Norman Nevills: Whitewater Man of the West," *Utah Historical Quarterly* 55 (Spring 1987): 181–200.
19. Earl Pomeroy, *In Search of the Golden West: The Tourist in Western America* (New York: Alfred A. Knopf, 1957), 225; Hal K. Rothman, *Devil's Bargains: Tourism in the Twentieth-Century American West* (Lawrence: University of Kansas Press, 1998), 27–32.
20. See Wild Rivers promotional brochures and videos from 1957 to the present. Yet even as early as 1948, Nevills sent his customers a reading list of over seventy books and articles, arranged by categories like "The San Juan and Navajo Country," "Indians, Ancient and Modern," and so on. See the Nevills Collection, Special Collections, Marriott Library, University of Utah, Salt Lake City.
21. Tony Hillerman, *A Thief of Time* (New York: Harper and Row, 1988), 85.
22. In addition to the three works cited below, see also Neil M. Clark, "Fast Water Man," *Saturday Evening Post,* 18 May 1946, 28–30; Mildred E. Baker, "Rough Water," *American Forests* 50 (1 November 1944): 520–29; Randall Henderson, "Floating down the San Juan and the Colorado," *Sierra Club Bulletin* 30 (December 1945): 63–70; and Weldon F. Heald, "The Canyon

Wilderness," in *The Inverted Mountains: Canyons of the West,* ed. Roderick Peattie (New York: Vanguard Press, 1948), 209–52.

23. Alfred M. Bailey, "Desert River through Navajoland," *National Geographic* 62 (August 1947): 149–72.
24. Wallace Stegner, "San Juan and Glen Canyon," in *The Sound of Mountain Water: The Changing American West* (New York: E. P. Dutton, 1980), 102–20.
25. Stegner's biographer discusses Turner's influence on the "dean of western writers," and Stegner acknowledged it in a 1986 interview. See Jackson J. Benson, *Wallace Stegner: His Life and Work* (New York: Penguin Books, 1996), 63–64.
26. Stegner, "Coda: Wilderness Letter," in *The Sound of Mountain Water,* 145–53.
27. Edward Abbey, "Thus I Reply to Rene Dubos," in *Down the River* (New York: E. P. Dutton, 1982), 119–20.
28. Abbey, "Running the San Juan," in *Down the River,* 127,135.
29. Paul W. Rea, "Blissed and Blasted on the San Juan," in *Canyon Interludes: Between White Water and Red Rock* (Salt Lake City: Signature Books, 1996), 233–54.
30. Abbey, "Down the River with Henry Thoreau," in *Down the River,* 13–48; Ann Zwinger and Edwin Way Teale, *A Conscious Stillness:Two Naturalists on Thoreau's Rivers* (New York: Harper and Row, 1982).
31. Ann Zwinger, *The Wind in the Rock: The Canyonlands of Southeastern Utah* (New York: Harper and Row, 1978), 127.
32. Most of the story takes place around the San Juan between No Man's Mesa, south of the river, and the Mike's Mesa-Nokai Dome area north of the river. The so-called Haunted Mesa is between Mike's Mesa and Nokai Dome. Louis L'Amour, *Haunted Mesa* (New York: Bantam, 1987). The book contains a detailed map of the area at the front.
33. Ibid., 264–65. Lest the reader think that L'Amour, through his depiction of the model farming village of the Third World, is advocating a world of peaceful nonviolence, consider that all conflict is resolved through fighting with weapons. Moreover, L'Amour makes a kind of NRA pitch against gun control on page 259, where he states the need for an armed populace to fend off enemy paratroopers who might drop down into a relatively uninhabited region like San Juan country. This rhetoric sounds frighteningly like certain paramilitary, right-wing, antigovernment groups currently operating on the fringes of American life.
34. Hillerman, *A Thief of Time,* 185.
35. Richard Benke, "Vast Landscape, Sun Inspire Novelist," *The Spectrum,* 20 September 1998, sec. C, p. 9. A Mancos, Colorado, company called Time Travelers Maps offers a "Tony Hillerman Indian Country Map and Guide."
36. Information on Goodman comes from three articles in the centennial issue of *Blue Mountain Shadows* 17, (Summer 1996): Drew Ross, "'I Have Struck It Rich at Last,'" 30–37; LaVerne Tate, "The Real Gold Mine of San Juan," 27–29; and "Historical Moments: Preserved by Charles Goodman's Camera," 38–45. Ross's article has been reprinted as "'I Have Struck It Rich at Last": Charles Goodman, Traveling Photographer," *Utah Historical Quarterly* 66 (Winter 1998): 65–83.
37. Peterson, "A Utah Moon," 10.
38. Thomas R. Vale and Geraldine R. Vale, *Western Landscapes: Travels along U.S. 89* (Tucson: University of Arizona Press, 1989), 8.
39. Ellen Meloy, "Long Body, Sinuous Gifts, *Orion* 15 (Summer 1996): 12.
40. Ellen Meloy, "Geese, River Ghost, the Holy Ghost—A Field Report," *Canyon Echo* 2 (June 1995): 10, 18. Meloy has also written a personal memoir of her attempts to come to terms with the uranium industry and nuclear testing, which took place in and around the Colorado Plateau, and link those concerns with her own journey to find and build a home on the banks of the San Juan in *The Last Cheater's Waltz: Beauty and Violence in the Desert Southwest* (New York: Henry Holt, 1999). The book contains some fine descriptions of the river area as well as an account of her experiment to fence out cattle from her property.
41. Ann Weiler Walka, "Preface," in *Waterlines: Journeys on a Desert River* (Flagstaff, Ariz.: Red Lake Books, 1993).
42. Ibid.
43. See *Blue Mountain Shadows* 12 (Summer 1993).
44. See *Canyon Echo* 2 (June 1995) and 4 (June 1997).
45. For a discussion of the effects of overphotography, see Susan Sontag, *On Photography* (1977; reprint, New York: Doubleday, 1989), 19, 109–10, 180; and Daniel Dancer, "Over-Glossied and Imaged Out: Toward a Deep Photographic Ethic," *Wild Earth* (Spring 1996): 81–87.

Epilogue

1. Laurie Goering, "U.S. Signs Global-Warming Treaty," *Salt Lake Tribune,* 13 November 1998, sec. A, pp. 1, 5. Although the Clinton administration

signed this treaty, it faces stiff opposition in the U.S. Senate. Moreover, China and India, two of the largest greenhouse-gas contributors, have refused to participate in these U. N.–sponsored negotiations.

2. Michael Collier, Robert H. Webb, and John C. Schmidt, *Dams and Rivers: A Primer on the Downstream Effects of Dams,* U.S. Geological Survey Circular 1126 (Washington, D.C.: GPO, 1996). For a popular account of this issue, see George Sibley, "Glen Canyon: Using a Dam to Heal a River," *High Country News,* 22 July 1996, pp. 1, 8–12.
3. Philip Shabecoff, "U.S. Bureau for Water Projects Shifts Focus to Conservation," *The New York Times,* 20 October 1987, p. 11. See the Bureau of Reclamation's upper Colorado region website, available at <www.uc.usbr.gov>
4. Dave Wegner, "Restoring Glen Canyon: Linking Our Future to the Importance of the Past," *Hidden Passage: The Journal of the Glen Canyon Institute* 1 (1998): 5–7; Dave Wegner, "Sediment, Water, and Erosion: The First Technical Study Focuses on Sediment," *Hidden Passage* 2 (1999): 13.
5. Tom Wharton, "A Dam Shame? Effort to Drain Powell May Be a Fight against the Currents," *Salt Lake Tribune,* 3 August 1997, sec. A, pp. 1, 13. Jared Farmer believes that mourning Glen Canyon is wasted emotion and environmentalists should be more concerned about the health of Lake Powell. He thinks that the dam will be here for some time. See Farmer, *Glen Canyon Damned: Inventing Lake Powell and the Canyon Country* (Tucson: University of Arizona Press, 1999), 192–201. Marc Reisner, author of *Cadillac Desert* (New York: Viking, 1986) and a critic of the bureau and dams, believes many western dams like Flaming Gorge could and should be dismantled but not Glen Canyon. He feels downstream dams depend too much on it for river control and local economies are too heavily invested in Lake Powell recreation. Marc Reisner, "The Cadillac Desert" (lecture presented at Southern Utah University's Convocation Series, Cedar City, Utah, 19 November 1999).
6. Ellen Meloy, interview by James Aton, 3 November 1996, Bluff, Utah. Nature writer and Bluff resident Meloy believes in the studies and would love for them to succeed in restoring native fish. But she also feels too many factors are working against them.
7. "Endangered Condors Give Flight to Curiosity, Soar More Than 250 Miles to Grand Mesa," *Salt Lake Tribune,* 28 August 1998, sec. A, p. 11.
8. See two articles by Thomas Michael Power: "The Economic Importance of Federal Grazing to the Economies of the West," *Southern Utah Wilderness Alliance Newsletter,* Spring 1995, insert; "The Economics of Wilderness Preservation in Utah," *Southern Utah Wilderness Newsletter,* Winter 1995, insert; and his book, *Lost Landscapes and Failed Economies: The Search for the Value of Place* (Washington, D.C.: Island Press, 1996), 177–86.
9. Linda Richmond, river ranger, Bureau of Land Management, San Juan Resource Area, telephone conversation with James Aton, 30 October 1998.
10. Land planners have been busy in Bluff since the mid-1990s. See Richard E. Toth et al., *Bluff, Utah: A Time for Change/A Time for Choice* (Logan: Utah State University, 1994); Craig Johnson, *Bluff Floodplain Nature Park,* part 2 of *Bluff, Utah: Study in Rural Community Planning* (Logan: Utah State University Press, 1995); *The Bluff Nature Preserve: An Innovative Oasis Where Community and Nature Meet,* (Bluff, Utah: Bluff Historical Preservation Association, 1996); and Four Corners Planning, Inc., *Community of Bluff: General Plan* (Bluff, Utah: Author, 1996). As of spring 1998, the nature preserve is moving ahead, although it faces a court challenge from the Navajo Tribe south of the river. The general plan was adopted by the country commissioners but awaits final passage of a zoning ordinance.
11. See Jim Woolf, "Wasatch Front Residents Know What They Want: More Wilderness in Utah," *Salt Lake Tribune,* 2 February 1995, sec. A, pp. 1, 13; Valley Research, Inc., "2 News Statewide Poll for Coalition for Utah's Future," 18 May 1995 (press release, mimeographed); and Brent Israelsen, "Utahns Want More Wilderness Survey Shows," *Salt Lake Tribune,* 7 July 1998, sec. B, p. 1.
12. Frank Waters, *Masked Gods: Navaho and Pueblo Ceremonialism* (1950; reprint New York: Ballantine Books, 1970), 5.
13. Larry Rogers, *Chapter Images: 1996: Profiles of 110 Navajo Nation Chapters* (Window Rock, Ariz.: Navajo Nation, 1997), 174–75.
14. Department of the Interior, Bureau of Land Management, *Grand Gulch Plateau Cultural and Recreational Area Management Plan* (Monticello, Utah: Author, 1993), 27.

Bibliography

Manuscripts

Amoss, Harold Lindsay, Jr. "Ute Mountain Utes." Ph.D. diss., University of California, 1951.

Billings, Alfred N. "Account Book and Diary of Alfred N. Billings." Utah State Historical Society, Salt Lake City.

Blake, H. Elwyn. "Boating the Wild Rivers." Marston Collection, Huntington Library, San Marino, California.

Brugge, David M. "Navajo Use and Occupation of Lands North of the San Juan River in Present-day Utah." Unpublished manuscript. In possession of Robert McPherson.

Crampton, Gregory, and Steven K. Madsen. "Boating on the Upper Colorado: A History of the Navigational Use of the Green, Colorado and San Juan Rivers and Their Major Tributaries" (1975). Utah State Historical Society, Salt Lake City.

Cummings, Byron R. "Field Notes of 1909." Arizona State Museum, University of Arizona, Tucson.

Eastwood, Alice. "Memoirs." Alice Eastwood Archives, California Academy of Sciences, San Francisco.

Estep, Evan W. to Colonel Dorrington, 3 September 1922. Navajo Archives, Edge of the Cedars Museum, Blanding, Utah.

———. To Commissioner of Indian Affairs, 25 January 1923. Navajo Archives, Edge of the Cedars Museum, Blanding, Utah.

Hewitt, E. L. "Field Notes 1906–09." Edge of the Cedars Museum, Blanding, Utah. Unpublished typescript. Photocopy on file.

Hunter, Lyman. "San Juan Remembered." Special Collections, Marriott Library, University of Utah, Salt Lake City.

Jensen, Bryant L. "An Historical Study of Bluff City, Utah, from 1878 to 1906." Master's thesis, Brigham Young University, 1966.

Jenson, Andrew. "History of San Juan Stake." Special Collections, Harold B. Lee Library, Brigham Young University, Provo, Utah.

Jones, Kumen. "San Juan Mission to the Indian." Joel E. Ricks Collection, Utah State Historical Society, Salt Lake City.

———. "The Writings of Kumen Jones." Special Collections, Harold B. Lee Library, Brigham Young University, Provo, Utah; and Joel E. Ricks Collection, Utah State Historical Society, Salt Lake City.

Kane, Francis F., and Frank M. Riter. "A Further Report to the Indian Rights Association on the Proposed Removal of the Southern Utes," 20 January 1892. Utah State Historical Society, Salt Lake City.

Kearns, Timothy M. "Aceramic Sites and the Archaic Occupation along the Middle San Juan River, Southeast Utah." Paper presented at the 54th Meeting of the Society for American Archaeology, Las Vegas, Nevada, 1990. Part of session "Sampling, Time, and Population along the San Juan River."

Kelly, Charles. "Aneth." Charles Kelly Papers, Special Collections, Marriott Library, University of Utah, Salt Lake City.

Leiby, Austin Nelson. "Borderland Pathfinders: The Diaries of Juan Maria Antonio de Rivera." Ph.D. diss., Northern Arizona University, 1985.

Loper, Bert. "This Is to Be an Attempt to Describe My Second San Juan Voyage." Marston Collection, Huntington Library, San Marino, California.

———. "U.S.G.S. Survey of the San Juan River, 1921." Marston Collection, Huntington Library, San Marino, California.

Lyman, Albert R. "History of San Juan County, 1879–1917." Special Collections, Harold B. Lee Library, Brigham Young University, Provo, Utah; Utah State Historical Society, Salt Lake City; and Bureau of Land Management Library, Monticello, Utah.

———. Journals (1897). Special Collections, Harold B. Lee Library, Brigham Young University, Provo, Utah.

Lyman, Platte D. Diary. Special Collections, Harold B. Lee Library, Brigham Young University, Provo, Utah.

———. Journal. Special Collections, Harold B. Lee Library, Brigham Young University, Provo, Utah.

"Minutes of County Commission of San Juan County, Utah, from April 26, 1880 to March 5, 1900." Recorder's Office, County Courthouse, Monticello, Utah.

Miser, Hugh D. to Heber Christensen, 16 March 1925. Marston Collection, Huntington Library, San Marino, California.

Muhlestein, Daniel K. "The Rise and Fall of the Cattle Companies in San Juan, 1880–1900," n. d. Utah State Historical Society, Salt Lake City.

National Wildlife Federation, Southern Utah Wilderness Alliance and Joseph Feller v. Bureau of Land Management, 20 December 1993. Bureau of Land Management Library, Monticello, Utah, on file.

Nevills Collection. Special Collections, Marriott Library, University of Utah, Salt Lake City.

Notes from the 1991 legislative audit of the Utah Navajo Development Council and the Utah Division of Indian Affairs, summary in possession of Robert McPherson.

Ohmart, Robert D., Wayne O. Deason, and Constance Burke. "A Riparian Case History: The Colorado River." Paper presented at symposium on Importance, Preservation, and Management of Riparian Habitat, Tucson, Arizona, 9 July 1977.

"Overall Economic Development Plan 1992–1993." Navajo Nation Division of Economic Development, Window Rock, Arizona.

Pettit, Ethan. "Diary of Ethan Pettit, 1855–1881." Utah State Historical Society, Salt Lake City.

Reisner, Marc. "The Cadillac Desert." Lecture presented at Southern Utah University's Convocation Series, Cedar City, Utah, 19 November 1999.

Riordan, D. M. to Commissioner of Indian Affairs, 31 December 1883. Record Group 75, Letters Received by the Office of Indian Affairs 1881–1907, National Archives, Washington, D.C.

Rowley, Samuel. "Autobiography." In "San Juan Stake History." LDS Church History Archives, Church of Jesus Christ of Latter-day Saints, Salt Lake City.

"San Juan Stake History." LDS Church History Archives, Church of Jesus Christ of Latter-day Saints, Salt Lake City.

Schroedl, Alan R. "Archaic of the Northern Colorado Plateau." Ph.D. diss., University of Utah, 1976.

Shelton, William T. to Commissioner of Indian Affairs, 12 August 1905. J. Lee Correll Collection, Navajo Archives, Navajo Nation, Window Rock, Arizona.

———. To Commissioner of Indian Affairs, 11 July 1906. J. Lee Correll Collection, Navajo Archives, Navajo Nation, Window Rock, Arizona.

———. To Commissioner of Indian Affairs, 18 March 1907. J. Lee Correll Collection, Navajo Archives, Navajo Nation, Window Rock, Arizona.

———. To Commissioner of Indian Affairs, 10 June 1907. J. Lee Correll Collection, Navajo Archives, Navajo Nation, Window Rock, Arizona.

———. To Commissioner of Indian Affairs, 24 June 1912. J. Lee Correll Collection, Navajo Archives, Navajo Nation, Window Rock, Arizona.

Sholly, Debra Jane. "A Study of Campsite Impacts on the San Juan River, Utah." Master's thesis, University of Montana, 1991.

Silvey, Frank. "Information on Indians" (1936). Utah State Historical Society, Salt Lake City.

Snow, Erastus, and Brigham Young to President Silas S. Smith, Platte D. Lyman, and the Saints, 5 September 1880. Special Collections, Harold B. Lee Library, Brigham Young University, Provo, Utah.

United States v. Utah (1931), Colorado River Bed Case. Utah State Historical Society, Salt Lake City.
Bowen, Cord C., testimony, September 1929.
Hyde, Ernest B., testimony, October 1929.
Hyde, Frank H., testimony, September 1929.
Jones, Kumen, testimony, September 1929.
Karnell, Frank H., testimony, November 1929.
Kroeger, A. L., testimony, September 1929.
Loper, Bert, testimony, September 1929.
Mendenhall, Walter E., testimony, October 1929.
Miser, Hugh D., testimony, November 1929.
Nielsen, Jens, testimony, August 1929.
Raplee, A. L., testimony, September 1929.
Wetherill, John, testimony, September 1929.
Wetherill, Louisa, testimony, October 1929.
Zahn, Otto J., testimony, September 1929.

"The Ute Indians of Southwestern Colorado" (1941). Compiled by Helen Sloan Daniels. Unpublished manuscript.

"U-262 Report" (1992). Abajo Archeology, Bluff, Utah.

Valley Research, Inc. "2 News Statewide Poll for Coalition for Utah's Future," 18 May 1995. Press release. Mimeographed.

Yost, Mrs. E. J. (Billie), to Otis Marston, 5 December 1955. Unpublished letter. In possession of Robert McPherson.

Interviews

Begay, Cyrus. Interview by Baxter Benally and Robert McPherson, 14 May 1991.

Begay, Florence. Interview by Robert McPherson, 30 January 1991.

Begay, Jerry. Interview by Baxter Benally and Robert McPherson, 16 January 1991.
Begay, John Knot. Interview with Baxter Benally and Robert McPherson, 7 May 1991.
Bellagamba, Sue. Telephone conversation with James Aton, February 1996, Moab, Utah.
Benally, Clyde. Telephone conversation with Robert McPherson, 30 April 1992.
Benally, Slim. Interview by Robert McPherson, 8 July 1988.
Bitsili, Mary. Interview by Aubrey Williams and Maxwell Yazzie, 19 January 1961. Doris Duke 710, Special Collections, Marriott Library, University of Utah, Salt Lake City.
Black, Ada. Interview by Robert McPherson, 11 October 1991.
Blueeyes, Charlie. Interviews by Robert McPherson, 7 June 1988, 8 July 1988, and 28 August 1988.
Bowns, Dr. James E. Interview by James Aton, Spring 1996, Cedar City, Utah.
Cantsee, Chester, Sr. Interview by Aldean Ketchum and Robert McPherson, 6 September 1994.
Canyon, Betty. Interview by Robert McPherson, 10 September 1991.
Curtis, Paul. Personal communication with Robert McPherson, 9 November 1999.
Cly, Guy. Interview by Robert McPherson, 7 August 1991.
Deeter, Steven. Personal communication with Robert McPherson, 12 November 1999.
Dougi, Tom. Interview by Robert McPherson, 8 April 1992.
Dutchie, Edward, Sr. Interview by Robert McPherson, 7 May 1996.
Everitt, Benjamin L. Telephone conversation with Robert McPherson, 16 September 1996.
Eyetoo, Stella. Interview by Aldean Ketchum and Robert McPherson, 21 December 1994.
Gezon, Phil. Interview by James Aton, 18 January 1996, Monticello, Utah.
Holiday, Tallis. Interview by Robert McPherson, 3 November 1987.
Howard, Mamie. Interview by Robert McPherson, 2 August 1988.
Howell, Kay. Interview by Robert McPherson, 14 May 1991.
Howell, Robert. Interview by Robert McPherson, 14 May 1991.
Hunt, Ray. Interviews by Robert McPherson, 21 January and 5 September 1991.
Hurst, Winston B. Interview by James Aton, 6 June 1994, Blanding, Utah.
Jay, Mary. Interview by Robert McPherson, 22 February 1991.
Karpowitz, Jim. Telephone conversation with James Aton, February 1996.
Knight, Terry. Interview by Mary Jane Yazzie and Robert McPherson, 19 December 1994.
Lameman, Tully. Telephone conversation with Robert McPherson, 13 August 1996.
Lee, Isabelle. Interview by Robert McPherson, 13 February 1991.
McCoy, Kathleen. Telephone conversation with James Aton, 1 February 1996.
Meadows, John. Interview by Robert McPherson, 30 May 1991.
Meloy, Ellen. Interview by James Aton, 3 November 1996.
Mike, Billy. Interview by Aldean Ketchum and Robert McPherson, 13 October 1993.
Monument Valley and Navajo Mountain residents, interview by Jim Dandy, n.d.
Nez, Martha. Interviews by Robert McPherson, 2 August and 10 August 1988.
Norton, John. Interview by Baxter Benally and Robert McPherson, 16 January 1991.
Oliver, Harvey. Interviews by Baxter Benally and Robert McPherson, 6 March and 7 May 1991.
Palmer, Curtis. Conversation with Robert McPherson, 25 July 1997.
Pinnoch, Clive. Interview by James Aton, 16 January 1996, Page, Arizona.
Redshaw, Helen. Interview by Robert McPherson, 16 May 1991.
Richmond, Linda. Interviews by James Aton, 17 January 1996, 9 July 1997, and 30 October 1998, Bluff, Utah.
Sakizzie, Ella. Interview by Baxter Benally and Robert McPherson, 14 May 1991.
Sandberg, Nick. Personal communication with Robert McPherson, 9 November 1999.
Seltzer, Pat. Telephone conversation with Robert McPherson, 12 June 1996.
Silas, Jane. Interview by Robert McPherson, 27 January 1991.
Silas, Jane. Interview by Baxter Benally and Robert McPherson, 27 February 1991.
Sims, Mike. Telephone conversation with James Aton, 22 October 1998, Farmington, New Mexico.
Smiley, Billy. Interview by Robert McPherson, 14 January 1991.
Stevenson, Gene. Interviews by James Aton, 16 May 1996 and 29 October 1998.
Sweetwater, Old Lady. Interview,11 June 1958. Doris Duke Oral History 13, Special Collections, Marriott Library, University of Utah, Salt Lake City.
Valdez, Rick. Telephone conversation with Robert McPherson, 13 September 1996.

Weston, Margaret. Interview by Robert McPherson, 13 February 1991.

Whitehorse, Ben. Interview by Baxter Benally and Robert McPherson, 30 January 1991.

Woyewodzic, Bob. Interview by James Aton, 17 January 1996; telephone conversation with James Aton, 26 January 1996, Monticello, Utah.

Yazzie, Fred. Interview by Robert McPherson, 5 November 1987.

Government Documents
Published

Annual Report. Bureau of Indian Affairs, 1930. Navajo Archives, Navajo Nation, Window Rock, Arizona.

Annual Report. Bureau of Indian Affairs, 1934. Navajo Archives, Navajo Nation, Window Rock, Arizona.

Atkins, Virginia M., ed. *Anasazi Basketmaker: Papers from the 1990 Wetherill-Grand Gulch Symposium.* BLM Cultural Resource Series, no. 24. Salt Lake City: Department of the Interior, Bureau of Land Management, 1993.

Burkham, D. E. *Hydraulic Effects of Changes in Bottom-Land Vegetation on Three Major Floods, Gila River in Southeastern Arizona.* U.S. Geological Survey Professional Paper 655-J. Washington, D.C.: GPO, 1976.

Christenson, G. E., et. al., eds. *Contributions to Quaternary Geology of the Colorado Plateau.* Special Studies 64. Salt Lake City: Utah Geological and Mineral Survey, 1985.

Collier, Michael, Robert H. Webb, and John C. Schmidt. *Dams and Rivers: A Primer on the Downstream Effects of Dams.* U.S. Geological Survey Circular 1126. Washington, D.C.: GPO, 1996.

Follansbee, Robert. *Years of Increasing Cooperation, July 1, 1919 to June 30, 1978.* Vol. 2 of *A History of the Water Resources Branch of the United States Geological Survey.* U.S. Geological Survey Archives, Denver.

Gregory, Herbert E. *Geology of the Navajo Country: A Reconnaissance of Parts of Arizona, New Mexico, and Utah.* U.S. Geological Survey Professional Paper 93. Washington, D.C.: GPO, 1917.

———. *The Navajo Country: A Geographic and Hydrographic Reconnaissance of Parts of Arizona, New Mexico, and Utah.* U.S. Geological Survey Water Supply Paper 380. Washington, D.C.: GPO, 1916.

———. *The San Juan Country: A Geographic and Geologic Reconnaissance of Southeastern Utah.* U.S. Geological Survey Professional Paper 188. Washington, D.C.: GPO, 1938.

———. "The San Juan Oil Field, San Juan County, Utah." In *Contributions to Economic Geology, 1909.* Part 2, Bulletin no. 431. Washington, D.C.: GPO, 1911.

Hayden, F. V. *Ninth Annual Report of the United States Geological and Geographical Survey for the Year 1875.* Washington, D.C.: GPO, 1877.

———. *Tenth Annual Report of the United States Biological and Geological Survey for the Year 1875.* Washington, D.C.: GPO, 1878.

Hereford, Richard, G. C. Jacoby, and V. A. S. McCord. *Geomorphic History of the Virgin River in Zion National Park Area, Southwestern Utah.* U.S. Geological Survey Open File Report 95-515. Washington, D.C.: GPO, 1995.

Hereford, Richard, and R. H. Webb. "Historic Variation of Warm-Season Rainfall, Southern Colorado Plateau, Southwestern USA." In *Climate Change.* Vol. 22. Washington, D.C.: GPO, 1992.

Horton, Jerome S. *Notes on the Introduction of Deciduous Tamarisk.* U.S. Forest Service Research Note RM-16. Fort Collins, Colo.: Rocky Mountain Forest and Range Experiment Station, 1964.

Irvine, Charles A. *The Desert Bighorn Sheep of Southeastern Utah.* Publication no. 69-12. Salt Lake City: Utah State Division of Fish and Game, 1969.

Kappler, Charles J. *Indian Affairs—Laws and Treaties.* Vol. 3. Washington, D.C.: GPO, 1913.

Kasprzyk, M. J., and G. L. Bryant. *Results of Biological Investigations from the Lower Virgin River Vegetation Management Study.* Boulder City, Nev.: Department of the Interior, Bureau of Reclamation, 1989.

LaRue, E. C. *Colorado River and its Utilization.* U.S. Geological Survey Water Supply Paper 395. Washington, D.C.: GPO, 1916.

———. *Water Power and Flood Control of Colorado River below Green River, Utah.* U.S. Geological Water Supply Paper 556. Washington, D.C.: GPO, 1925.

Lipe, William D., et al., eds. *Dolores Archaeological Program: Anasazi Communities at Dolores: Grass Mesa Village.* Denver: Department of the Interior, Bureau of Land Management, 1988.

Macomb, J. N. *Report of the Exploring Expedition from Santa Fe, New Mexico, to the Junction of the Grand and Green Rivers of the Great Colorado of the West in 1859.* Washington, D.C.: GPO, 1876.

Miser, Hugh D. *The San Juan Canyon, Southeastern Utah: A Geographic and Hydrographic Reconnaissance.* U.S. Geological Survey Water Supply Paper 538. Washington, D.C.: GPO, 1924.

Muhn, James, and Hanson R. Stuart. *Opportunity and Challenge—The Story of BLM.* Washington, D.C.: Department of the Interior, Bureau of Land Management, 1988.

Patterson, James L., and William P. Somers. "Part 9. Colorado Basin." In *Magnitude and Frequency of Floods in the United States.* U.S. Geological Survey Water Supply Paper 1683. Washington, D.C.: GPO, 1966.

Peterson, Kenneth L., and Janet D. Orcutt, eds. *Dolores Archaeological Program: Supporting Studies: Settlement and Environment.* Denver: Department of the Interior, Bureau of Land Management, 1987.

Quartarone, Fred. *Historical Accounts of Upper Colorado River Basin Endangered Fish.* Denver: Department of the Interior, Fish and Wildlife Service, 1995.

Report of the Commissioner of Indian Affairs. New Mexico Superintendency. Washington. D.C.: GPO, 1892.

Report of the Commissioner of Indian Affairs. Washington, D.C.: GPO, 1897.

Report of the Commissioner of Indian Affairs. Washington, D.C.: GPO, 1901.

Report of the Commissioner of Indian Affairs. Washington, D.C.: GPO, 1906.

Robinson, T. W. *Introduction Spread and Areal Extent of Salt Cedar (Tamarix) in the Western United States.* U.S. Geological Survey Professional Paper 491-A. Washington, D.C.: GPO, 1965.

Spangler, L. E., D. L. Naftz, and Z. E. Peterman. *Hydrology, Chemical Quality, and Characterization of Salinity in the Navajo Aquifer in and Near the Greater Aneth Oil Field, San Juan County, Utah.* U.S. Geological Survey Report 96-4155. Salt Lake City: U.S. Geological Survey, 1996.

Thompson, Kendall R. *Characteristics of Suspended Sediment in the San Juan River Near Bluff, Utah.* Water Resources Investigation Report no. 82-4104. Salt Lake City: U.S. Geological Survey, 1982.

Turner, Raymond M. *Quantitative and Historical Evidence of Vegetation Changes along the Upper Gila River, Arizona.* U.S. Geological Survey Professional Paper 655-H. Washington, D.C.: GPO, 1974.

U.S. Congress. House. *Amending the Act of March 1, 1933 . . . as an Addition to the Navajo Indian Reservation.* 90th Cong., 2d sess., 1968, S. Report 1324.

U.S. Congress. House. Subcommittee on Irrigation and Reclamation. *Colorado River Storage Project: Hearings on H. R. 4449, H. R. 4443, and H. R. 4463.* 83d Cong., 2d sess., 1954.

———. *Colorado River Storage Project: Hearings on H. R. 270, H. R. 2836, H. R. 3383, H. R. 3384, and H. R. 4488.* 84th Cong., 1st sess., 1955.

U.S. Congress. Senate. *Navajo Indians in New Mexico and Arizona.* 52d Cong., 1st sess., 4 August 1892, Ex. Doc. 156.

———. *The Western Range—Letter of Secretary of Agriculture.* 74th Cong., 2d sess., 24 April 1936, S. Doc. 199.

U.S. Congress. Senate. Subcommittee on Irrigation and Reclamation. *Colorado River Storage Project: Hearings on S.1555.* 83d Cong. 2d sess., 1954.

———. *Colorado River Storage Project: Hearings on S.500.* 84th Cong., 1st sess., 1955.

U.S. Department of the Interior. Bureau of Land Management. *Grand Gulch Plateau Cultural and Recreational Area Management Plan.* Monticello, Utah: Author, 1993.

———. *Proposed Resource Management Plan and Final Environmental Impact Statement for the San Juan Resource Area, Moab District.* Vol. 1. Washington, D.C.: GPO, 1987.

———. *Utah Statewide Wilderness Study Report.* Salt Lake City: Author, 1991.

U.S. Department of the Interior. Bureau of Reclamation. *The Colorado River: "A Natural Menace Becomes a National Resource," A Comprehensive Report on the Development of the Water Resources of the Colorado River Basin for Irrigation, Power Production, and Other Beneficial Uses in Arizona, California, Colorado, Nevada, New Mexico, Utah, and Wyoming.* Washington, D.C.: GPO, 1946.

———. *Colorado River Storage Project and Participating Projects.* Salt Lake City: Author, 1949.

———. *Colorado River Storage Project and Participating Projects.* Salt Lake City: Author, 1950.

———. *Operation of Glen Canyon Dam: Final Environmental Impact Statement.* Salt Lake City: Author, 1995.

———. *San Juan Investigation: Utah and Colorado.* Washington, D.C.: GPO, 1969.

U.S. Department of the Interior. Fish and Wildlife Service. *Colorado River Squawfish Recovery Plan.* Denver: Author, 1991.

———. "Endangered and Threatened Wildlife and Plants: Determination of Critical Habitat for Four Colorado River Endangered Fishes; Final Rule," 13374–13400. *Federal Register* 59, no. 54 (21 March 1994).

U.S. Department of the Interior, National Park Service. *Glen Canyon National Recreation Area Annual Report.* San Juan County Tourist Information Bureau, Monticello, Utah, 1999, on file.

Using Geochemical Data to Identify Sources of Salinity to the Freshwater Navajo Aquifer in Southeastern Utah. Salt Lake City: U.S. Geological Survey, 1995.

"Utah Stream Water Quality." In *National Water Summary, 1990–91.* U.S. Geological Survey Water Supply Paper 2400. Washington, D.C.: GPO, 1991.

"Utah Surface-Water Resources." In *National Water Summary, 1985.* U.S. Geological Survey Water Supply Paper 2300. Washington, D.C.: GPO, 1985.

van Hylckama, T. E. A. *Weather and Evapo-transpiration Studies in a Salt Cedar Thicket, Arizona.* Washington, D.C.: GPO, 1974.

Wilson, Lanny O. *Distribution and Ecology of the Desert Bighorn Sheep in Southeastern Utah.* Publication no.

68-5. Salt Lake City: Utah State Division of Fish and Game, 1968.
Woolley, Ralf R. *Cloudburst Floods in Utah, 1850–1938.* U.S. Geological Survey Paper 994. Washington, D.C.: GPO, 1946.

Government Documents Unpublished

Ballard, Jerry. "San Juan River Management Plan—Staff Report," 18 August 1979. Bureau of Land Management Library, Monticello, Utah.
Gregory, Herbert E. "Field Notes," Book 8 (1913). U.S. Geological Survey Archives, Denver.
Kendrick, Major Henry L. "Report to Maj. Wm. A Nichols," 15 August 1853. Old Military Records, National Archives, Washington, D.C.
Klessert, Anthony L. "Inventory Report— BIA," 17 August 1989. Report no. BIA— NAO NTM 89-205, Navajo Archives, Navajo Nation, Window Rock, Arizona.
Miser, Hugh D. "Field Records," (1921). Accession no. 4124-A, U.S. Geological Survey Archives, Denver.
Mitchell, Henry L., et. al. "Evaluation of Property Destroyed in Kane County, Utah," 24 December 1879. Record Group 75, Bureau of Indian Affairs, Consolidated Ute Agency Records, Federal Records Center, Denver.
Sturgis, Captain E. A. to Adjutant, Fort Wingate, 12 August 1908. Record Group 75, Letters Received by the Office of Indian Affairs 1881–1907, National Archives, Washington, D.C.
U.S. Department of the Interior. Bureau of Land Management. "Supplement No. 6 to Memorandum of Understanding between National Park Service, Utah State Office and Bureau of Land Management, Utah State Office," 1 January 1 1979. Bureau of Land Management Library, Monticello, Utah.
U.S. Department of the Interior, Fish and Wildlife Service. "Memorandum: Biological Opinion on the Effects of Water Well Depletion," 8 December 1994. Fish and Wildlife Service Office, Denver.
———. "Memorandum: Formal Section 7 Consultation, Biological Opinion for the Proposed Permitting of Oil and Gas Activities," 20 July 1993. Fish and Wildlife Service, Albuquerque, New Mexico. Bureau of Land Management Office, Monticello, Utah. Photocopy.
———. "Staff Report on the December 13, 1994 Meeting between the U.S. Fish and Wildlife Service and Bureau of Land Management to Discuss Oil and Gas, Water Quality and Threatened and Endangered Fish Habitats of the San Juan River," 16 December 1994. Bureau of Land Management Office, Monticello, Utah. Photocopy.

Published Books and Articles

Abbey, Edward. *Down the River.* New York: E. P. Dutton, 1982.
Agenbroad, Larry D., Jim I. Mead, and Lisa W. Nelson, eds. *Megafauna and Man: Discovery of America.* Scientific Papers, vol. 1. Hot Springs, S. Dak.: The Mammoth Site of Hot Springs, South Dakota, 1990.
Ambler, J. Richard, and Mark Q. Sutton. "The Anasazi Abandonment of the San Juan Drainage and the Numic Expansion." *North American Anthropologist* 10 (1989): 39–57.
Anderson, Donna G. "Trials and Triumphs: Bluff's Struggle with Water." *Blue Mountain Shadows* 2 (Fall 1988): 18–26.
Aton, James. "The River, the Ditch and the Volcano—Bluff, 1879–1884." *Blue Mountain Shadows* 14 (Summer 1993): 15–23.
———. "Lake Bluff?" *Canyon Echo* 4 (June 1997): 1, 14.
Baars, Donald L. *Navajo Country: A Geological and Natural History of the Four Corners Region.* Albuquerque: University of New Mexico Press, 1995.
———. *Redrock Country: The Geologic History of the Colorado Plateau.* Garden City, N.Y.: Doubleday/Natural History Press, 1971.
———, ed. *Geology of the Canyons of the San Juan River.* Durango, Colo.: Four Corners Geological Society, 1974.
Baars, Donald L., and Rex C. Buchanan. *The Canyon Revisited: A Rephotography of the Grand Canyon, 1923/1991.* Salt Lake City: University of Utah Press, 1994.
Bailey, Alfred M. "Desert River through Navajoland." *National Geographic* 62 (August 1947): 149–72.
Bailey, Garrick, and Roberta Glenn Bailey. *A History of the Navajos—The Reservation Years.* Santa Fe: School of American Research Press, 1986.
Baker, Mildred E. "Rough Water." *American Forests* 50 (1 November 1944): 520–29.
Baker, Richard Allan. *Conservation Politics: The Senate Career of Clinton P. Anderson.* Albuquerque: University of New Mexico Press, 1985.
Barnes, F. A. *Hiking the Historic Route of the 1859 Macomb Expedition.* Moab, Utah: Canyon Country Publications, 1989.
Bartlett, Richard A. *Great Surveys of the West.* Norman: University of Oklahoma Press, 1962.
Behnke, R. J., and D. E. Benson. *Endangered and Threatened Fishes of the Upper Colorado River Basin.* Cooperative Extension Service, Bulletin 503A. Fort Collins, Colo.: Colorado State University, 1983.
Benson, Jackson J. *Wallace Stegner: His Life and Work.* New York: Penguin Books, 1996.

Betancourt, Julio L. "Late Quaternary Plant Zonation and Climate in Southeastern Utah." *The Great Basin Naturalist* 44 (1984): 1–35.

Betancourt, Julio L., and Thomas R. Van Devender. "Holocene Vegetation in Chaco Canyon." *Science* 214 (1981): 656–58.

Blue Mountain Shadows 12 (Summer 1993).

The Bluff Nature Preserve: An Innovative Oasis Where Community and Nature Meet. Bluff, Utah: Bluff Historical Preservation Association, 1996.

Briggs, Walter. *Without Noise of Arms: The 1776 Dominguez-Escalante Search for a Route from Santa Fe to Monterey.* Flagstaff, Ariz.: Northland Press, 1976.

Brightman, Robert. *Grateful Prey: Rock Cree Human-Animal Relationships.* Los Angeles: University of California Press, 1993.

Brough, R. Clayton, Dale L. Jones, and Dale J. Stevens. *Utah's Comprehensive Weather Almanac.* Salt Lake City: Publisher's Press, 1987.

Brugge, David M. "Vizcarra's Navajo Campaign of 1823." *Arizona and the West* 6 (1964): 223–44.

Bunte, Pamela A., and Robert J. Franklin. *From the Sands to the Mountain: Change and Persistence in a Southern Paiute Community.* Lincoln: University of Nebraska Press, 1987.

Canyon Echo 2 (June 1995).

Carlisle, Ronald C., ed. *Ice-Age Origins.* Ethnography Monograph, no. 12. Pittsburgh: University of Pittsburgh Press, 1988.

Carothers, Steven W., and Bryan T. Brown. *The Colorado River through Grand Canyon: Natural History and Human Change.* Vol. 2. Tucson: University of Arizona Press, 1991.

Chamberlain, Ralph V. "Some Plant Names of the Ute Indians." *American Anthropologist* 2, no. 1 (January–March 1909): 27–40.

Chavez, Fray Angelico, and Ted Warner, eds. *The Dominguez-Escalante Journal.* Provo, Utah: Brigham Young University Press, 1976.

Christenson, Andrew L. "The Last of the Great Expeditions: The Rainbow Bridge/Monument Valley Expeditions 1933–1938." *Plateau* 58 (1987): 1–32.

Clark, Ira G. *Water in New Mexico: A History of Its Management and Use.* Albuquerque: University of New Mexico Press, 1987.

Clark, Neil M. "Fast Water Man." *Saturday Evening Post,* 18 May 1946, 28–30.

Cole, Frank. *Well Spacing in the Aneth Reservoir.* Norman: University of Oklahoma Press, 1962.

Conetah, Fred A. *A History of the Northern Utes.* Edited by Kathryn L. McKay and Floyd A. O'Neil. Salt Lake City: Uintah Ouray Ute Tribe, 1982.

Cordell, Linda S. *Prehistory of the Southwest.* San Diego: Academic Press, 1984.

Correll, J. Lee. *Through White Men's Eyes—A Contribution to Navajo History.* Vol. 6. Window Rock, Ariz.: Navajo Heritage Center, 1979.

Crampton, C. Gregory. *Ghosts of Glen Canyon: History beneath Lake Powell.* St. George, Utah: Publishers Place, 1986.

———. *Historical Sites in Glen Canyon Mouth of San Juan River to Lee's Ferry.* University of Utah Anthropological Papers, no. 46. Salt Lake City: University of Utah Press, 1960.

———. *Outline History of the Glen Canyon.* University of Utah Anthropological Papers, no. 42. Salt Lake City: University of Utah Press, 1959.

———. *The San Juan Canyon Historical Sites.* University of Utah Anthropological Papers, no. 70. Salt Lake City: University of Utah Press, 1964.

———. *Standing Up Country: The Canyon Lands of Utah and Arizona.* 1969. Reprint, Salt Lake City: Peregrine Smith Books, 1983.

Cutter, Donald C. "Prelude to a Pageant in the Wilderness." *Western Historical Quarterly* 8 (January 1977): 4–14.

Dancer, Daniel. "Over-Glossied and Imaged Out: Toward a Deep Photographic Ethic." *Wild Earth* (Spring 1996): 81–87.

Daniels, Helen Sloan. *Adventures with the Anasazi of Falls Creek.* Occasional Papers of the Center for Southwest Studies, no. 3. Durango, Colo.: Fort Lewis College, 1954.

Davidson, Dale, et al. "San Juan County Almost 8000 Years Ago: Ongoing Excavations at Old Man Cave." *Blue Mountain Shadows* 13 (1994): 7–12.

Davis, Owen, et al. "Riparian Plants Were a Major Component of the Diet of Mammoths of Southern Utah." *Current Research in the Pleistocene* 2 (1985): 81–82.

Davis, William E. "The Lime Ridge Clovis Site." *Utah Archaeology* 2, no. 1 (1989): 66–76.

Davis, William E., and Gary M. Brown. "The Lime Ridge Clovis Site." *Current Research in the Pleistocene* 3 (1986): 1–3.

D'Azevedo, Warren, ed. *Handbook of North American Indians.* Vol. 11. Washington, D.C.: Smithsonian Institution, 1986.

deBuys, William. *Enchantment and Exploitation: The Life and Hard Times of a New Mexico Mountain Range.* Albuquerque: University of New Mexico Press, 1988.

Dobyns, Henry F. *From Fire to Flood: Historic Human Destruction of Sonoran Riverine Oases.* Ballena Press Anthropological Papers, no. 20. Edited by Lowell John Bean and Thomas C. Blackburn. Socorro, N. Mex.: Ballena Press, 1981.

Dunmire, William W., and Gail D. Tierney. *Wild Plants and Native Peoples of the Four Corners.* Santa Fe: Museum of New Mexico Press, 1997.

Dyk, Walter. *A Navaho Autobiography.* New York: Viking Fund, 1947.

Dyk, Walter, and Ruth Dyk. *Left Handed: A Navajo Autobiography.* New York: Columbia University Press, 1980.

Eastwood, Alice. "General Notes of a Trip through Southeastern Utah." *Zoe* 3 (January 1893): 354–61.

———. "Notes on the Cliff Dwellers." *Zoe* 3 (January 1893): 375–76.

———. "Report on a Collection of Plants from San Juan County, in Southeastern Utah." *California Academy of Sciences Proceedings* 6 (August 1896): 271–329.

Ellis, Fern D. *Come Back to My Valley.* Cortez, Colo.: Cortez Printers, 1976.

Engstrand, Iris H. W. *Spanish Scientists in the New World: The Eighteenth- Century Expeditions.* Seattle: University of Washington Press, 1981.

Euler, Robert C. "The Canyon Dwellers. "*American West* 4 (1967): 22–27, 67–71.

Fagan, Brian M. *The Great Journey: The Peopling of North America.* New York: Thomas and Hudson, 1987.

Farmer, Jared. *Glen Canyon Damned: Inventing Lake Powell and the Canyon Country.* Tucson: University of Arizona Press, 1999.

Fenner, Patty, Ward W. Brady, and David R. Patton. "Effects of Regulated Water Flows on Regeneration of Fremont Cottonwood." *Journal of Range Management* 38, no. 2 (March 1985): 135–38.

Ferris, Robert G., ed. *The American West: An Appraisal.* Santa Fe: Museum of New Mexico, 1963.

Fletcher, Maurine S., ed. *The Wetherills of the Mesa Verde: Autobiography of Benjamin Alfred Wetherill.* 1977. Reprint, Lincoln: University of Nebraska Press, 1987.

Foster, Mike. *Strange Genius: The Life of Ferdinand Vandeveer Hayden.* Niwot, Colo.: Roberts Rinehart Publishers, 1994.

Foster, Robert J. *General Geology.* Columbus, Ohio: Merrill Publishing, 1988.

Four Corners Planning. *Community of Bluff: General Plan.* Bluff Utah: Author, 1996.

Fradkin, Philip. *A River No More: The Colorado River and the West.* Tucson: University of Arizona Press, 1984.

Franciscan Fathers. *An Ethnologic Dictionary of the Navajo Language.* Saint Michaels, Ariz.: Saint Michaels Press, 1910.

Freeman, Ira. *A History of Montezuma County, Colorado.* Boulder, Colo.: Johnson Publishing, 1958.

Genoways, Hugh H., and Mary R. Dawson, eds. *Contributions to Quaternary Vertebrate Paleontology.* Special Publication of Carnegie Museum of Natural History, no. 8. Pittsburgh: Carnegie Museum, 1984.

Gillmor, Francis, and Louisa Wade Wetherill. *Traders to the Navajos.* Albuquerque: University of New Mexico Press, 1952.

Glausiusz, Josie. "Trees of Salt." *Discover* 17 (March 1996): 30–32.

Goetzmann, William H. *Army Exploration in the American West, 1803–1863.* 1959. Reprint, Lincoln: University of Nebraska Press, 1979.

———. *New Men, New Lands: America and the Second Great Age of Discovery.* New York: Viking, 1986.

Graf, William L. "Fluvial Adjustments to the Spread of Tamarisk in the Colorado Plateau Region." *Geological Society of America Bulletin* 89 (October 1978): 1491–1501.

Grant, U. S. "The Dinosaur Dam Sites Are Not Needed." *Living Wilderness* 34 (Autumn 1950): 17–24.

Gregory, Herbert E. "The Navajo Country." *American Biographical Society Bulletin* 8 (1915): 561–672.

———. "Scientific Explorations in Southern Utah." *American Journal of Science* 243 (October 1945): 527–49.

Hafen, Leroy R., ed. *The Mountain Men and the Fur Trade of the Far West.* Vols. 2 and 5. Glendale, Calif.: Arthur H. Clark, 1965, 1966.

Hafen, Leroy R., and Ann Hafen. *Old Spanish Trail: Santa Fe to Los Angeles.* Glendale, Calif.: Arthur H. Clark, 1965.

Hale, Peter B. *William Henry Jackson and the Transformation of the American Landscape.* Philadelphia: Temple University Press, 1988.

Hansen, Adrian N. "The Endangered Species Act and Extinction of Reserved Indian Water Rights on the San Juan River." *Arizona Law Review* 37 (1995): 1305–44.

Hargrave, Lyndon Lane. *Report on Archaeological Reconnaissance in the Rainbow Plateau Area of Northern Arizona and Southern Utah.* Berkeley: University of California Press, 1935.

Harvey, Mark W. T. "Defending the Park System: The Controversy over Rainbow Bridge." *New Mexico Historical Review* 73 (January 1998): 45–67.

———. "Echo Park, Glen Canyon, and the Postwar Wilderness Movement." *Pacific Historical Review* 60 (February 1991): 43–67.

———. *Symbol of Wilderness: Echo Park and the American Conservation Movement.* Albuquerque: University of New Mexico Press, 1994.

Hausman, Gerald. *The Gift of the Gila Monster.* New York: Simon and Schuster, 1993.

Hawley, Florence, et al. "Culture Process and Change in Ute Adaptation—Part 1," *El Palacio* (October 1950): 311–61.

Haynes, Gary. *Mammoth, Mastodonts, and Elephants: Biology, Behavior, and the Fossil Record.* Cambridge: Cambridge University Press, 1991.

Hays, Samuel P. *Beauty, Health and Permanence: Environmental Politics in the United States, 1955–1985.* New York: Cambridge University Press, 1987.

———. *Conservation and the Gospel of Efficiency: The Progressive Conservation Movement, 1890–1920.* Cambridge: Harvard University Press, 1959.

Henderson, Randall. "Floating down the San Juan and the Colorado." *Sierra Club Bulletin* 30 (December 1945): 63–70.

Hill, Joseph J. "Spanish and Mexican Exploration and Trade Northwest from New Mexico into the Great Basin, 1765–1853." *Utah Historical Quarterly* 3 (January 1930): 2–23.

Hill, W. W. *The Agricultural and Hunting Methods of the Navaho Indians.* New Haven: Yale University Press, 1938.

Hillerman, Tony. *Hunting Badger.* New York: HarperCollins, 1999.

———. *A Thief of Time.* New York: Harper and Row, 1988.

Hogan, Patrick, and Joseph C. Winter, eds. *Economy and Interaction along the Lower Chaco River.* Albuquerque: Maxwell Museum, 1983.

Holden, Paul B., and William Masslich. *San Juan River Recovery Implementation Program: Summary Report, 1991–1996.* Logan, Utah: Bio-West, 1997.

Howell, James Thomas. "'I Remember When I Think. . . .'" *Leaflet of Western Botany* 2 (August 1954): 153–64.

Hundley, Norris, Jr. *Water and the West: The Colorado River Compact and the Politics of Water in the American West.* Berkeley: University of California Press, 1975.

Hunter, Christopher J. *Better Trout Habitat—A Guide to Stream Restoration and Management.* Washington, D.C.: Island Press, 1991.

Hurst, Winston B. "The Mysterious Telluride Blanket." *Blue Mountain Shadows* 13 (1994): 68–69.

Hurst, Winston B., and Joe Pachak. *Spirit Windows: Native American Rock Art of Southeastern Utah.* Blanding, Utah: Edge of the Cedars Museum, 1992.

Iverson, Peter. *The Navajo Nation.* Albuquerque: University of New Mexico Press, 1981.

Jacobs, Clell G. "The Phantom Pathfinder: Juan Maria Antonio de Rivera and His Expedition." *Utah Historical Quarterly* 60 (Summer 1992): 200–23.

Jefferson, James, Robert W. Delaney, and Gregory Thompson. *The Southern Utes: A Tribal History.* Ignacio, Colo.: Southern Ute Tribe, 1972.

Jennings, Jesse D. *Accidental Archaeologist: Memoirs of Jesse D. Jennings.* Salt Lake City: University of Utah Press, 1994.

———. *Glen Canyon: A Summary.* University of Utah Anthropological Papers, no. 81. Salt Lake City: University of Utah Press, 1966.

———, ed. *Sudden Shelter.* University of Utah Anthropological Papers, no. 103. Salt Lake City: University of Utah Press, 1980.

Jennings, Jesse D., and Floyd W. Sharrock. "The Glen Canyon: A Multi-Discipline Project." *Utah Historical Quarterly* 33 (Winter 1965): 34–50.

Johnson, Craig. *Bluff Floodplain Nature Park.* Part 2 of *Bluff, Utah: Study in Rural Community Planning.* Logan: Utah State University Press, 1995.

Jones, Sondra. "'Redeeming' the Indian: The Enslavement of Indian Children in New Mexico and Utah." *Utah Historical Quarterly* 67 (Summer 1999): 220–41.

———. *The Trial of Don Pedro Leon Lujan: The Attack against Indian Slavery and Mexican Traders in Utah.* Salt Lake City: University of Utah Press, 1999.

Judd, Neil M. "The Discovery of Rainbow Bridge." In *The Discovery of Rainbow Bridge,* 8–13. Bulletin l. Tucson: Cummings Publication Council, 1959.

———. *Men Met along the Trail: Adventures in Archaeology.* Norman: University of Oklahoma Press, 1968.

Kelley, Klara Bonsack, and Harris Francis. *Navajo Sacred Places.* Indianapolis: Indiana University Press, 1994.

Kittridge, William. *Owning It All.* St. Paul: Graywolf Press, 1987.

Klett, Mark, et al. *Second View: The Rephotographic Survey Project.* Albuquerque: University of New Mexico Press, 1984.

Klinck, Richard E. *Land of Room Enough and Time Enough.* Salt Lake City: Peregrine Smith Books, 1984.

Kohler, Timothy A., and Meredith H. Matthews. "Long-Term Anasazi Land Use and Forest Reduction: A Case Study from Southwest Colorado." *American Antiquity* 53, no. 3 (1988): 537–64.

Kricher, John C. "Needs of Weeds." *Natural History* 89, no. 12 (December 1980): 36–45.

Kroeber, Clifton, ed. "The Route of James O. Pattie on the Colorado in 1826: A Reappraisal by A. L. Kroeber." *Arizona and the West* 4 (Summer 1964): 130–31.

Kurten, Bjorn, and Elaine Anderson. *Pleistocene Animals of North America.* New York: Columbia University Press, 1980.

Kyokai, Bukkyo Dendo. *The Teaching of Buddha.* Tokyo: Kosaido Printing Company, 1966.

Lambert, Neal. "Al Scorup: Cattleman of the Canyons." *Utah Historical Quarterly* 32, no. 3 (Summer 1964): 301–20.

L'Amour, Louis. *Haunted Mesa.* New York: Bantam, 1987.

Lavendar, David. *Colorado River Country.* New York: E. P. Dutton, 1982.

———. *River Runners of the Grand Canyon.* Grand Canyon, Ariz.: Grand Canyon Natural History Association, 1985.

Leblanc, Steven A. *Prehistoric Warfare in the American Southwest.* Salt Lake City: University of Utah Press, 1999.

Levin, Harold L. *Contemporary Physical Geology.* 2d ed. Philadelphia: Saunders College Publishing, 1986.

Limerick, Patricia Nelson. *The Legacy of Conquest: The Unbroken Past of the American West.* New York: Norton, 1987.

Lindsay, Alexander J., Jr. "The Beaver Creek Agricultural Community on the San Juan River, Utah." *American Antiquity* 27 (1961): 174–87.

Lindsay, Alexander J., Jr., et al. *Survey and Excavations North and East of Navajo Mountain, Utah, 1959–1962.* Museum of Northern Arizona Bulletin, no. 45. Glen Canyon Series, no. 8. Flagstaff: Museum of Northern Arizona, 1968.

Luckert, Karl. *Navajo Mountain and Rainbow Bridge Religion.* Flagstaff: Museum of Northern Arizona, 1977.

Lupo, Karen D. "The Historical Occurrence and Demise of Bison in Northern Utah." *Utah Historical Quarterly* 64 (Spring 1996): 168–80.

Lyman, Albert R. "Fort on the Firing Line." Serialized in *Improvement Era,* 1948–49.

———. *Indians and Outlaws: Settling the San Juan Frontier.* 1962. Reprint, Salt Lake City: Publisher's Press, 1980.

———. *The Outlaw of Navajo Mountain.* Salt Lake City: Deseret Books, 1963.

———. *The Trail of the Ancients.* Blanding, Utah: The Trail of the Ancients Association, 1972.

McCool, Daniel C., ed. *Waters of Zion: The Politics of Water in Utah.* Salt Lake City: University of Utah Press, 1995.

McKenney, Rose, Robert B. Jacobson, and Robert C. Wertheimer. "Woody Vegetation and Channel Morphogenesis in Low-Gradient, Gravel-Bed Streams in the Ozark Plateaus, Missouri and Arkansas." In *Biogeomorphology, Terrestrial and Freshwater Aquatic Systems: Geomorphology,* vol. 13, edited by C. R. Hupp, W. R. Ostertramp, and A. D. Howard (Binghamton: State University of New York, 1995), 175–98.

McNamee, Gregory. *Gila: The Life and Death of an American River.* New York: Orion Books, 1994.

McNitt, Frank. *The Indian Traders.* Albuquerque: University of New Mexico Press, 1962.

———. *Richard Wetherill: Anasazi: Pioneer Explorer of Southwestern Ruins.* Rev. ed. Albuquerque: University of New Mexico Press, 1966.

McPherson, Robert S. *A History of San Juan County: In the Palm of Time.* Utah Centennial County History Series. Salt Lake City: Utah State Historical Society, 1995.

———. "Howard Antes and the Navajo Faith Mission: Evangelist of Southeastern Utah." *Blue Mountain Shadows* 17 (Summer 1996): 14–24.

———. *Navajo Land, Navajo Culture: Persistence and Change in Southeastern Utah.* Norman: University of Oklahoma Press, forthcoming.

———. *The Northern Navajo Frontier, 1860–1900: Expansion through Adversity.* Albuquerque: University of New Mexico Press, 1988.

———. *Sacred Land, Sacred View.* Provo, Utah: Brigham Young University Press, 1992.

Marshall, Eliot. "Clovis Counterrevolution." *Science* 249 (1990): 738–41.

Martin, Paul S. "The Discovery of America." *Science* 179 (1973): 969–74.

Martin, Paul S., and Richard G. Klein, eds. *Quaternary Extinctions: A Prehistoric Revolution.* Tucson: University of Arizona Press, 1984.

Martin, Russell. *A Story That Stands Like a Dam: Glen Canyon and the Struggle for the Soul of the West.* New York: Henry Holt, 1989.

Matson, R. G., William D. Lipe, and William R. Haase, III. "Adaptational Continuities and Occupational Discontinuities: The Cedar Mesa Anasazi." *Journal of Field Archaeology* 15 (1980): 245–64.

Matthews, Washington. *Navaho Legends.* New York: Houghton, Mifflin and Company, 1897.

Matthews, Washington. *Navaho Legends.* Reprint, Salt Lake City: University of Utah Press, 1994.

May, Dave. "Maligned or Malignant; Tamarisk: The Plant We Love to Hate," *Canyon Legacy* (Winter 1989): 3–7.

Mayes, Vernon O., and Barbara Lacy. *Nanise': A Navajo Herbal.* Tsaile, Ariz.: Navajo Community College, 1989.

Meloy, Ellen. "Geese, River Ghost, the Holy Ghost—A Field Report." *Canyon Echo* 2 (June 1995): 10–19.

———. *The Last Cheater's Waltz: Beauty and Violence in the Desert Southwest.* New York: Henry Holt, 1999.

———. "Long Body, Sinuous Gifts." *Orion* 15 (Summer 1996):12.

Miller, Angela. *The Empire of the Eye: Landscape Representation and American Cultural Politics,*

1825–1875. Ithaca, N. Y.: Cornell University Press, 1993.

Miller, David E. *Hole-in-the-Rock: An Epic in the Colonization of the Great American West.* Salt Lake City: University of Utah Press, 1959.

———. *Hole-in-the-Rock: An Epic in the Colonization of the Great American West.* 2d ed. Salt Lake City: University of Utah Press, 1966.

Milner, Clyde A., II. "The Shared Memory of Montana's Pioneers." *Montana* 37 (Winter 1987): 2–13.

Monroe, James S., and Reed Wicander. *Physical Geology: Exploring the Earth.* St. Paul: West Publishing, 1992.

Monson, Gale, and Lowell Summer, eds. *The Desert Bighorn: Its Life History, Ecology and Management.* Tucson: University of Arizona Press, 1980.

Montgomery, Carla W. *Physical Geology.* 2d ed. Dubuque: Wm. C. Brown, 1988.

Moorhead, Warren K. "In Search of a Lost Race." *Illustrated American,* August 1892.

———. "Across the Desert." *Illustrated American,* 16 July 1892, 410.

Muhlestein, Harold, and Fay Muhlestein. *Monticello Journal—A History of Monticello until 1937.* Monticello, Utah: Authors, 1988.

Myers, William A. *Iron Men and Copper Wires: A Centennial History of the Southern California Edison Company.* Glendale, Calif.: Trans-Anglo Books, 1983.

Navajo Stock Reduction Interviews. Compiled by Utah State Historical Society and California State University at Fullerton. Fullerton, Calif.: California State University at Fullerton Press, 1984.

Neilly, Robert B., ed. *Basketmaker Settlement and Subsistence along the San Juan River, Utah.* Salt Lake City: Utah Division of State History, 1982.

Nelson, Lisa. *Ice Age Mammals of the Colorado Plateau.* Flagstaff: Northern Arizona University Press, 1990.

Nelson, Nancy. *Any Time, Any Place, Any River: The Nevills of Mexican Hat.* Flagstaff, Ariz.: Red Lake Books, 1991.

Newcomb, Franc Johnson. *Navaho Folk Tales.* Albuquerque: University of New Mexico Press, 1967.

Nickens, Paul R., ed. *An Archaeology of the Eastern Ute: A Symposium.* Occasional Papers, no. 1. Denver: Colorado Council of Professional Archaeology, 1988.

Novak, Barbara. *Nature and Culture: American Landscape Painting, 1825–1875.* New York: Oxford University Press, 1980.

O'Bryan, Aileen. *Navaho Indian Myths.* New York: Dover Publications, 1993.

Office of Program Development. *Navajo Nation Overall Economic Development Program.* Window Rock, Ariz.: Navajo Nation, 1980.

Olsen, Nancy H. *Hovenweep Rock Art: An Anasazi Visual Communication System.* Occasional Paper, no. 14. Los Angeles: UCLA Institute of Archaeology, 1985.

Olson, A. P. "Split-Twig Figurines from NA5607, Northern Arizona." *Plateau* 38 (1966): 55–64.

O'Neil, Floyd, ed. *Southern Ute: A Tribal History.* Ignacio, Colo.: Southern Ute Tribe, 1972.

Opler, Marvin Kaufman. *The Southern Ute of Colorado.* New York: D. Appleton-Century, 1940.

Palmer, Tim. *Endangered Rivers and the Conservation Movement.* Berkeley: University of California Press, 1989.

———. *Wild and Scenic Rivers of America.* Washington, D.C.: Island Press, 1993.

Patterson, Alex. *A Field Guide to Rock Art Symbols of the Greater Southwest.* Boulder, Colo.: Johnson Books, 1992.

Pattie, James O. *The Personal Narrative of James O. Pattie.* Edited by William H. Goetzmann. Philadelphia: J. B. Lippincott, 1962.

Peattie, Roderick, ed. *The Inverted Mountains: Canyons of the West.* New York: Vanguard Press, 1948.

Perkins, Cornelia Adams, Marian Gardner Nielson, and Lenora Butt Jones. *Saga of San Juan.* Monticello, Utah: San Juan Chapter, Daughters of Utah Pioneers, 1957.

Peterson, Charles S. *Look to the Mountain: Southeastern Utah and the La Sal National Forest.* Provo, Utah: Brigham Young University Press, 1975.

———. "'A Utah Moon': Perceptions of Southern Utah." The Juanita Brooks Lecture at Dixie College, St. George, Utah, 1984.

———. *Water Rights on the Little Colorado River—First Draft.* Missoula, Mont.: History Research Associates, 1986.

Pisani, Donald J. *To Reclaim a Divided West: Water, Law, and Public Policy, 1848-1902.* Albuquerque: University of New Mexico Press, 1992.

Pomeroy, Earl. *In Search of the Golden West: The Tourist in Western America.* New York: Alfred A. Knopf, 1957.

Potter, Loren D., and Charles L. Drake. *Lake Powell: Virgin Flow to Dynamo.* Albuquerque: University of New Mexico Press, 1989.

Powell, Allan Kent, ed. *San Juan County, Utah: People, Resources, and History.* Salt Lake City: Utah State Historical Society, 1982.

Power, Thomas Michael. "The Economic Importance of Federal Grazing to the Economies of the West." *Southern Utah Wilderness Alliance Newsletter,* Spring 1995, insert.

———. "The Economics of Wilderness Preservation in Utah." *Southern Utah Wilderness Alliance Newsletter,* Winter 1995, insert.

———. *Lost Landscapes and Failed Economies: The Search for the Value of Place.* Washington, D.C.: Island Press, 1996.

Preston, Don, ed. *A Symposium of the Oil and Gas Fields in Utah.* Salt Lake City: Intermountain Association of Petroleum Geologists, 1961.

Prudden, T. Mitchell. *Biographical Sketches and Letters of T. Mitchell Prudden, M. D.* New Haven: Yale University Press, 1927.

———. "An Elder Brother to the Cliff Dwellers." *Harper's New Monthly Magazine* 95 (1897): 55–62.

———. *On the Great American Plateau.* New York: G. P. Putnam's, 1907.

———. "The Prehistoric Ruins of the San Juan Watershed." *American Anthropologist* 5 (1903): 224–88.

Pyne, Stephen J. *How the Canyon Became Grand: A Short History.* New York: Viking, 1998.

Rea, Paul W. "Blissed and Blasted on the San Juan." In *Canyon Interludes: Between White Water and Red Rock.* Salt Lake City: Signature Books, 1996.

Reay, Lee. *Incredible Passage through the Hole-in-the-Rock.* Salt Lake City: Publisher's Press, 1980.

Reichard, Gladys. *Navaho Religion—A Study of Symbolism.* Princeton: Princeton University Press, 1963.

Reilly, P. T. *Lee's Ferry: From Mormon Crossing to National Park.* Edited by Robert H. Webb. Logan: Utah State University Press, 1999.

———. "Norman Nevills: Whitewater Man of the West." *Utah Historical Quarterly* 55 (Spring 1987): 181–200.

Reisner, Marc. *Cadillac Desert: The American West and Its Disappearing Water.* New York: Viking, 1986.

Reno, Philip. *Mother Earth, Father Sky, and Economic Development: Navajo Resources and Their Use.* Albuquerque: University of New Mexico Press, 1981.

"A Review of Animas–La Plata: The West's last big water project." *High Country News Reports* (1996).

Riley, Carroll L., and Walter W. Taylor, eds. *American Historical Anthropology: Essays in Honor of Leslie Spier.* Carbondale: Southern Illinois University Press, 1967.

Roessel, Ruth, and Broderick H. Johnson, comps. *Navajo Livestock Reduction: A National Disgrace.* Tsaile, Ariz.: Navajo Community College Press, 1974.

Rogers, Larry. *Chapter Images: 1996: Profiles of 110 Navajo Nation Chapters.* Window Rock, Ariz.: Navajo Nation, 1997.

Rose, Eben. "In Kinship with Tamarisk." *The News: The Journal of Grand Canyon River Guides* [now *Boatman's Quarterly Review*] 6 (Winter 1992–93): 20–29.

Ross, Drew. "I Have Struck It Rich At Last." *Blue Mountain Shadows* 17 (Summer 1996): 30–37.

———. "'I Have Struck It Rich at Last': Charles Goodman, Traveling Photographer." *Utah Historical Quarterly* 66 (Winter 1998): 65–83.

Rothman, Hal K. *Devil's Bargains: Tourism in the Twentieth-Century American West.* Lawrence: University of Kansas Press, 1998.

Runte, Alfred. *National Parks: The American Experience.* 3rd ed. Lincoln: University of Nebraska Press, 1997.

Sanchez, Joseph P. *Explorers, Traders, and Slavers: Forging the Old Spanish Trail, 1678–1850.* Salt Lake City: University of Utah Press, 1997.

Schaafsma, Polly. *Indian Rock Art of the Southwest.* Santa Fe: School of American Research, 1980.

Schmidt, John L., and Douglas L. Olbert, eds. *Big Game of North America.* Harrisburg, Pa.: Stockpole Books, 1978.

Schroedl, Alan R. "The Grand Canyon Figurine Complex." *American Antiquity* 42 (1977): 254–65.

———. *Kayenta Anasazi Archeology and Navajo Ethnohistory on the Northwestern Shonto Plateau: The N-16 Project.* Salt Lake City: P-III Associates, 1989.

Schwartz, D. L., et al. "Split-Twig Figurines in the Grand Canyon. *American Antiquity* 23 (1958): 264–74.

Smith, Anne M. *Ethnography of the Northern Utes.* Albuquerque: Museum of New Mexico Press, 1974.

———. *Ute Tales.* Salt Lake City: University of Utah Press, 1992.

Sontag, Susan. *On Photography.* New York: Doubleday, 1989.

Spaulding, Geoffrey W. *Paleoecological Investigations at the Coombs Site (42 GA 34) Megafauna and Man.* Las Vegas: Dames and Moore, 1994.

Stegner, Wallace. *The Sound of Mountain Water: The Changing American West.* New York: E. P. Dutton, 1980.

Stevens, Joseph E. *Hoover Dam: An American Adventure.* Norman: University of Oklahoma Press, 1988.

Stevens, Larry. "Scourge of the West: The Natural History of Tamarisk in the Grand Canyon." *The News: The Journal of Grand Canyon River Guides* [now *Boatman's Quarterly Review*] 6 (Summer 1993): 14–15.

Stokes, William Lee. *Geology of Utah.* Salt Lake City: Utah Museum of Natural History, 1986.

Stowe, Carlton. *Utah's Oil and Gas Industry: Past, Present, and Future.* Salt Lake City: Utah Engineering Experiment Station, University of Utah, 1979.

Tanner, Faun McConkie. *The Far Country: A Regional History of Moab and La Sal.* Salt Lake City: Olympus Publishing, 1976.

Tate, LaVerne. "Historical Moments: Preserved by Charles Goodman's Camera." *Blue Mountain Shadows* 17 (Summer 1996): 38–45

———. "The Real Gold Mine of San Juan." *Blue Mountain Shadows* 17 (Summer 1996): 27–29.

Taylor, J. Golden, et. al., eds. *A Literary History of the American West.* Fort Worth: Texas Christian University Press, 1987.

Teal, Louise. *Breaking into the Current: Boatwomen of the Grand Canyon.* Tucson: University of Arizona Press, 1994.

Thomas, Keith. *Man and the Natural World: A History of the Modern Sensibility.* New York: Pantheon, 1983.

Thompson, Gregory C. "The Unwanted Utes: The Southern Utes in Southeastern Utah." *Utah Historical Quarterly* 49 (Spring 1981): 189–203.

"Tony Hillerman Indian Country Map and Guide." Mancos, Colo.: Time Travelers Maps, 1999.

Topping, Gary. *Glen Canyon and the San Juan Country.* Moscow: University of Idaho Press, 1997.

Toth, Richard E., et al. *Bluff, Utah: A Time for Change/A Time for Choice.* Logan: Utah State University, 1994.

Trimble, Stanley W., and Alexandra C. Mendel. "The Cow as a Geomorphic Agent —A Critical Review." *Geomorphology* 13 (1995): 233–53.

Turner, Christy G., II. *Petroglyphs of the Glen Canyon Region.* Museum of Northern Arizona Bulletin, no. 48. Flagstaff: Museum of Northern Arizona, 1963.

———. "Revised Dating for Early Rock Art of the Glen Canyon Region." *American Antiquity* 36 (1971): 469–71.

Turner, Christy G., II, and Jacqueline A. Turner. *Man Corn: Cannibalism and Violence in the Prehistoric American Southwest.* Salt Lake City: University of Utah Press, 1999.

Utah Wilderness Coalition. *Wilderness at the Edge.* Salt Lake City: Author, 1990.

Utley, Robert M. *A Life Wild and Perilous: Mountain Men and the Paths to the Pacific.* New York: Henry Holt, 1997.

Vale, Thomas R., and Geraldine R. Vale. *Western Landscapes: Travels along U.S. 89.* Tucson: University of Arizona Press, 1989.

Valle, Doris. *Looking Back around the Hat—A History of Mexican Hat.* Mexican Hat, Utah: Author, 1986.

Van Cott, John W. *Utah Place Names.* Salt Lake City: University of Utah Press, 1990.

Van Devender, T. R., and W. G. Spaulding. "Development of Vegetation and Climate in the Southwestern United States." *Science* 204 (1979): 701–10.

Walka, Ann Weiler. *Waterlines: Journeys on a Desert River.* Flagstaff, Ariz.: Red Lake Books, 1993.

Waters, Frank. *The Colorado.* 1950. Reprint, New York: Rinehart and Company, 1946.

———. *Masked Gods: Navaho and Pueblo Ceremonialism.* New York: Ballantine Books, 1970.

Weatherford, Gary B., and Phillip Nichols. *Legal-Political History of Water Resource Development in the Upper Colorado River Basin.* Lake Powell Research Bulletin, no. 4, part 1. Los Angeles: Lake Powell Research Project, 1974.

Webb, Roy. *Call of the Colorado.* Moscow: University of Idaho Press, 1994.

———. "'Never Was Anything So Heavenly': Nevills' Expeditions on the San Juan River." *Blue Mountain Shadows* 12 (Summer 1993): 35–50.

Weber, David J. *The Taos Trappers: The Fur Trade in the Far Southwest, 1540–1846.* Norman: University of Oklahoma Press, 1968.

Wegner, Dave. "Restoring Glen Canyon: Linking Our Future to the Importance of the Past." *Hidden Passage: The Journal of the Glen Canyon Institute* 1 (1998): 5–7.

———. "Sediment, Water, and Erosion: The First Technical Study Focuses on Sediment." *Hidden Passage* 2 (1999): 13.

Wellman, Klaus F. "Kokopelli of Indian Paleology: Humpbacked Rain Priest, Hunting Magician, and Don Juan of the Old Southwest." *Journal of the American Medical Association* 212 (1970): 1678–82.

Westwood, Richard E. *Rough Water Man: Elwyn Blake's Colorado River Expeditions.* Reno: University of Nevada Press, 1992.

White, Richard. "Historiographic Essay, American Environmental History: The Development of a New Field." *Pacific Historical Review* 54 (1985): 297–335.

———. *"It's Your Misfortune and None of My Own": A History of the American West.* Norman: University of Oklahoma Press, 1991.

———. *The Organic Machine: The Remaking of the Columbia River.* New York: Hill and Wang, 1995.

———. *The Roots of Dependency: Subsistence, Environment, and Social Change among the Choctaws, Pawnees and Navajos.* Lincoln: University of Nebraska Press, 1983.

Whitson, Tom D., and Roy Riechenbeck. *Weeds and Poisonous Plants of Wyoming and Utah.* Cooperative Extension Series. Laramie: University of Wyoming, 1987.

Widdison, Jerold G., ed. *Anasazi: Why Did They Leave? Where Did They Go?.* Albuquerque: Southwest Cultural Heritage Association, 1991.

Wilcox, D. R., and W. B. Masse, eds. *The Protohistoric Period in the North American Southwest, A. D. 1350–1700.* Archaeological Research Papers, no. 24. Tempe: Arizona State University Press, 1981.

Wilkinson, Todd. "Homecoming." *National Parks* (May–June 1996): 40–45.

Wilson, Carol Green. *Alice Eastwood's Wonderland: The Adventures of a Botanist.* San Francisco: California Academy of Sciences, 1955.

Wishart, David J. *The Fur Trade of the American West, 1807–1840: A Geographical Synthesis.* Lincoln: University of Nebraska Press, 1979.

Woodbury, Angus. *Notes on the Human Ecology of Glen Canyon.* University of Utah Anthropological Papers, no. 74. Salt Lake City: University of Utah Press, 1965.

Woodbury, Angus, et al. *Preliminary Report on Biological Resources of the Glen Canyon Reservoir.* University of Utah Anthropological Papers, no. 31. Salt Lake City: University of Utah Press, 1958.

Woodbury, Angus, et al., eds. *Ecological Studies of the Flora and Fauna in Glen Canyon.* University of Utah Anthropological Papers, no. 40. Salt Lake City: University of Utah Press, 1959.

Woodbury, Angus M., and Henry N. Russell, Jr. *Birds of the Navajo Country.* University of Utah Bulletin, Biological Series, no. 9. Salt Lake City: University of Utah Press, 1945.

Woodbury, Angus, Stephen D. Durrant, and Seville Flowers. *Survey of Vegetation in the Glen Canyon Reservoir Basin.* University of Utah Anthropological Papers, no. 36. Salt Lake City: University of Utah Press, 1959.

Worster, Donald. "History as Natural History: An Essay on Theory and Method." *Pacific Historical Review* 53 (1984): 1–19.

———. *Rivers of Empire: Water, Aridity, and the Growth of the American West.* New York: Pantheon Books, 1985.

———. *An Unsettled Country: Changing Landscapes of the American West.* Albuquerque: University of New Mexico Press, 1994.

—— ed. *The Ends of the Earth: Perspectives on Modern Environmental History.* New York: Cambridge University Press, 1988.

Worster, Donald, et al. "A Roundtable: Environmental History." *Journal of American History* 76, no. 4 (March 1990): 1087–1147.

Young, Norma Perkins. *Anchored Lariats on the San Juan Frontier.* Provo, Utah: Community Press, 1985.

Young, Robert W., and William Morgan. *Navajo Historical Selections.* Lawrence, Kans.: Bureau of Indian Affairs, 1954.

Zwinger, Ann. *The Wind in the Rock: The Canyonlands of Southeastern Utah.* New York: Harper and Row, 1978.

Zwinger, Ann, and Edwin Way Teale. *A Conscious Stillness:Two Naturalists on Thoreau's Rivers.* New York: Harper and Row, 1982.

Newspapers

Cortez Journal-Herald, 1930.

Denver Post, 1907.

Deseret News, 1854–1998.

Deseret Weekly, 1882.

Grand Valley Times, 1904–10.

High Country News, 1996–98.

Mancos Times, 1893–99.

Mancos Times-Tribune, 1904–16.

Montezuma Journal, 1903–14.

Navajo Times, 1978–97.

The New York Times, 1987.

Salt Lake Herald, 1892–93.

Salt Lake Mining Review, 1910.

Salt Lake Tribune, 1892–1998.

San Juan Record, 1941–74.

The Spectrum, 1998.

The Times-Independent, 1922.

Western Oil Reporter, 1978

Websites

Bureau of Reclamation., upper Colorado region. Available at <www.uc.usbr.gov>

Southern Utah Wilderness Alliance, Salt Lake City. Available at <www.suwa.org>

U.S. Geological Survey, Washington, D.C. Available at <h2o.usgs.gov/swr/>

INDEX